# COLLEGE CHEMISTRY IN THE LABORATORY

FIFTH EDITION

Morris Hein
Leo R. Best
Robert L. Miner

Mt. San Antonio College

James M. Ritchey

California State University at Sacramento

BROOKS/COLE PUBLISHING COMPANY
Pacific Grove, California

I(T)P ™
The trademark ITP is used under license.

Brooks/Cole Publishing Company
A Division of Wadsworth, Inc.



Printed in the United States of America

10 9 8 7 6 5 4 3

ISBN 0-534-17529-5

Sponsoring Editor: Faith B. Stoddard
Editorial Assistant: Beth Wilbur
Production Editor: Joan Marsh
Cover Design: Vernon T. Boes
Cover Photo: Herb Charles Ohlmeyer, Fran Heyl Associates
Printing and Binding: Malloy Lithographing, Inc.

# Contents


Preface .... v
To the Student .... vi

**EXPERIMENTS**

1. Laboratory Techniques .... 1
2. Measurements .... 11
3. Preparation and Properties of Oxygen .... 21
4. Preparation and Properties of Hydrogen .... 31
5. Water in Hydrates .... 39
6. Freezing Points—Graphing of Data .... 45
7. Properties of Solutions .... 53
8. Composition of Potassium Chlorate .... 61
9. Double Displacement Reactions .... 69
10. Single Displacement Reactions .... 75
11. Ionization—Acids, Bases, and Salts .... 79
12. Identification of Selected Anions .... 93
13. Identification of Halide Ions in Mixtures .... 101
14. Properties of Lead(II), Silver, and Mercury(I) Ions .... 107
15. Quantitative Precipitation of Chromate Ion .... 115
16. Boyle's Law .... 123
17. Charles' Law .... 131
18. Gaseous Diffusion .... 141
19. Liquids—Vapor Pressure and Boiling Points .... 147
20. Neutralization—Titration I .... 155
21. Neutralization—Titration II .... 163
22. Chloride Content of Salts .... 171
23. Oxidation-Reduction—Sulfur Dioxide .... 177
24. Halogens .... 185
25. Chemical Equilibrium—Reversible Reactions .... 193
26. Heat of Reaction .... 201
27. Boiling Points and Melting Points .... 209
28. Hydrocarbons .... 215
29. Alcohols, Esters, Aldehydes, and Ketones .... 223
30. Esterification—Distillation: Synthesis of n-Butyl Acetate .... 231
31. Synthesis of Aspirin .... 239
32. Polymers—Macromolecules .... 245
33. Carbohydrates .... 253
34. Glucose Levels in Whole Blood .... 263
35. Amino Acids and Proteins .... 269
36. Paper Chromatography .... 279
37. Ion-Exchange Chromatography of Amino Acids .... 287
38. Identification of an Unknown Aminc Acid by Titration .... 295
39. Enzymatic Catalysis—Catalase .... 305
40. A Partial Chemical Analysis of Urine .... 311
41. Lipids .... 317
42. Blood Cholesterol Levels .... 327

## STUDY AIDS

1. Significant Figures ... 333
2. Formulas and Chemical Equations ... 335
3. Graphs ... 339
4. Use of the Mole in Chemical Calculations ... 343
5. Organic Chemistry—An Introduction ... 347

## EXERCISES

1. Significant Figures and Exponential Notation ... 351
2. Measurements ... 353
3. Names and Formulas I ... 355
4. Names and Formulas II ... 357
5. Names and Formulas III ... 359
6. Equation Writing and Balancing I ... 361
7. Equation Writing and Balancing II ... 363
8. Equation Writing and Balancing III ... 365
9. Graphical Representation of Data ... 367
10. Moles ... 369
11. Stoichiometry I ... 371
12. Gas Laws ... 373
13. Solution Concentrations ... 375
14. Stoichiometry II ... 377
15. Chemical Equilibrium ... 379
16. Oxidation-Reduction Equations I ... 381
17. Oxidation-Reduction Equations II ... 383
18. Hydrocarbons ... 385
19. Alcohols, Esters, Aldehydes, and Ketones ... 387
20. Functional Groups ... 389
21. Synthetic Polymers ... 391
22. Carbohydrates ... 393
23. Amino Acids and Polypeptides ... 395
24. Lipids ... 397
25. Molecular Models and Isomerism ... 399
26. Stereoisomerism—Optical Isomers ... 403

## APPENDICES

1. Suggested List of Equipment ... 409
2. List of Reagents Required and Preparation of Solutions ... 411
3. Special Equipment or Preparations Needed ... 418
4. Units of Measurements ... 422
5. Solubility Table ... 423
6. Vapor Pressure of Water ... 424
7. Boiling Points of Liquids ... 425

Periodic Table of the Elements ... 426
Table of Atomic Masses (Weights) Based on Carbon 12 ... 427
Names, Formulas, and Charges of Common Ions ... 428

# Preface

This manual is intended for the student who has not had a course in chemistry and needs a basic one-year course that includes the fundamentals of organic and biochemistry. The experiments are designed to be challenging but understandable to the student. Experimentation begins with simple laboratory techniques and measurements and progresses to relatively complex biochemical procedures. The forty-two experiments are graded in difficulty to keep pace with the expanding capability of the student. This number of experiments is sufficient for a full academic year and is also sufficient to allow the instructor reasonable flexibility in preparing the laboratory schedule.

Our major objectives are to afford the student 1) an introduction to laboratory experimentation, 2) increased knowledge of the capabilities and limitations of measurements, 3) familiarity with a variety of chemical reactions and the equations used to describe them, 4) opportunities to use instruments, 5) experience in collecting and processing numerical data, and 6) practice in drawing valid conclusions from experimental evidence. We have sought to establish a balance between descriptive and quantitative experiments. Eleven of the experiments include unknowns for student analysis.

This edition includes two new Exercises (Molecular Models and Isomerism, and Stereoisomerism—Optical Isomers). These Exercises also make good Experiments to introduce the student to the subject of molecular models and isomerism.

We have tried to devise a helpful and convenient format for both student and instructor. This format includes these special features.

1. A concise discussion of the important underlying principles is provided for each experiment. (This discussion deals with immediately pertinent material and is intended to supplement, not replace, the textbook.)
2. Experimental procedures are outlined in detail sufficient to permit the student to work with only general supervision.
3. The names and, wherever practicable, the formulas of all reagents needed are listed at the beginning of each experiment.
4. **Special safety precautions, when necessary, are indicated at the point where they are required in an experiment.**
5. For convenience in cross referencing, the letters and subtitles in the Procedure section correspond to those in the Report section of each experiment.
6. Five Study Aids deal with the important topics of a) significant figures, b) chemical formulas and equations, c) graphs and graphical presentation of data, d) the mole in chemical calculations, and e) an introduction to organic chemistry. Twenty-six Exercises afford practice in dealing with these topics.
7. For the convenience of the instructor and stockroom personnel, the Appendix provides an experiment-by-experiment listing of the special equipment and preparations needed. The Appendix also contains a list of suggested equipment for student lockers, suggested auxiliary equipment and a complete list of reagents and details on the concentration and preparation of needed solutions.

We are especially indebted to our colleagues in the chemistry departments of Mt. San Antonio College and California State University, Sacramento, for their many helpful comments and contributions—and to the many students of both institutions who provided such helpful feedback in the development of this manual.

M. Hein L.R. Best R.L. Miner J.M. Ritchey

# To the Student

Since your laboratory time is limited, it is important to carefully study and clearly understand the experiment scheduled for a laboratory period **before** you come to the lab. Without at least 1 hour of preparation before each lab—and this should be considered as a standing homework assignment—it is doubtful that the experiment and its report form can be satisfactorily completed in the allotted time.

Each of the experiments in this manual is composed of four parts:

1. Materials and Equipment—a list that includes the formulas of all compounds used.

2. Discussion—a brief discussion of the principles underlying the experiment.

3. Procedure—detailed directions for performing the experiment with safety precautions.

4. Report for Experiment—a form for recording data and observations, performing calculations, and answering questions.

Follow the directions in the procedure carefully, and consult your instructor if you have any questions. For convenience, the letters and subtitles in the report form have been set up to correspond with those in the procedure section of each experiment.

As you make your observations and obtain your data, record them on the report form. Try to use your time efficiently; when a reaction or process is occurring that takes considerable time and requires little watching, start working on other parts of the experiment, perform calculations, answer questions on the report form, or clean up your equipment.

Except when your instructor directs otherwise, you should do all the work individually. You may profit by discussing experimental results with your classmates, but in the final analysis you must rely on your own judgment in completing the report form.

## PREVENTING ACCIDENTS

Accidents in beginning chemistry laboratories are most frequently caused by:

1. Using improper techniques to insert and remove glass tubing from rubber stoppers;

2. Putting too much pressure on glassware;

3. Handling hot glassware and other equipment improperly;

4. Mixing incorrect reagents;

5. Using improper amounts of reagents;

6. Using improper techniques to dispose of laboratory wastes.

Accidents are mainly caused by carelessness, but you can help avoid them by closely following the precautions printed in this manual and those given by your instructor. At all times be concerned for your neighbors' safety as well as your own. You should be aware that chemical reactions may be dangerous, but if the proper precautions and techniques are used, none of the experiments in this book will be hazardous.

⚠ This symbol is to remind you to pay special attention to the precautions.

## LABORATORY RULES AND SAFETY PROCEDURES

1. Wear protective goggles or glasses at all times in the laboratory work areas.

2. Wear a laboratory apron to protect your clothing.

3. After completing the experiment, clean and put away your glassware and equipment. Clean your work area and make sure the gas and water are turned off.

4. Dispose of insoluble wastes such as filter paper, litmus paper, matches, and broken glass in the wastebasket, not in the sink. Dispose of solid chemicals and liquids as directed by your instructor. When liquids are poured into the sink, wash them down with plenty of water.

5. Do not take reagent bottles to your laboratory work area. Use test tubes, beakers, or paper to obtain chemicals from the dispensing area. Take small quantities of reagents. You can always get more if you run short.

6. Check carefully the label on each reagent bottle to be sure you have the correct reagent. The names of many substances appear similar at first glance.

7. To avoid possible contamination, never return unused chemicals to the reagent bottles. (See Laboratory Rule No. 4.)

8. Do not insert medicine droppers into reagent bottles. Instead pour a little of the liquid into a small beaker.

9. Be neat in your work; if you spill something, clean it up immediately.

10. Wash your hands anytime you get chemicals on them and at the end of the laboratory period.

11. Keep the balance and the area around it clean. Do not place chemicals directly on the balance pans; place a piece of weighing paper or a small container on the pan first, and then weigh your material. Never weigh an object while it is hot.

12. Do not heat graduated cylinders, burets, pipets, or bottles with a burner flame.

13. Do not look down into the open end of a test tube in which the contents are being heated or in which a reaction is being conducted.

14. Do not perform unauthorized experiments.

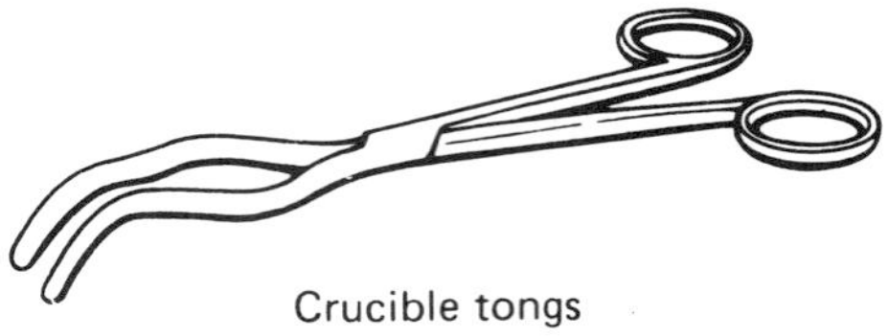

Crucible tongs

Ring support

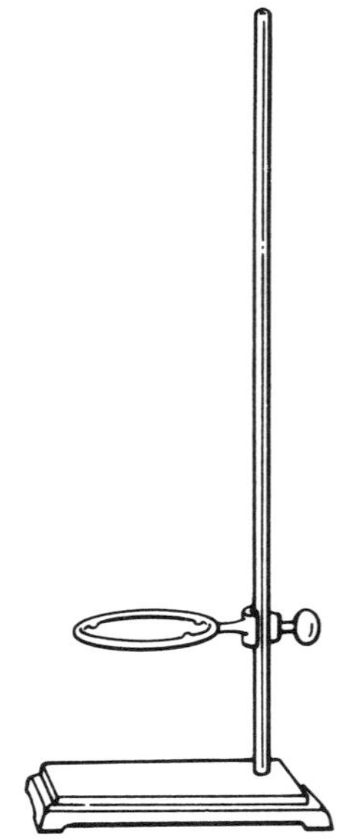

Ring stand

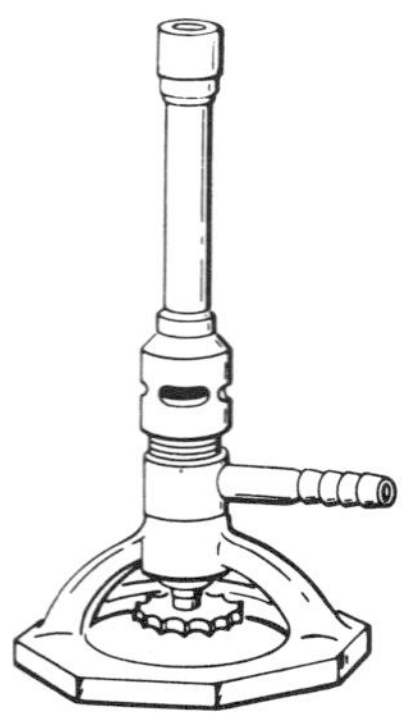

Bunsen burner
(Tirrill type)

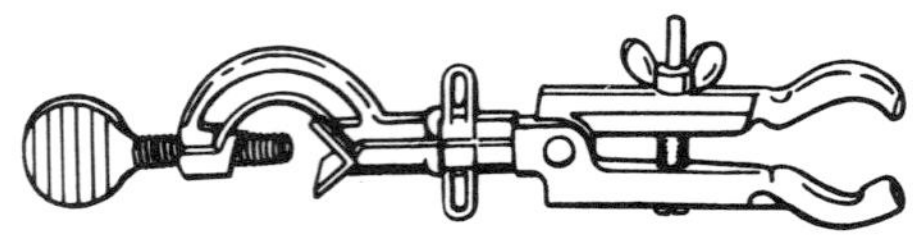

Utility clamp

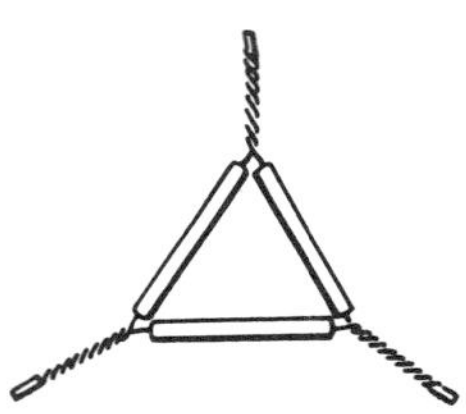

Clay triangle

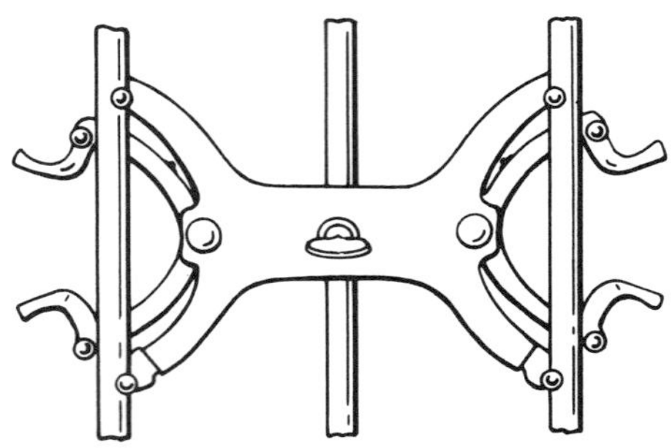

Buret clamp

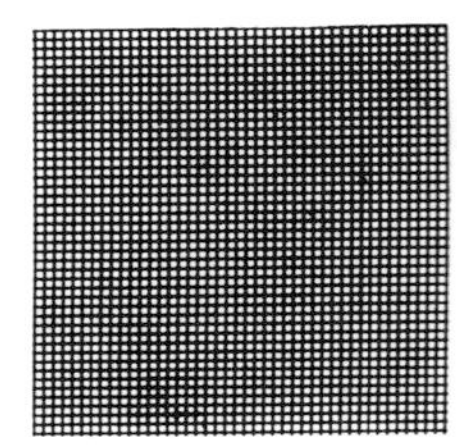

Wire gauze

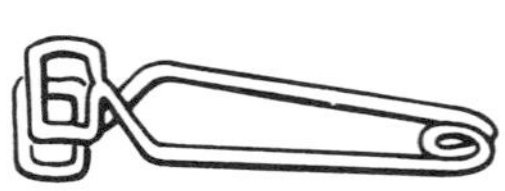

Test tube holder

Wing top
(flame spreader)

Evaporating dish

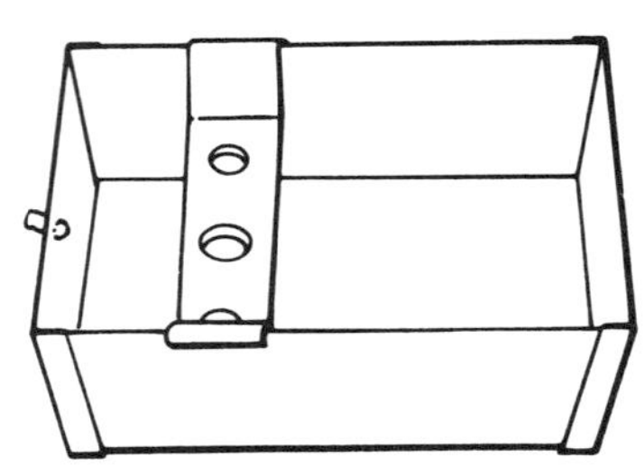

Pneumatic-trough

Watch glass

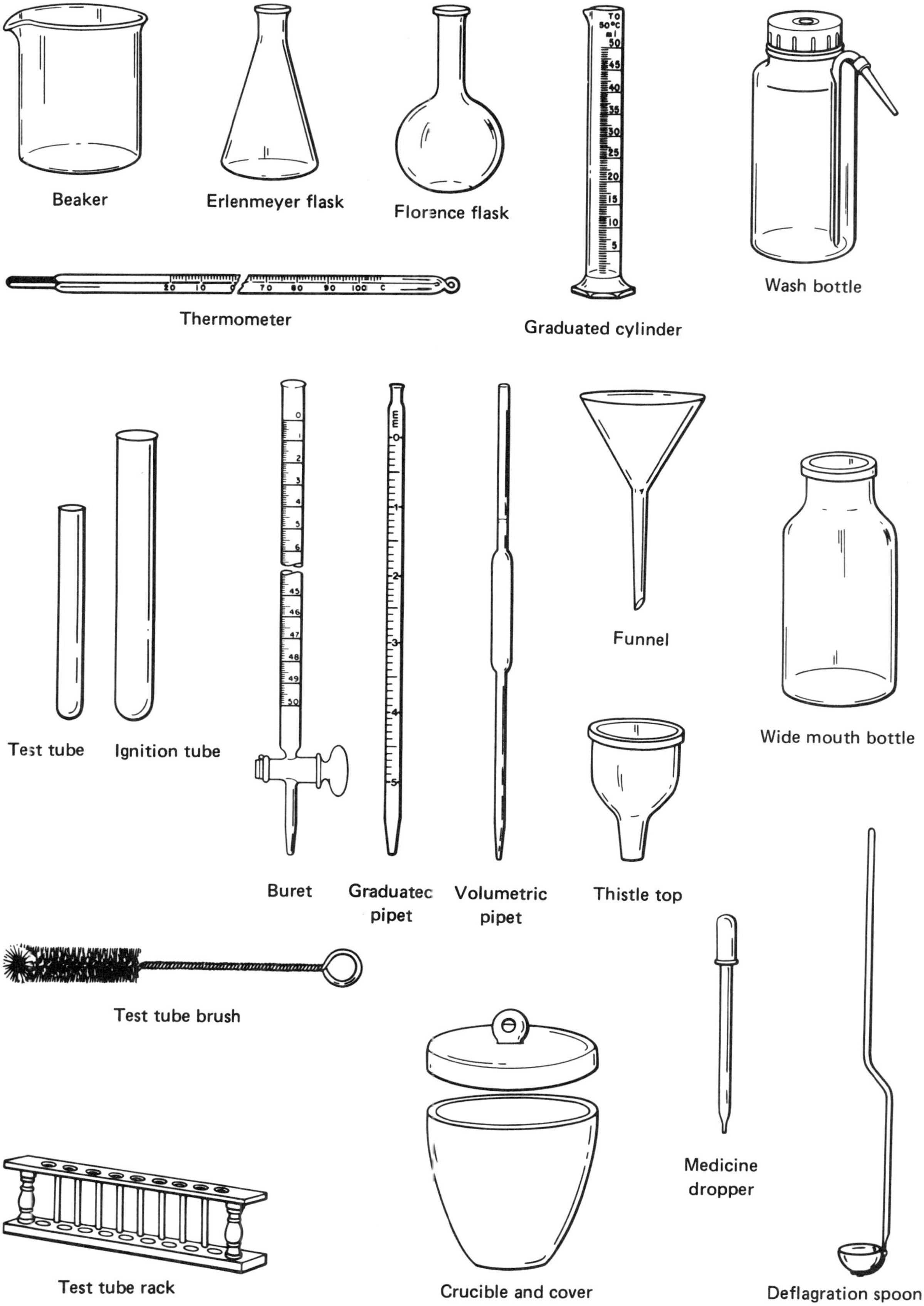
Beaker
Erlenmeyer flask
Florence flask
Graduated cylinder
Wash bottle
Thermometer
Test tube
Ignition tube
Buret
Graduated pipet
Volumetric pipet
Funnel
Thistle top
Wide mouth bottle
Test tube brush
Crucible and cover
Medicine dropper
Deflagration spoon
Test tube rack

# EXPERIMENT 1

# Laboratory Techniques

## MATERIALS AND EQUIPMENT

Solids: lead(II) chromate ($PbCrO_4$), potassium nitrate ($KNO_3$), and sodium chloride (NaCl). Liquid: glycerol. Solutions: 0.1 M lead(II) nitrate [$Pb(NO_3)_2$] and 0.1 M potassium chromate ($K_2CrO_4$). Ceramfab pad, 100 mL and 400 mL beakers, Bunsen burner, No. 1 evaporating dish, triangular file, funnel, wire gauze, filter paper, glass rod, clay triangle, 6 mm glass tubing.

## DISCUSSION AND PROCEDURE

**Wear protective glasses.**

### A. Laboratory Burners

Almost all laboratory burners used today are modifications of a design by the German chemist Robert Bunsen. In Bunsen's fundamental design, also widely used in domestic and industrial gas burners, gas and air are premixed by admitting the gas at relatively high velocity from a jet in the base of the burner. This rapidly moving stream of gas causes air to be drawn into the barrel from side ports and to mix with the gas before entering the combustion zone at the top of the burner.

The burner is connected to a gas cock by a short length of rubber or plastic tubing. With some burners the gas cock is turned to the **fully on** position when the burner is in use, and the amount of gas admitted to the burner is controlled by adjusting a needle valve in the base of the burner. In burners that do not have this needle valve, the gas flow is regulated by partly opening or closing the gas cock. With either type of burner **the gas should always be turned off at the gas cock when the burner is not in use** (to avoid possible dangerous leakage).

1. **Operation of the Burner.** Examine the construction of your burner (Figure 1.1) and familiarize yourself with its operation. A burner is usually lighted with the air inlet ports nearly closed. The ports are closed by rotating the barrel of the burner in a clockwise direction. After the gas has been turned on and lighted, the size and quality of the flame is adjusted by admitting air and regulating the flow of gas. Air is admitted by rotating the barrel; gas is regulated with the needle valve, if present, or the gas cock. Insufficient air will cause a luminous yellow, smoky flame; too much air will cause the flame to be noisy and possibly blow out. A Bunsen burner flame that is satisfactory for most purposes is shown in Figure 1.2; such a flame is said to be "nonluminous." Note that the hottest region is immediately above the bright blue cone of a well-adjusted flame.

### B. Glassworking

In laboratory work it is often necessary to fabricate simple items of equipment, making use of glass tubing and rubber stoppers. In working with glass tubing, improper techniques may result not only in an unsatisfactory apparatus but also in severe cuts and burns. Therefore the numbered instructions below should be studied carefully. Prepare the following list of items (illustrated in

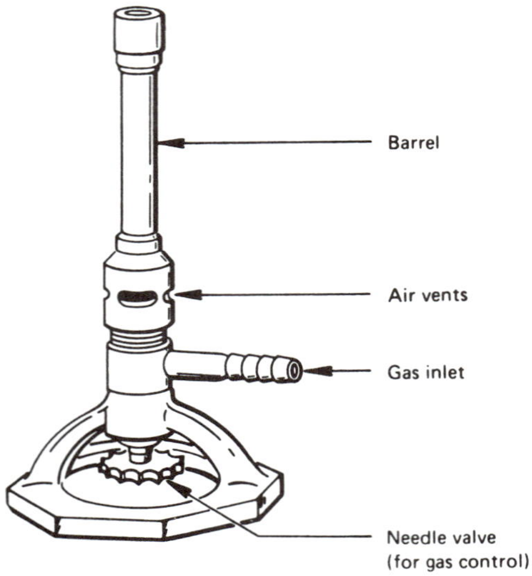


Figure 1.1. Bunsen burner (Tirrill type)

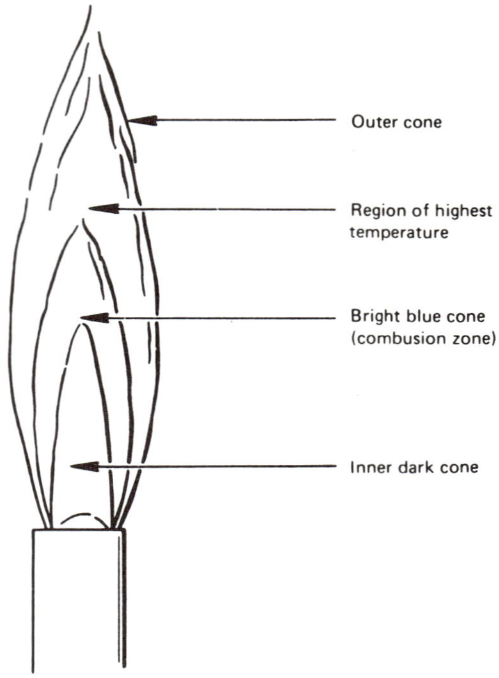


Figure 1.2. Bunsen burner flame

Figure 1.3), using 6 millimeter (mm) glass tubing and rod.

**Two** straight tubes, one 24 centimeters (cm) long, the other 12 cm (Figure 1.3A).

**Two** right-angle bends (Figure 1.3B).

**One** delivery tube (Figure 1.3C).

**Two** buret tips (Figure 1.3D). (Optional)

**One** stirring rod if there is none in your locker (Figure 1.3E).

This equipment will be used in future experiments. After it has been completed and approved by your instructor, store it in your locker.

1. **Cutting Glass Tubing.** (See Figure 1.4) Mark the tube with a pencil or ball-point pen at the point where it is to be cut. Grasp the tubing about 1 cm from the mark and hold it in position on the laboratory table. Hold the file by the tang (or handle) end and, pressing the edge of the file firmly against the glass at right angles to the tubing, make a scratch on the tubing by pushing the file away from you. If the file is in good condition a single stroke should suffice. Several strokes may be required if the file is dull, but if more than one stroke is needed, all must follow the same path so that only one scratch mark is present on the tubing. The scratch need not be very deep or very long, but it should be clearly defined.

Grasp the tubing with your thumbs together directly opposite the scratch mark (see Figure 1.4). Now apply pressure with the thumbs as though bending the ends toward your body while at the same time exerting a slight pull on the tubing. A straight, clean break should result. Use the flat side of your file to remove any sharp projections from the ends of the cut tubing.

2. **Fire-Polishing Glass. Fire polishing** is the process of removing the sharp edges of glass by heating the tubing in a burner flame.

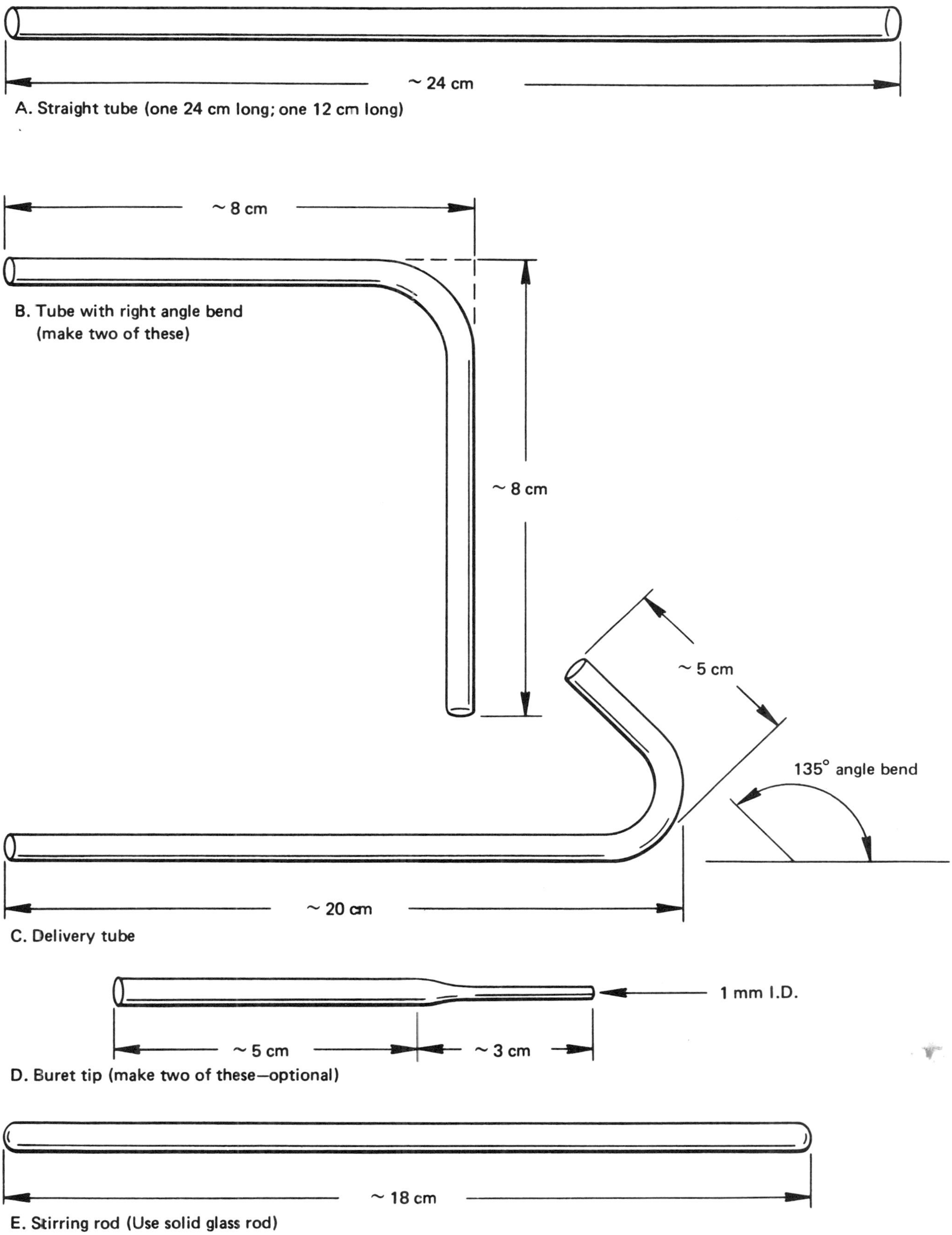


**Figure 1.3. Glassware (Illustrations are not to scale)**

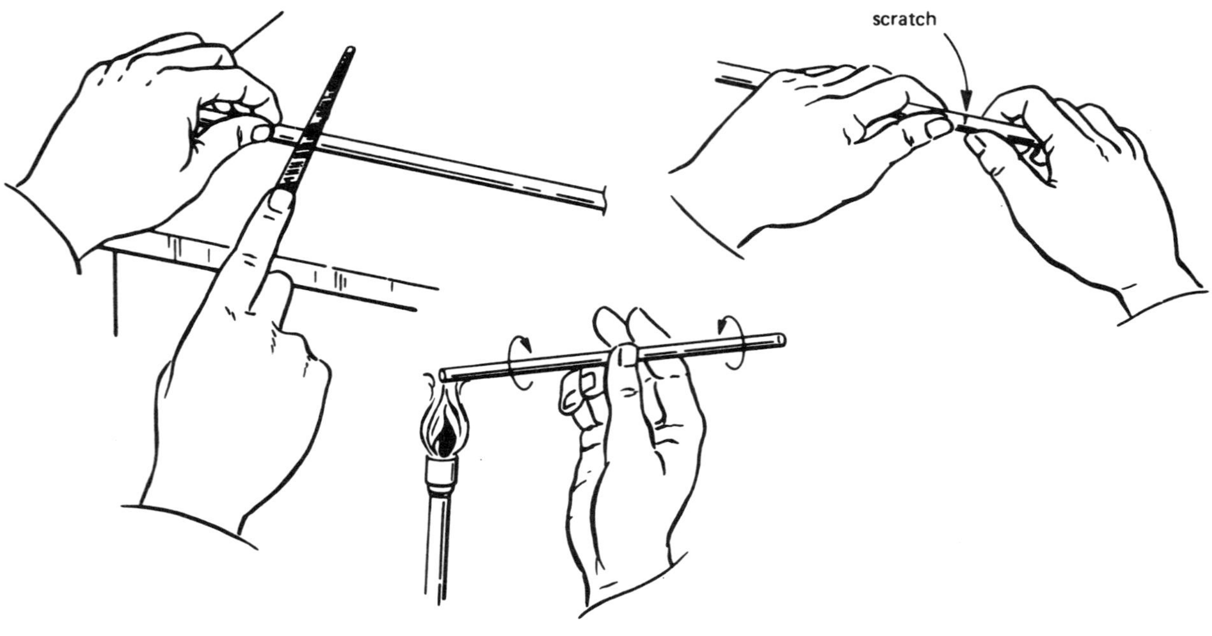


**Figure 1.4. Cutting and fire-polishing glass tubing**

While continuously rotating the tubing, heat the end in the hottest part of the flame until the sharp edges are smooth. Be careful not to heat the tubing too much lest the bore be constricted. When the fire polishing is completed, put the tubing on a Ceramfab pad to cool. Your instructor will have some examples of properly fire-polished tubing available for your inspection.

> NOTE: The Ceramfab pad protects the hot glass from sudden chilling (thermal shock) and the table top from injury. More important, it is an excellent safety device. If hot objects are always placed on the pad and allowed to cool, then picked up with caution, one is less likely to get burned.

Laboratory stirring rods are easily made by cutting glass rod in the same way described for tubing and fire-polishing the ends until they are smooth and rounded.

**Whenever glass is cut it must be fire-polished in order to avoid personal injury.**

3. **Bending Glass Tubing.** Put the wing top (flame spreader) on your burner and adjust the flame so that a sharply defined region of intense blue color is visible. Grasp the tubing to be bent at both ends and hold it in the flame lengthwise just above the zone of intense blue color. Continuously rotate the tubing in the flame until it has softened enough to bend easily (Figure 1.5). Remove the tubing from the flame, bend to the desired shape, and set aside to cool on the Ceramfab pad. If the bend is not satisfactory, discard it and repeat the work with a new piece.

If a bend is to be made where one arm of tubing is too short to hold in the hand while heating (Figure 1.3C), follow one of two procedures: (1) Proceed as in the above paragraph, using enough tubing to handle it from both ends; then cut to size and fire-polish it after the bend is completed. (2) Heat a piece of tubing of proper size, holding it at one end and rotating it until it is soft in the region to be bent; then remove it from the flame and bend by grasping with tongs or by inserting the tang of a file into the hot end of the tubing.

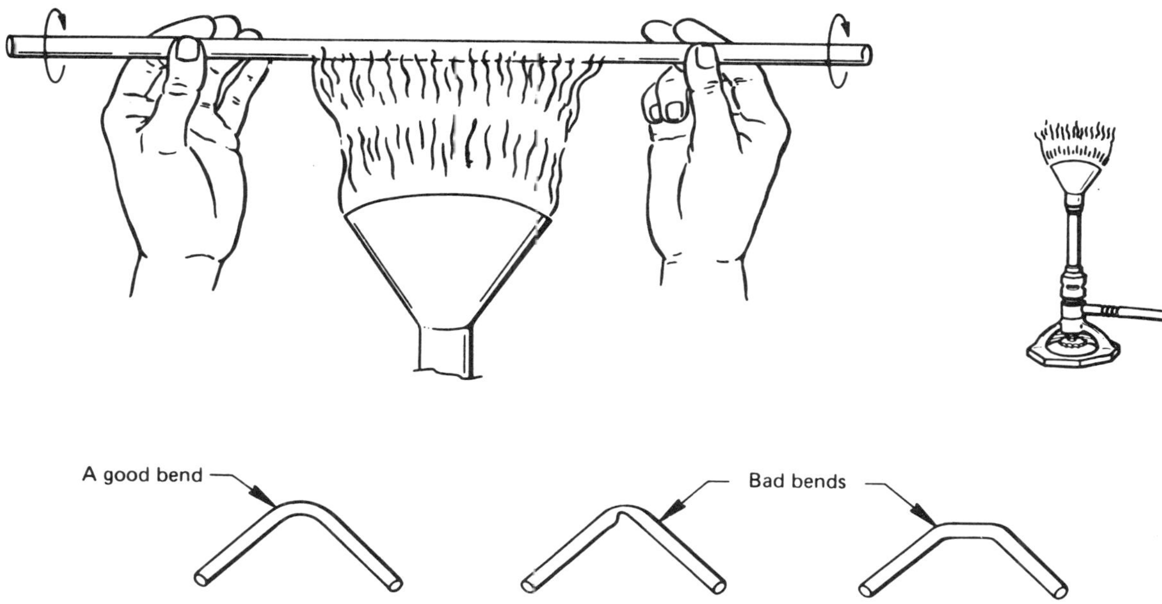


**Figure 1.5. Heating glass tubing with a wing top**

4. **Preparing Buret Tips (Jets).** A buret tip (or jet) (Figure 1.3D) is prepared as follows: Heat a small section in the center of a 14 cm length of tubing while rotating it in the hottest portion of the burner flame (without wing top) until the tubing is **very** soft. Remove from the flame and **slowly** pull the ends away from each other while holding the tubing in a **vertical** position. After cooling, cut the tips to the desired dimensions. Fire-polish all edges.

5. **Inserting Glass Tubing into a Stopper.**

> ⚠ CAUTION: To avoid personal injury it is important when inserting glass tubing into a rubber stopper to hold the tubing close to the end to be inserted. It is also good safety practice to protect your hands with a towel when inserting or removing glass tubing from rubber stoppers.

Lubricate the hole in the stopper with glycerol (glycerine), using a stirring rod to make sure that the lubricant actually gets into the hole. Grasp the **fire-polished** tubing about **1 cm** from the end to be inserted. Holding the **lubricated** stopper in the other hand, start the tubing into the hole by gently twisting it, and gradually work it all the way through the stopper. Be sure to grip the tubing at a point **not more than 1 cm** from the stopper at all times when making the insertion. Gripping at greater distances and twisting will break the tubing and probably cause personal injury.

The end of the tube should protrude at least 5 mm from the stopper, so that the free passage of fluids is not prevented by flaps of rubber. After insertion make sure that the tube is not plugged, and remove the excess glycerol either by washing or with a bit of filter paper. Note the lubricating properties of glycerol by rubbing a drop between your thumb and forefinger.

6. **Removing Glass Tubing from Stoppers.** Tubing should be removed from stoppers at the end of the laboratory period. Be careful to grip the tubing at a point **very close** to the stopper when twisting. If allowed to stand for several days the stopper may stick to the glass and be difficult to remove. In the event of sticking, do not use "strong arm" methods. A cork borer (No. 3 for 6 mm

tubing) may be inserted between the rubber and glass to help remove the tubing from the stopper but consult your instructor before attempting this procedure.

## C. Evaporation

**Evaporation** is one of the processes used to separate a dissolved solid from a liquid.

1. **Preparing a Solution.** Cover the bottom of a test tube with a small quantity of sodium chloride. Add distilled water until the tube is about one-quarter full and shake until the salt is dissolved.

2. **Evaporating the Solution.** Pour the salt solution into an evaporating dish. Place the evaporating dish on a beaker of boiling water, as illustrated in Figure 1.6. Continue boiling the water in the beaker—replenishing, if necessary—until all of the water has evaporated from the solution in the evaporating dish, leaving the original salt as the residue.

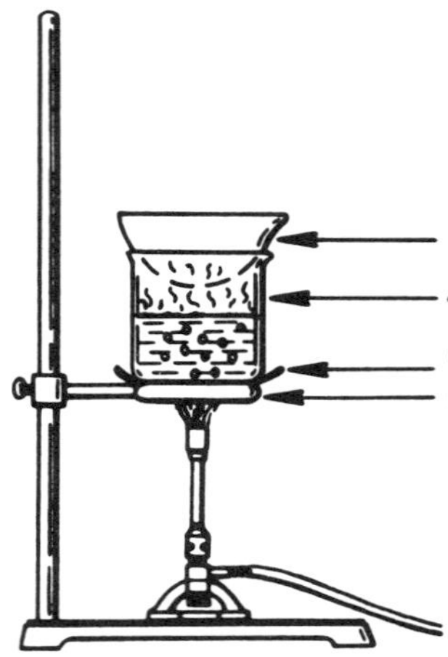


**Figure 1.6. Evaporation on a simple steam bath**

## D. Filtration

The process of separating suspended solids from liquids by means of filters is called **filtration.** Certain chemical reactions occurring in solution yield insoluble solids called **precipitates.** In the laboratory these precipitates are generally separated from the mixture by filtering them out on filter paper. The liquid that passes through the paper is the **filtrate**; the solid remaining on the filter paper is the **residue** or **precipitate.**

1. **Forming a Precipitate.** At the reagent shelf fill a test tube about one-half full of lead(II) nitrate solution. Fill a second tube about one-quarter full of potassium chromate solution. These solutions contain lead(II) nitrate and potassium chromate, each dissolved in water. Pour the lead(II) nitrate solution into a 100 milliliter (mL) beaker. Now slowly pour the potassium chromate solution into the beaker, stir, and observe the results. The chemical reaction that occurred formed potassium nitrate and lead(II) chromate, one of which is a yellow precipitate.

2. **Filtering the Precipitate.** Prepare a filter as follows (see Figure 1.7):

   (a) Fold a circle of filter paper in half. Fold in half again and tear off one corner of one folded edge.

(b) Fit the opened filter cone into a funnel, placing the torn edge next to the glass. Wet with distilled water and press the top edge of the paper against the funnel, forming a seal.

Set up the funnel for filtering as shown in Figure 1.8.

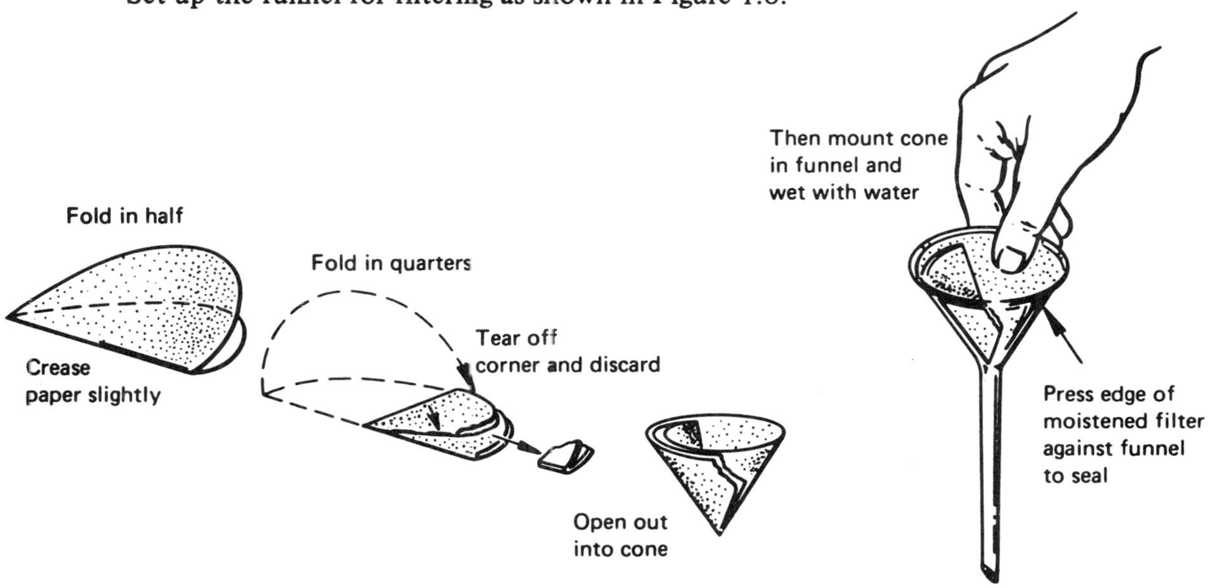


**Figure 1.7. Folding and mounting filter paper**

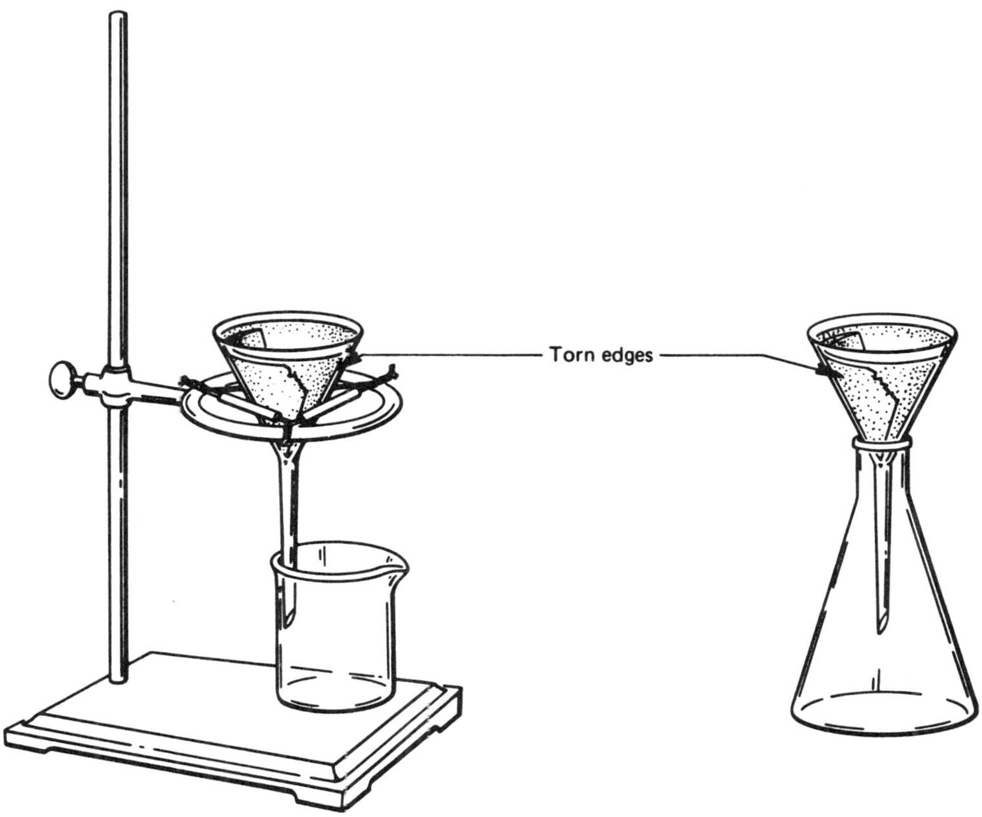


**Figure 1.8. Set ups for filtration**

Stir the mixture of potassium nitrate and lead(II) chromate and slowly pour it down the stirring rod into the funnel (Figure 1.9). Be careful not to overfill the paper filter cone. After the filtration is completed take the filter cone to the reagent shelf and determine whether the precipitate is potassium nitrate or lead(II) chromate by comparing it with solid samples of these two substances. Dispose of the precipitate and the filter paper in the special container provided by your instructor.

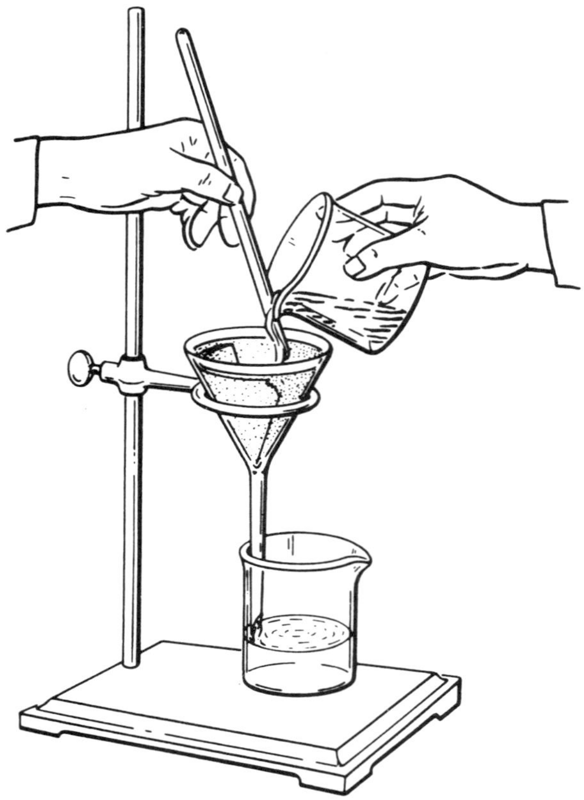

**Figure 1.9. Pouring a solution down a stirring rod**

# *EXPERIMENT 2*

# Measurements

## MATERIALS AND EQUIPMENT

Solids: sodium chloride (NaCl) and ice. Balance, ruler, thermometer, solid object for density determination.

## DISCUSSION

Chemistry is an experimental science, and measurements are fundamental to most of the experiments. It is important to learn how to make and use these measurements properly.

### Metric System

In science the metric system is used almost exclusively. The unit of length is the meter; of mass, the gram; and of volume, the liter. The meter, the gram, and the liter are related to larger and smaller units by multiples of 10. "Deci," "centi," and "milli" indicate units that are 1/10, 1/100, 1/1000, respectively, of the original unit. The prefix "kilo" is used to designate a unit 1000 times larger than the fundamental unit.

Deci = 0.1
Centi = 0.01
Milli = 0.001
Kilo = 1000.

Examples: decigram, centimeter, milliliter, milligram, kilometer, kilogram. It is valuable to know that 1 milliliter (mL) equals 1 cubic centimeter ($cm^3$ or cc).

It will at times be necessary to convert from the American system to the metric system or vice versa. Many conversion factors are available but the following are usually sufficient:

1 pound (lb) = 453.6 grams (g)
1 inch (in.) = 2.54 centimeters (cm)
1 quart (qt) = 946 milliliters (mL)

The Celsius (or centigrade) temperature scale is commonly used in scientific work. On this scale the freezing point of water is designated 0°C, the boiling point 100°C.

### Precision of Measurements

Scientific measurements must be as precise as possible, which often means estimating between the smallest scale divisions on the instrument being used. Suppose we are measuring a piece of wire, using the metric scale on a ruler that is calibrated in tenths of centimeters. One end of the wire is placed at exactly zero cm and the other end falls somewhere between 6.3 cm and 6.4 cm. Since the distance between 6.3 cm and 6.4 cm is very small, it is difficult to determine the next digit exactly. We might estimate the length of the wire as 6.34 cm, though a more precise instrument might show it was 6.36 cm. Despite the error, 6.34 cm is a more precise, and thus more valuable, measurement than 6.3 cm. If the wire had come to exactly 6 cm, reporting the length as 6 cm would be an error, for it would indicate only that the length is closer to 6 cm than to 5 or 7 cm. What we really mean is that, as closely as we can read it, it is exactly 6 cm. But "exactly" implies perfection; that is, 6.000. . . cm. So we must write the number in such a way that it tells how closely we can read it. On this scale we can estimate to 0.01 cm, so our length should be reported as 6.00 cm.

### Significant Figures

The result of multiplication, division, or other mathematical manipulation cannot be more precise than the least precise measurement used in the calculation. For instance, suppose we have an object that weighs 3.62 lb and we want to calculate the mass in grams. Longhand multiplication of 3.62 lb times 453.6 g/lb gives 1,642.032 g. To report 1,642.032 g as the mass is absurd, for it implies a precision far beyond that of the original measurement. Although the conversion factor has four significant figures, the **mass** in pounds has only three significant figures. Therefore the answer should have only three significant figures; that is, 1,640 g. In this case the zero cannot be considered significant. This value can be more properly expressed as $1.64 \times 10^3$ g. For a more comprehensive discussion of significant figures see Study Aid 1 on page 333.

### Precise Quantities versus Approximate Quantities

In conducting an experiment it is often unnecessary to measure an exact quantity of material. For instance, directions might state, "Weigh about 2 g of sodium sulfite." This instruction indicates that the measured quantity of salt should be approximately 2 g; for example, somewhere between 1.8 and 2.2 g. To weigh exactly 2.00 g only wastes time, since the directions call for weighing "about 2 g."

Suppose the directions read, "Weigh about 2 g of sodium sulfite **to the nearest 0.001 g.**" This instruction does not imply that the amount is 2.000 g but only **about** 2 g and that it should be weighed **accurately** to 0.001 g. Therefore four different students might weigh their samples and obtain 2.141 g, 2.034 g, 1.812 g, and 1.937 g, respectively, and each would have satisfactorily followed the directions.

### Temperature

The simple act of measuring a temperature with a thermometer can easily involve errors. When measuring the temperature of a liquid, one can minimize possible error by observing the following procedures: Hold the thermometer away from walls of the container; allow sufficient time for the thermometer to reach equilibrium with the liquid; be sure the liquid is adequately mixed.

When converting from degrees Celsius to Fahrenheit or vice versa, we make use of the following formulas:

$$°C = \frac{(°F - 32)}{1.8} \quad \text{or} \quad °F = (1.8 \times °C) + 32$$

**Example Problem:** Convert 70.0° F to degrees Celsius:

$$°C = \frac{(70.0°F - 32)}{1.8}$$

$$°C = \frac{38.0}{1.8} = 21.11°C = 21.1°C$$

This example shows not only how the formula is used but also a typical setup of the way chemistry problems should be written. It shows how the numbers are used, but does not show the longhand multiplication and division, which should be worked out on scratch paper or by calculator. Note that the answer was changed from 21.11°C to 21.1°C because the initial temperature, 70.0°F, had only three significant figures. The 1.8 and 32 in the formulas are exact numbers and have no effect on the number of significant figures.

### Mass (Weight)

The directions in this manual are written for a 0.001 gram precision balance, but all the experiments can be performed satisfactorily using a 0.01 gram or 0.0001 gram precision balance. Your instructor will give specific directions on how to use the balance, but the following precautions should be observed:

1. Never place chemicals directly on the balance pan; first place them on a weighing paper or in a container.
2. Clean up any materials you spill on or around the balance.
3. Before moving objects on and off the pan, be sure the balance is in the "arrest" position.
4. When you leave the balance, return the balance to the "arrest" position.
5. Never try to make adjustments on a balance. If it seems out of order, tell your instructor.

### Volume

Beakers and flasks are marked to indicate only approximate volumes. You will usually make measurements of volume in a graduated cylinder. When observing a volume in a graduated cylinder, read the point on the graduated scale that coincides with the bottom of the curved surface—called the **meniscus**—of the liquid (see Figure 2.1). Volumes measured in your graduated cylinder should be estimated and recorded to the nearest 0.1 mL.

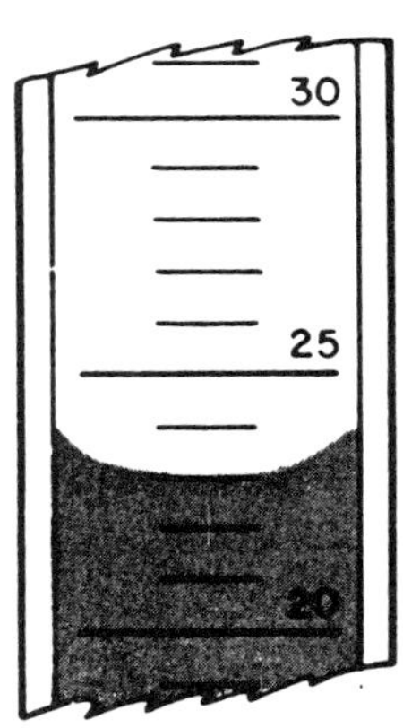


**Figure 2.1. Read the bottom of the meniscus. The volume is 23.0 mL.**

**Density**

Density is an inherent property of a substance and is useful in identifying the substance. **Density** is the ratio of the mass of a substance to the volume occupied by that mass; it is the mass per unit volume and is given by the equations

$$\text{Density} = d = \frac{\text{Mass}}{\text{Volume}} = \frac{m}{V} = \frac{g}{mL} \quad \text{or} \quad \frac{g}{cm^3}$$

In calculating density it is important to make correct use of units and mathematical setups.

**Example Problem:** An object weighs 283.5 g and occupies a volume of 14.6 mL. What is its density?

$$d = \frac{m}{V} = \frac{283.5\ g}{14.6\ mL} = 19.4\ g/mL$$

Note that all the operations involved in the calculation are properly indicated and that all units are shown. If we divide grams by milliliters, we get an answer in grams per milliliter.

The volume of an irregularly shaped object is usually measured by the displacement of a liquid. An object completely submerged in a liquid displaces a volume of the liquid equal to the volume of the object.

Measurement data and calculations must always be accompanied by appropriate units.

## PROCEDURE

**Wear protective glasses.**

Record your data on the report sheet as soon as obtained.

### A. Temperature

Record all temperatures to the **nearest 0.1°C.**

1. Fill a 400 mL beaker half full of tap water. Place your thermometer in the beaker. Give it a minute to reach thermal equilibrium. Keeping the thermometer in the water and holding the tip of the thermometer away from the glass, read and record the temperature.

2. Fill a 150 mL beaker half full of tap water. Set up a ring stand, with the ring 25 to 30 cm above the base. Support the beaker on the ring with a wire gauze. Heat the water to boiling. Read and record the temperature of the boiling water, being sure to hold the thermometer away from the bottom of the beaker.

3. Fill a 250 mL beaker one-fourth full of tap water and add a 100 mL beaker of crushed ice. Without stirring, place the thermometer in the beaker, resting it on the bottom. Wait at least 1 minute, then read and record the temperature. Now stir the mixture for about 1 minute. If almost all the ice melts, add more. Holding the thermometer off the bottom, read and record the temperature. Save the mixture for Part 4.

4. Weigh approximately 4 to 6 g of sodium chloride and add it to the ice-water mixture. Stir for 1 minute, adding more ice if needed. Read and record the temperature.

## B. Mass

Using the balance provided, do the following, recording all the masses **to the precision specified.**

1. Weigh a 400 mL beaker to the nearest 0.1 g.

2. Weigh a 250 mL Erlenmeyer flask to the nearest 0.1 g.

3. Weigh a piece of weighing paper (or filter paper if no weighing paper is available) to the nearest 0.01 g.

4. Use the weighing paper from B.3 and weigh approximately 2 g of sodium chloride to the nearest 0.01 g. Remember that chemicals are never placed directly on the balance pan, so a weighing paper must be used.

## C. Length

Using a ruler, make the following measurements in both inches and centimeters; measure to the nearest 1/16th inch and to the nearest 0.1 centimeter:

1. Measure the external height of a 400 mL beaker.

2. Measure the total length of a test tube.

## D. Volume

Using the graduated cylinder most appropriate, measure the following volumes to the nearest 0.1 mL:

1. Fill a test tube to the brim with water and measure the volume of the water.

2. Fill a 125 mL Erlenmeyer flask to the brim with water and measure the volume of the water.

3. Pour 5.0 mL of water into a test tube. With a ruler, measure the height (in centimeters) of this water. In the future you will often find it convenient to estimate this volume simply by observing the height of the liquid in the test tube.

4. Pour 10. mL of water into a test tube. Measure the height of this volume of water also.

## E. Density

Estimate and record all volumes to the nearest 0.1 mL. Make all weighings to the highest precision of the balance. Note that you must supply the units for the measurements and calculations in this section.

1. **Density of Water.** Weigh a clean, dry 50 mL graduated cylinder and record its mass. (Graduated cylinders should never be dried over a flame.) Fill with distilled water to 50.0 mL. Use a medicine dropper to adjust the meniscus to the 50.0 mL mark. Record the volume. Reweigh and calculate the density of water.

2. **Density of a Rubber Stopper.** Weigh a solid No. 1 or No. 2 stopper. Fill the 50 mL graduated cylinder with tap water to approximately 25 mL; read and record the exact volume. Carefully

place the rubber stopper into the graduated cylinder so that it is submerged. Read and record the volume. Calculate the density of the rubber stopper.

3. **Density of a Solid Object.** Obtain a solid object from your instructor. Record the object's number (and letter, if it has one). Determine the density of your solid by following the procedure given in Part 2 for the rubber stopper. To avoid the possibility of breakage, incline the graduated cylinder at an angle and slide, rather than drop, the solid into it.

Return the solid object to your instructor.

# EXPERIMENT 3

# Preparation and Properties of Oxygen

## MATERIALS AND EQUIPMENT

Solids: candles, magnesium (Mg) strips, manganese dioxide ($MnO_2$), fine steel wool (Fe), roll sulfur (S), wood splints. Solution: 9 percent hydrogen peroxide ($H_2O_2$)*. Deflagration spoon, pneumatic trough, 20 to 25 cm length rubber tubing, 25 × 200 mm ignition tube, five wide-mouth (gas-collecting) bottles, five glass cover-plates. Demonstration supplies: (1) cotton, sodium peroxide ($Na_2O_2$); (2) steel wool, 25 × 200 mm test tube; (3) Hoffman electrolysis apparatus.

*If Procedure A-1 is used, solid reagent-grade potassium chlorate ($KClO_3$) is needed.

## DISCUSSION

Oxygen is the most abundant and widespread of all the elements in the earth's crust. It occurs both as free oxygen gas and combined in compounds with other elements. Free oxygen gas is **diatomic** and has the formula $O_2$. Oxygen is found combined with more elements than any other single element, and it will combine with all the elements except some of the noble gases. Water is 88.9 percent oxygen by mass. The atmosphere is about 21 percent oxygen by volume. Oxygen gas is colorless and odorless, and is only very slightly soluble in water, a property important to its collection in this experiment.

Oxygen may be obtained by decomposing a variety of oxygen-containing compounds. Some of these are mercury(II) oxide (HgO, mercuric oxide), lead(IV) oxide ($PbO_2$, lead dioxide), potassium chlorate ($KClO_3$), potassium nitrate ($KNO_3$), hydrogen peroxide ($H_2O_2$), and water.

In this experiment several bottles of oxygen will be collected. The oxygen is produced either by decomposing hydrogen peroxide (Procedure A) or by decomposing potassium chlorate (Procedure A-1). Your instructor will tell you which procedure to use.

### A. Decomposing Hydrogen Peroxide

Hydrogen peroxide decomposes very slowly at room temperature. The rate of decomposition is greatly increased by adding a catalyst, manganese dioxide. These equations represent the chemical changes that occur.

**Word Equation:** Hydrogen peroxide ⟶ Water + Oxygen

**Formula Equation:** $2\,H_2O_2 \xrightarrow{MnO_2} 2\,H_2O + O_2\uparrow$

### A-1. Decomposing Potassium Chlorate

Potassium chlorate melts at 368°C and decomposes at temperatures above 400°C. The rate of decomposition is rather slow but can be accelerated by adding a catalyst—manganese dioxide.

Although manganese dioxide contains oxygen, it is not decomposed under the conditions of this experiment. The following equations show the chemical changes that occur.

**Word Equation:** Potassium chlorate $\xrightarrow{\text{Heat}}$ Potassium chloride + Oxygen

**Formula Equation:** $2\,KClO_3 \xrightarrow[MnO_2]{\Delta} 2\,KCl + 3\,O_2\uparrow$

The oxygen is collected by a method known as the **downward displacement of water.** The gas is conducted from a generator to a bottle of water inverted in a pneumatic trough. The oxygen, which is only very slightly soluble in water, rises in the bottle and forces the water out.

One outstanding and important property of oxygen is its ability to support combustion. During combustion oxygen is consumed but does not burn, and this ability to support combustion is one test for oxygen. Compounds containing oxygen and one other element are known as **oxides.** Thus when elements such as sulfur, hydrogen, carbon, and magnesium burn in air or oxygen, they form sulfur dioxide, hydrogen oxide (water), carbon dioxide, and magnesium oxide, respectively. These chemical reactions may be represented by equations; for example:

**Word Equation:** Sulfur + Oxygen $\longrightarrow$ Sulfur dioxide

**Formula Equation:** $S + O_2 \longrightarrow SO_2$

See Study Aid 2, page 335, for a discussion of writing formulas and chemical equations.

## PROCEDURE

### A. Generation of Oxygen from Hydrogen Peroxide

⚠ PRECAUTIONS:

1. **Wear protective glasses.**
2. If hydrogen peroxide solution gets on your skin, wash it off with water.

Assemble the apparatus shown in Figure 3.1. It consists of a 250 mL Erlenmeyer flask, two-hole stopper, thistle tube, glass right-angle bend (Figure 1.3B), glass delivery tube with 135 degree

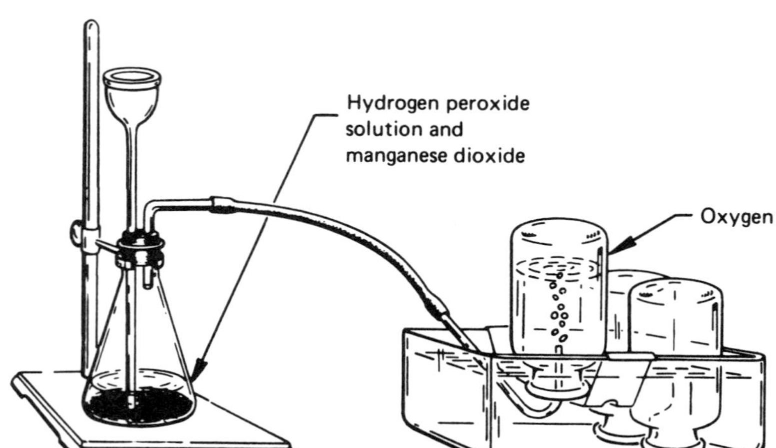


**Figure 3.1. Preparing oxygen by decomposing hydrogen peroxide.**

bend (Figure 1.3C), and a 20-25 cm length of rubber tubing. The thistle tube should be at least 24 cm (~10 in.) long and be inserted in the rubber stopper so that there is about 3 mm (1/8 in.) clearance between the end of the tube and the bottom of the flask with the stopper in place. **Remember to use glycerol when inserting the glass tubing into the rubber stopper.**

Fill a pneumatic trough with water until the water level is just above the removable shelf. Then completely fill five wide-mouth bottles with water. Transfer each bottle to the pneumatic trough by covering its mouth with a glass plate, inverting it, and lowering it into the water. Remove the glass plate below the water level. Place two bottles on the shelf in the trough leaving the other three standing ready for transfer to the shelf when needed.

Using a spatula, put a pea-sized quantity of manganese dioxide ($MnO_2$) in the generator flask. Replace the stopper and make sure that all glass-rubber connections are tight. Add 25 mL of water to the flask through the thistle tube. **Make sure that the end of the thistle tube is covered with water (to prevent escape of oxygen gas through the thistle tube).** Using a 50 mL graduated cylinder, measure out 50 mL of 9 percent hydrogen peroxide solution.

To start the generation of oxygen, pour 5 to 10 mL of the peroxide solution into the thistle tube. If all of the peroxide solution does not run into the generator, momentarily lift the delivery tube from the water in the trough. Immediately replace the end of the delivery tube under water and in the mouth of the first bottle to collect the gas. Continue generating oxygen by adding an additional 5 to 10 mL portion of hydrogen peroxide solution whenever the rate of gas production slows down markedly. When one bottle is filled with gas, immediately start filling the next bottle.

Cover the mouth of each gas-filled bottle with a glass plate before removing it from the water. Store each bottle mouth upward without removing the glass plate; the oxygen will not readily escape since it is slightly more dense than air. Note which bottle of gas was collected first and continue until a total of five bottles of gas have been collected.

### A-1. Alternate Method. Generation of Oxygen from Potassium Chlorate

⚠ PRECAUTIONS:

1. **Wear protective glasses.**

2. Be certain to securely clamp the generator tube at an angle, as shown in Figure 3.2. **Hot potassium chlorate is an explosion hazard when in contact with the rubber stopper.**

3. Remove the delivery tube from the water before you stop heating the potassium chlorate mixture.

4. Potassium chlorate is a fire hazard. Dispose of any excess or spilled potassium chlorate as directed by your instructor. **Do not throw potassium chlorate in the trash cans.**

Assemble the apparatus shown in Figure 3.2. It consists of a clean, **dry 25 × 200 mm ignition tube** fitted with a rubber stopper containing a piece of glass tubing with a right-angle bend (Figure 1.3B). The right-angle tubing is attached to a 20 to 25 cm piece of rubber tubing which in turn is attached to a glass delivery tube with a 135-degree bend (Figure 1.3C). Be sure the rubber tubing fits tightly on the glass tubing.

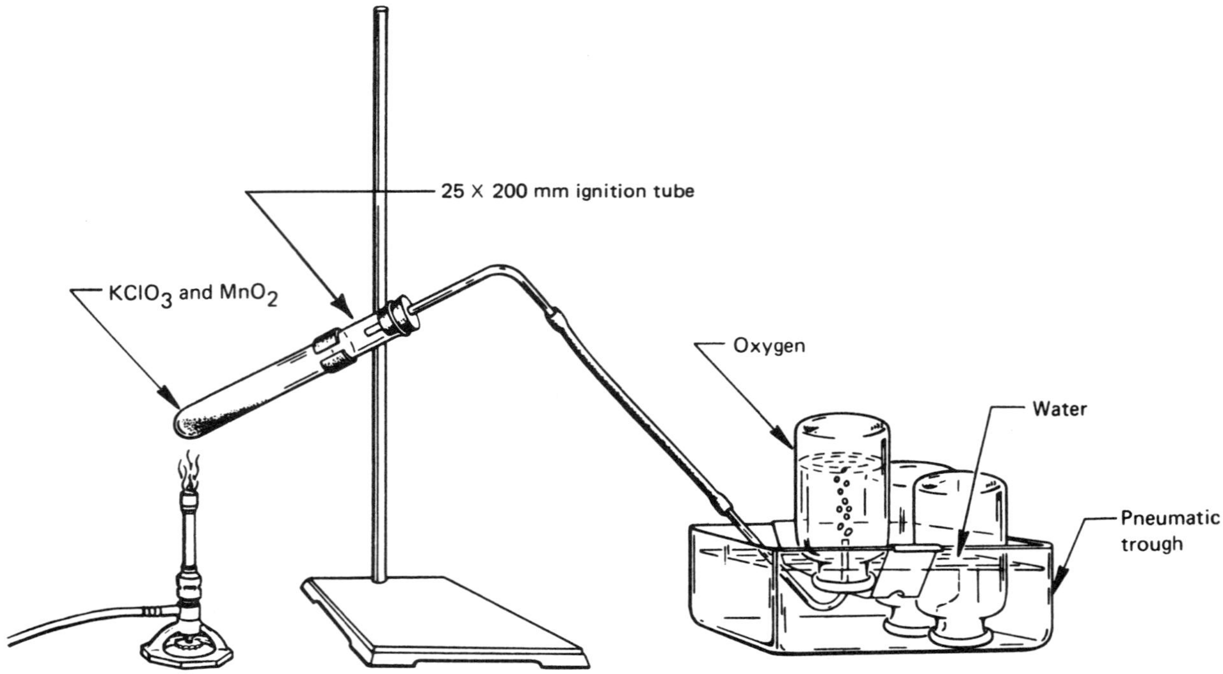


**Figure 3.2. Preparing oxygen by decomposing potassium chlorate.**

Fill a pneumatic trough with water until the water level is just above the removable shelf. Then completely fill five wide-mouth bottles with water. Transfer each bottle to the pneumatic trough by covering its mouth with a glass plate, inverting it, and lowering it into the water. Remove the glass plate below the water level. Place two bottles on the shelf in the trough leaving the other three standing ready for transfer to the shelf when needed.

Weigh approximately 12 g of potassium chlorate and 2 g of manganese dioxide into a small beaker. Mix them together, transfer to the ignition tube, and spread the mixture out so that it covers about the lower one-third of the tube. Insert the stopper securely into the tube and clamp the tube to the ring stand at about a 30-degree angle (see Figure 3.2). Insert the end of the delivery tube into one of the gas-collecting bottles on the shelf. Adjust the burner so that the **tip of the flame just touches the ignition tube,** and start heating the mixture. After oxygen has started coming from the delivery tube, the rate may be controlled by regulating the amount of heat. When one bottle is filled with gas, immediately start filling the next bottle. Then cover the mouth of the gas-filled bottle with a glass plate and remove it from the water. Store the bottles mouth upward without removing the glass plates; the oxygen will not readily escape since it is slightly heavier than air. Note which of the bottles was the first one filled with gas and continue until a total of five bottles of gas have been collected.

## B. Properties of Oxygen

Each of the following tests (except B.6) is conducted with a bottle of oxygen and, for comparison, with a bottle of air. Record your observations on the report form.

1. Ignite a wood splint, blow out the flame, and insert the still-glowing splint into the first bottle of oxygen collected. Repeat with a bottle of air. To ensure having a bottle of air, fill the bottle with water and then empty it, thus washing out other gases that may be present.

2. Place a small lump of sulfur in a deflagration spoon and start it burning by heating it in the burner flame. Lower the burning sulfur alternately into a bottle of oxygen and a bottle of air and compare combustions. **Do this part of the experiment in the hood and quench the excess burning sulfur in a beaker of water.**

3. Stand a small candle (no longer than 5 cm) on a glass plate and light it. Lower a bottle of oxygen over the burning candle, placing the mouth of the bottle on the glass plate. Measure and record the time, in seconds, that the candle continues to burn. Repeat with a bottle of air. Note also the difference in the brilliance of the candle flame in oxygen and in air. Return the unused portion of the candle to the reagent shelf.

4. Invert a bottle of oxygen, covered with glass plate, and place it mouth to mouth over a bottle of air. Then remove the glass plate from between the bottles and allow them to stand mouth to mouth for 3 minutes. Cover each bottle with a glass plate and set the bottles down, mouths upward. Test the contents of each bottle with a glowing splint.

5. Pour 25 mL of water into the fifth bottle of oxygen and replace the cover. Place the bottle close to (within 5 or 6 cm) the burner. Take a **loose,** 4 or 5 cm wad of steel wool (iron) in the crucible tongs and **momentarily** heat in the burner flame until some of the steel wool first begins to glow. **Immediately** lower the glowing metal into the bottle of oxygen. (It is **essential** that some of the steel wool be glowing when it goes into the oxygen.) Repeat, using a bottle of air.

> NOTE: The 25 mL of water is to prevent breakage if the glowing steel wool is accidentally dropped into the bottle.

6. Take a 2 to 5 cm strip of magnesium metal in the crucible tongs and ignite it by heating it in the burner flame.

⚠ PRECAUTIONS:

> 1. Do not look directly at the glowing magnesium, since considerable ultraviolet light is emitted during the reaction.
>
> 2. Do not put burning magnesium into a bottle of oxygen.

## C. Instructor Demonstrations (Optional)

1. **Sodium Peroxide as a Source of Oxygen.** Spread some cotton on the bottom of an evaporating dish and sprinkle a small amount (less than 1 g) of fresh sodium peroxide on it. Sprinkle a few drops of water on the peroxide. Spontaneous combustion of the cotton will occur.

2. **Approximate Percentage of Oxygen in the Air.** Push a small wad of steel wool to the bottom of a 25 X 200 mm test tube. Wet the steel wool by covering with water; pour out the surplus water; and place the tube, mouth downward, in a 400 mL beaker half full of water. After the oxygen in the trapped air has reacted with the steel wool—at least three days are needed for complete reaction—adjust the water levels inside and outside the tube to the same height. Cover the mouth of the tube, remove from the beaker, and measure the volume of water in the tube. Alternatively, the height of the water column may be measured (in millimeters) without removing the tube from the beaker. The volume of water in the tube is approximately equal to the volume of oxygen originally present in the tube of air.

$$\%\ \text{oxygen} = \frac{\text{Volume of water in tube}}{\text{Volume of tube}} \times 100\%$$

or

$$\%\ \text{oxygen} = \frac{\text{Height of water column}}{\text{Length of tube}} \times 100\%$$

3. **Decomposition of Water.** Set up the Hoffman electrolysis apparatus, as shown in Figure 3.3. The solution used in the apparatus should contain about 2 mL of sulfuric acid per 100 mL of water. Direct current may be obtained from several 1.5 volt type A cells connected in series or from some other D.C. source.

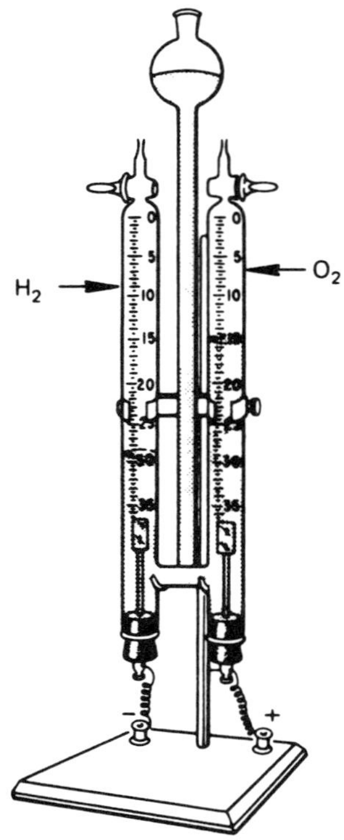


**Figure 3.3. Hoffman electrolysis apparatus.**

NAME ______________________

SECTION __________ DATE __________

INSTRUCTOR ______________________

## *REPORT FOR EXPERIMENT 3*

# Preparation and Properties of Oxygen

### A or A-1. Generation and Collection of Oxygen

1. What evidence did you observe that oxygen is not very soluble in water?

2. What is the source of oxygen in the procedure you used?

   Name ______________________ Formula ______________________

3. What purpose does the manganese dioxide serve in this preparation of oxygen?

4. What gas was in the apparatus before you started generating oxygen? Where did it go?

5. What is different about the composition of the first bottle of gas collected compared to the other four?

6. Why are the bottles of oxygen stored with the mouth up?

7. (a) What is the symbol of the element oxygen? __________

   (b) What is the formula for oxygen gas? __________

8. Which of the following formulas represent oxides? (Circle) $MgO$, $KClO_3$, $SO_2$, $MnO_2$, $O_2$, $NaOH$, $PbO_2$, $Na_2O_2$.

9. Write the word and formula equations for the preparation of oxygen from (a) hydrogen peroxide and from (b) potassium chlorate.

   (a) **Word Equation:**

   **Formula Equation:**

   (b) **Word Equation:**

   **Formula Equation:**

10. For the decomposition procedure that you used to generate oxygen, what substances, other than oxygen, are in the generator when the decomposition is complete?

**B. Properties of Oxygen**

1. Write word equations for the chemical reactions that occurred. (See Study Aid 2, page 227.)

   B.1. Combustion of wood. Assume carbon is the combustible material.

   B.2. Combustion of sulfur.

   B.5. Combustion of steel wool (iron). (Call the product iron oxide.)

   B.6. Combustion of magnesium.

2. Write formula equations for these same combustions.

   B.1. ($CO_2$ is the oxide of carbon that is formed.)

   B.2. ($SO_2$ is the oxide of sulfur that is formed.)

B.5. ($Fe_3O_4$ is the oxide of iron that is formed.)

B.6. (MgO is the oxide of magnesium that is formed.)

3. Combustion of a candle.

   (a) Number of seconds that the candle burned in the bottle of oxygen. ____________

   (b) Number of seconds that the candle burned in the bottle of air. ____________

   (c) Explain this difference in combustion time.

   (d) Is it scientifically sound to conclude that all the oxygen in the bottle was reacted when the candle stopped burning? Explain.

4. What were the results of the experiment in which a bottle of oxygen was placed over a bottle of air? Explain the results.

5. (a) Describe the material that is formed when magnesium is burned in air.

   (b) What elements are in this product?

6. (a) What is your conclusion about the rate or speed of a chemical reaction with respect to the concentration of the reactants—for example, a combustion in a high concentration of oxygen (pure oxygen) compared to a combustion in a low concentration of oxygen (air)?

(b) What evidence did you observe in the burning of sulfur to confirm your conclusion in 6(a)?

# EXPERIMENT 4

# Preparation and Properties of Hydrogen

## MATERIALS AND EQUIPMENT

Solids: strips of copper, magnesium, and zinc; sodium metal; steel wool; mossy zinc; wood splints. Liquid: concentrated sulfuric acid ($H_2SO_4$). Solutions: dilute (6 M) acetic acid ($HC_2H_3O_2$), 0.1 M copper(II) sulfate ($CuSO_4$), dilute (6 M) hydrochloric acid (HCl), dilute (3 M) phosphoric acid ($H_3PO_4$), and dilute (3 M) sulfuric acid ($H_2SO_4$). Pneumatic trough, five wide-mouth (gas-collecting) bottles.

## DISCUSSION

Hydrogen, having atomic number 1 and atomic weight 1.008, is the simplest element. It is the ninth most abundant element in the earth's crust (about 0.9 percent by mass). At ordinary temperatures and pressures it is a gas, composed of diatomic molecules, $H_2$, and is only very slightly soluble in water. Hydrogen is usually found combined with other elements. Water is the most common and probably most important compound of hydrogen. Hydrogen will not support combustion, but in the presence of oxygen it burns readily to form water:

$$2\,H_2 + O_2 \longrightarrow 2\,H_2O$$

This reaction is used as a simple test for hydrogen, for mixtures of hydrogen and air (or oxygen, burn explosively with a distinctive "popping" or "barking" sound.

### Methods of Preparation

There are several ways of producing hydrogen gas. Two methods are demonstrated in this experiment.

1. **Active Metal with Water.** Several of the most active metals—such as lithium, sodium, potassium, rubidium, cesium, and calcium—will react with cold water; magnesium will react with hot water. An example of such a reaction is

$$\text{Sodium} + \text{Water} \longrightarrow \text{Sodium hydroxide} + \text{Hydrogen}$$

$$2\,Na + 2\,H_2O \longrightarrow 2\,NaOH + H_2\uparrow$$

Besides hydrogen this reaction also produces sodium hydroxide, which causes the solution to become basic (alkaline).

2. **Active Metal with Dilute Acid.** In general, metals that are more active than hydrogen will react with **dilute** acids to produce hydrogen gas and a salt. For example, the salts produced in the following reactions are magnesium chloride, zinc sulfate, and zinc phosphate, respectively.

$$Mg + 2\,HCl \longrightarrow MgCl_2 + H_2\uparrow$$

$$Zn + H_2SO_4 \longrightarrow ZnSO_4 + H_2\uparrow$$

$$3\,Zn + 2\,H_3PO_4 \longrightarrow Zn_3(PO_4)_2 + 3\,H_2\uparrow$$

The strong oxidizing acids, such as nitric acid ($HNO_3$) and **concentrated** sulfuric acid, also react with metals but do not produce hydrogen gas.

### Properties of Hydrogen

In this experiment you will collect several bottles of hydrogen. In testing this hydrogen be sure to keep in mind that there is a distinction between combustion and support of combustion. We can observe three different situations: (1) A flame inserted into a bottle of pure hydrogen will go out, because there is no oxygen to support combustion. (2) A flame inserted into a mixture of air and hydrogen will set off a very rapid combustion (burning) of the hydrogen, causing a small explosion. (3) A flame brought to the mouth of a bottle of pure hydrogen will cause the hydrogen to burn, but only at the mouth, where the hydrogen is in contact with the air. Situation (3) is the hardest to detect because the reaction is not explosive, there is little noise, and hydrogen burns with a colorless flame.

Other tests involving the combustion of hydrogen demonstrate that hydrogen is lighter than air and diffuses rapidly (spontaneously mixes with other gases).

## PROCEDURE

⚠ PRECAUTIONS:

1. **Wear protective glasses.**

2. Since mixtures of hydrogen and air may explode when ignited, wrap the generator with a towel to prevent flying glass.

3. Keep burners away from the generator and the delivery tube.

4. Concentrated sulfuric acid is an extremely hazardous chemical. It attacks the skin rapidly. If you get it on your skin, wash with water immediately.

5. After pouring concentrated sulfuric acid, replace the stopper and rinse off the bottle with water from the shoulder of the bottle down. Do not allow water to get onto the top or into the bottle since it may cause serious spattering.

### A. Preparing Hydrogen from Water

Fill a test tube half full of water and place it in the test tube rack. Obtain a piece of sodium (no larger than a 4 mm cube) from your instructor and place it on a piece of filter paper; **do not touch the sodium with your fingers**. Fold the filter paper over the sodium and press out the kerosene, noting how soft the sodium metal is. (Sodium is kept in kerosene because it reacts rapidly with water or with the oxygen in air.) Pick up the sodium with tweezers or tongs and, holding it at arm's length, drop it into the test tube.

⚠ **CAUTION: Do not put your head over the tube while the sodium is reacting.**

Immediately bring a flame to the mouth of the test tube and observe the results. When the reaction has ceased, use a clean stirring rod to place drops of the solution on pieces of red and blue litmus paper to determine whether the solution is acidic or basic. Acids turn litmus red, bases turn litmus blue.

## B. Preparing Hydrogen from Acids

NOTE: Dispose of any unreacted metal strips by rinsing with water and putting them in the waste baskets or a container provided by your instructor, not in the sink!

1. **Using Various Metals.** Set up four test tubes in a rack. Place one small strip of zinc into the first tube. In like manner place samples of copper, steel wool (iron), and magnesium, in this order, into the other three tubes. In rapid succession add a few milliliters of dilute (6 M) hydrochloric acid to each test tube. Note whether gas is evolved in each case and also note its relative rate of evolution. Test any gas evolved for evidence of hydrogen by bringing a flame to the mouth of the test tube.

2. **Using Various Acids.** In the preceding section you found that zinc is one of the metals capable of releasing hydrogen from hydrochloric acid. In this section you will compare the relative ease with which zinc displaces hydrogen from a variety of acids. The stronger acids react at a faster rate than the weaker acids.

Set up four test tubes in a rack. Place several milliliters of dilute (6 M) hydrochloric acid into the first tube, dilute (6 M) acetic acid into the second, dilute (3 M) sulfuric acid into the third, and dilute (3 M) phosphoric acid into the fourth. Now drop a small strip of zinc into each of the four tubes. The variation in the rates of evolution of hydrogen gas is a measure of the relative strengths of the acids. Let the reactions proceed for three minutes before making your evaluation.

## C. Collecting Hydrogen

Assemble the generator shown in Figure 4.1, using a wide-mouth bottle or a 250 mL Erlenmeyer flask, a 2-hole rubber stopper equipped with a thistle tube reaching to within 1 cm of the bottom of the bottle, and a delivery tube. Clamp the generator to the ring stand. All connections must be airtight.

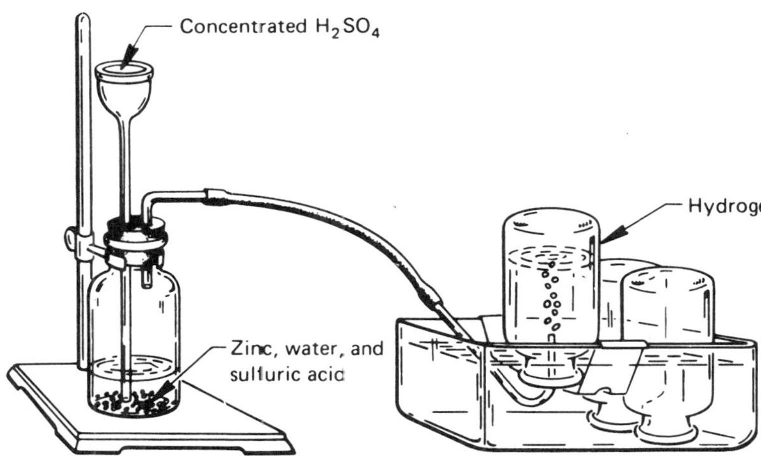


**Figure 4.1. Preparing hydrogen from zinc and sulfuric acid.**

Place approximately 10 g of mossy zinc in the bottle; add 2 mL of 0.1 M copper(II) sulfate solution (as a catalyst) and 50 mL of water and **make sure that the bottom end of the thistle tube is under water.** Fill four wide-mouth bottles with water and invert them in the pneumatic trough. When you are ready to collect the gas, pour about 2 mL of concentrated sulfuric acid through the thistle tube. Add more acid in 2 to 3 mL increments, as needed, to keep the reaction going. As each bottle becomes filled with gas, place a glass plate over its mouth while the bottle is still under water; then remove the bottle from the water and store it **mouth downward** without removing the glass plate. Set aside, but do not discard, the first bottle of gas collected. Fill the other three bottles with hydrogen. Now fill the generator with water to quench the reaction. Open the generator, rinse the remaining zinc thoroughly with water, and return the unreacted zinc to the container provided by the instructor.

### D. Reaction of Hydrogen

> NOTE: Unless otherwise directed, keep the bottles mouth downward while performing the following tests.

1. Raise the first bottle of gas collected a few inches straight up from the table, and immediately apply a burning splint to the mouth of the bottle.

2. Raise the second bottle straight up a few inches from the table. Then without delay slowly insert a burning splint halfway into the bottle. Continue with the insertion of the splint even though you hear a muffled report as the flame approaches the mouth of the bottle. **Slowly** withdraw the splint until the charred end is in the neck of the bottle; hold the splint there a few seconds, but do not withdraw it completely. Repeat inserting and withdrawing the splint several times.

3. Place the third bottle mouth upward and remove the glass plate. After one minute, bring a burning splint to its mouth.

4. Keeping the cover plate over its mouth, place the fourth bottle of hydrogen (still upside down) mouth to mouth over a bottle of air. Then remove the cover plate from between them. Let the two bottles remain in this position for three minutes, then replace the cover plate between them. Leaving the cover plate on the mouth of the lower bottle, raise the top bottle straight up, at least 6 inches, and immediately bring a burning splint to its mouth. Turn the lower bottle mouth downward, with the cover plate in place. Bring a burning splint to the mouth of this bottle as you lift it straight up. Compare the results.

NAME ____________________

SECTION __________ DATE __________

INSTRUCTOR ____________________

*REPORT FOR EXPERIMENT 4*

# Preparation and Properties of Hydrogen

## A. Preparing Hydrogen from Water

1. Describe what you observed when sodium was dropped into water.

2. Describe what you observed when a flame was brought to the mouth of the test tube.

3. What color did the litmus papers turn? __________

4. Did the reaction make the solution acidic or basic? __________

5. Complete and balance the following word and formula equations:

   Sodium + Water $\longrightarrow$

   $Na \quad + \quad H_2O \longrightarrow$

## B. Preparing Hydrogen from Acids

1. (a) Write the symbols of the metals that react with dilute hydrochloric acid.

   ____________________

   (b) Write the symbols of the four metals in order of decreasing ability to liberate hydrogen from hydrochloric acid.

   ________ ________ ________ ________

2. Write the formulas of the four acids in order of decreasing strength based on the rates at which they liberate hydrogen when reacting with zinc.

   ________ ________ ________ ________

## C. Collecting Hydrogen

1. Why was water added to cover the bottom of the thistle tube?

2. What do we call this method of collecting a gas?

3. What physical property of hydrogen, other than that it is less dense than water, allows it to be collected in this manner?

## D. Reaction of Hydrogen

1. What happened when the splint was brought to the mouth of the first bottle of gas collected?

2. Describe fully the results of testing the second bottle of gas.

3. (a) Is hydrogen combustible? ____________

   (b) What evidence do you have of this from testing the second bottle of gas?

4. (a) Does hydrogen support combustion? ____________

   (b) What evidence do you have of this from testing the second bottle of gas?

5. What compound was formed during the testing of the first and second bottles of gas?

   ____________

6. Why did the first bottle of gas behave differently from the second bottle?

7. (a) What happened when the splint was brought to the mouth of the third bottle of gas?

   (b) How do you account for this?

8. When testing the fourth bottle:

   (a) What was the result with the top bottle?

   (b) What was the result with the lower bottle?

   (c) How do you account for these results?

9. Complete the following word equations:

   Zinc + Sulfuric acid $\longrightarrow$

   Magnesium + Hydrochloric acid $\longrightarrow$

   Hydrogen + Oxygen $\longrightarrow$

10. Complete and balance the following corresponding formula equations:

   $Zn + H_2SO_4 \longrightarrow$

   $Mg + HCl \longrightarrow$

   $H_2 + O_2 \longrightarrow$

## EXPERIMENT 5

# Water in Hydrates

### MATERIALS AND EQUIPMENT

Solids: finely ground copper(II) sulfate pentahydrate ($CuSO_4 \cdot 5H_2O$), and unknown hydrate. Cobalt chloride test paper, clay triangle, crucible and cover, 25 × 200 mm hard-glass test tube.

### DISCUSSION

Many salts form compounds in which a definite number of moles of water are combined with each mole of the salt. Such compounds are called **hydrates.** The water that is chemically combined in a hydrate is referred to as **water of crystallization** or **water of hydration.** The following are representative examples:

$$CaSO_4 \cdot 2H_2O, \quad CoCl_2 \cdot 6H_2O, \quad MgSO_4 \cdot 7H_2O, \quad Na_2CO_3 \cdot 10H_2O.$$

In a hydrate the water molecules are distinct parts of the compound but are joined to it by bonds that are weaker than either those forming the anhydrous salt or those forming the water molecules. In the formula of a hydrate a dot is commonly used to separate the formula of the anhydrous salt from the number of molecules of water of crystallization. For example, the formula of calcium sulfate dihydrate is written $CaSO_4 \cdot 2H_2O$ rather than $H_4CaSO_6$.

Hydrated salts can usually be converted to the anhydrous form by heating:

$$\text{Hydrated salt} \xrightarrow{\Delta} \text{Anhydrous salt} + \text{Water}$$

Hence it is possible to determine the percentage of water present in a hydrated salt by determining the amount of mass lost (water driven off) when a known mass of the hydrate is heated:

$$\text{Percentage water} = \frac{\text{Mass lost}}{\text{Mass of sample}} \times 100$$

It is possible to condense the vapor driven off the hydrate and demonstrate that it is water by testing it with anhydrous cobalt(II) chloride ($CoCl_2$). Anhydrous cobalt(II) chloride is blue but reacts with water to form the red hexahydrate, $CoCl_2 \cdot 6H_2O$.

### PROCEDURE

**Wear protective glasses.**

#### A. Qualitative Determination of Water

Fold a 2.5 × 20 cm strip of paper lengthwise to form a V-shaped trough or chute. Load about 4 g of finely ground copper(II) sulfate pentahydrate in this trough, spreading it evenly along the length of the trough.

Clamp a **dry** 25 X 200 mm hard-glass test tube so that its mouth is 15–20 degrees **above the horizontal** (Figure 5.1a). Insert the loaded trough into the tube. Rotate the tube to a nearly vertical position (Figure 5.1b) to deposit the copper(II) sulfate in the bottom of the tube. Tap the paper chute gently if necessary, but make sure that no copper sulfate is spilled and adhering to the sides of the upper part of the tube. Remove the chute and turn the tube until it slants mouth downward at an angle of 15–20 degrees **below the horizontal** (Figure 5.1c). Make sure that **all** of the copper(II) sulfate remains at the bottom of the tube. To obtain a sample of the liquid that will condense in the cooler part of the tube, place a small clean, dry test tube, held in an upright position in either a rack or an Erlenmeyer flask, just below the mouth of the tube containing the hydrate.

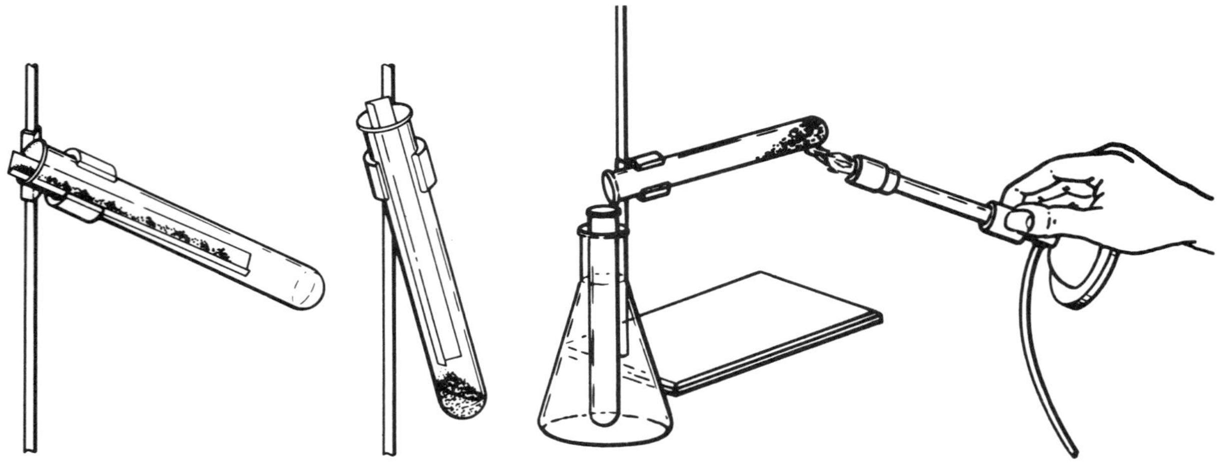

(a) (b) (c)

**Figure 5.1. Setup for dehydration of a hydrate.**

Heat the hydrate gently at first to avoid excessive spattering. Gradually increase the rate of heating, noting any changes that occur and collecting some of the liquid that condenses in the cooler part of the tube. Continue heating until the blue color of the hydrate has disappeared, but do not heat until the residue in the tube has turned black. Finally warm the tube over its entire length—without directly applying the flame to the clamp—for a minute or two to drive off most of the liquid that has condensed on the inner wall of the tube. Allow the tube and contents to cool.

> NOTE: At excessively high temperatures (above 600°C) copper(II) sulfate decomposes; sulfur trioxide is driven off and the black copper(II) oxide remains as a residue.

Observe the appearance and odor of the liquid that has been collected.

While the tube is cooling, dry a piece of cobalt chloride test paper by holding it about 20 to 25 cm above a burner flame; that is, close enough to heat but not close enough to char or ignite the paper. When properly dried, the test paper should be blue. Using a clean stirring rod, place a drop of the liquid collected from the hydrate on the dried cobalt chloride test paper. For comparison place a drop of distilled water on the cobalt chloride paper. Record your observations.

Empty the anhydrous salt residue in the tube onto a watch glass and divide it into two portions. Add 3 or 4 drops of the liquid collected from the hydrate to one portion and 3 or 4 drops of distilled water to the other. Compare and record the results of these tests.

## B. Quantitative Determination of Water in a Hydrate

NOTES:

1. Weigh crucible and contents to the highest precision possible with the balance available to you.

2. Since there is some inaccuracy in any balance, use the same balance for successive weighings of the same sample. When subtractions are made to give mass of sample and mass lost, the inaccuracy due to the balance should cancel out.

3. Handle crucibles and covers with tongs only, after the initial heating.

4. Be sure crucibles are at or near room temperature when weighed.

5. **Record all data directly on the report sheet as soon as you obtain them.**

Obtain a sample of an unknown hydrate, as directed by your instructor. Be sure to record the identifying number. Accurately weigh a clean, dry crucible and cover. Place between 2 and 3 g of

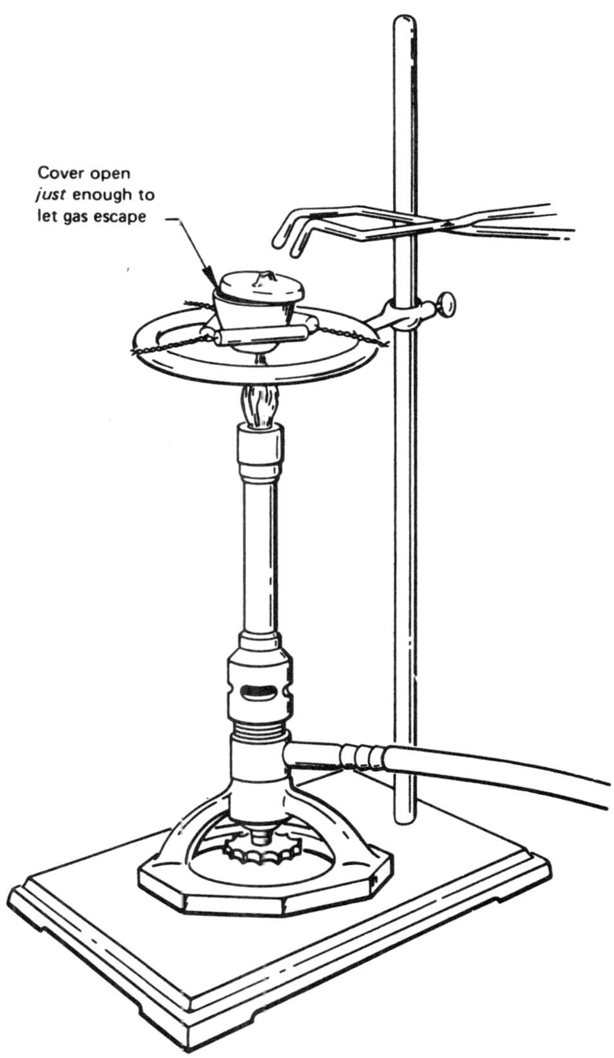


**Figure 5.2. Method of heating a crucible.**

the unknown in the weighed crucible, cover, and accurately weigh the crucible and contents. Place the covered crucible on a clay triangle; adjust the cover so that it is slightly ajar, to allow the water vapor to escape (see Figure 5.2); and **very gently** heat the crucible for about 5 minutes. Readjust the flame so that a sharp, inner-blue cone is formed. Heat for another 12 minutes, with the tip of the inner-blue cone just touching the bottom of the crucible. The crucible bottom should become dull red during this period. After this first heating is completed, close the cover, cool (about 10 minutes), and weigh. Heat the covered crucible and contents for an additional 6 minutes at maximum temperature; cool and reweigh. The results of the last two weighings should agree within 0.05 g. If the decrease in mass between the two weighings is greater than this amount, repeat the heating and weighing until the results of two successive weighings agree to within 0.05 g. Calculate the percentage of water in your sample on the basis of the final weighing.

## *EXPERIMENT 6*

# Freezing Points—Graphing of Data

### MATERIALS AND EQUIPMENT

Solids: benzoic acid ($C_6H_5COOH$) and crushed ice. Liquid: glacial acetic acid ($HC_2H_3O_2$). Thermometer, watch or clock with sweep second hand, slotted corks or stoppers.

### DISCUSSION

Each individual compound possesses a unique set of physical and chemical properties. Just as one human being can be distinguished from all others by certain characteristics–fingerprints, for example–it is also possible, through knowledge of its properties, to distinguish any given compound from among the many hundreds of thousands that are known.

The melting point and the boiling point are easily determined properties that are very useful in identifying a substance. Consequently, these properties are almost always recorded when a compound is described in the chemical literature (textbooks, handbooks, journal articles, etc.).

A pure substance will freeze (or melt) at a fixed temperature. This is the temperature at which the solid phase is in equilibrium with the liquid phase.

$$\text{Solid} \underset{\text{Freezing}}{\overset{\text{Melting}}{\rightleftarrows}} \text{Liquid}$$

Both melting and freezing occur at the same temperature; the terms "melting point" and "freezing point" are used to tell which process is occurring.

When a small amount of a compound (solute) is dissolved in another compound (solvent), the freezing point of the resulting solution will be lower than that of the solvent. For example, solutions of salt in water may freeze at temperatures as low as $-21°C$ ($21°$ below the freezing point of pure water).

Melting point data are of great value in determining the identity and/or purity of substances, especially in the field of organic chemistry. If a sample of a compound melts appreciably below the known melting point of the pure substance, we know that the sample contains impurities which have lowered the melting point. If the melting point of an unknown compound agrees with that of a known compound, the identity can often be confirmed by mixing the unknown compound with the known and determining the melting point of the mixture. If the melting point of the mixture is the same as that of the known compound, the compounds are identical. On the other hand, a lower melting point for the mixture indicates that the two compounds are not identical.

Frequently when a substance or a solution is being cooled, the temperature will fall below the true freezing point before crystals begin to form. This phenomenon is known as **supercooling.** Once crystallization has begun, the temperature will rise again because of the heat released by the crystallization process (heat of crystallization).

In this experiment you will study the cooling and freezing behavior of pure acetic acid (glacial) and of a solution of acetic acid and benzoic acid. Acetic acid is the solvent in the solution.

## PROCEDURE

**Wear protective glasses**

NOTES:

1. Since water and other foreign substances will affect the results of this experiment adversely, **use only clean, dry equipment.**

2. Read and record all temperatures to the **nearest 0.1°C.**

1. Fasten a utility clamp to the top of a clean, dry 18 × 150 mm test tube. Position this clamp-tube assembly on a ring stand so that the bottom of the tube is about 20 cm above the ring stand base.

2. Obtain a slotted one-hole cork (or stopper) to fit the test tube. Insert a thermometer in the cork and position in the test tube so that the end of the bulb is about 1.5 cm from the bottom of the test tube. Turn the thermometer so that the temperature scale can be read in the slot.

3. Measure 10.0 mL of glacial acetic acid in your smallest graduated cylinder and pour the acid into the test tube. Replace the thermometer and adjust the temperature of the acetic acid to approximately 25°C by warming or cooling the tube in a beaker of water.

4. Fill a 400 mL beaker about three-quarters full of crushed ice; add cold water until the ice is almost covered. Position the beaker of ice and water on the ring stand base under the clamped tube-thermometer assembly.

5. Read the temperature of the acetic acid and record as the 0.0 minute time reading in the Data Table. Now loosen the clamp on the ring stand and observe the second hand of your watch or clock. As the second hand crosses 12, lower the clamped tube-thermometer assembly so that **all of the acetic acid in the tube is below the surface of the ice water.** Fasten the clamp to hold the tube in this position.

6. Loosen the cork on the tube and stir the acid with the thermometer, **keeping the bulb of the thermometer completely immersed in the acid.** Take accurate temperature readings at 30-second intervals as the acid cools. (Zero Time was when the second hand crossed 12.) Stop stirring and center the thermometer bulb in the tube as soon as you are sure that crystals are forming in the acid (three to four minutes).

7. Continue to take accurate temperature readings at 30-second intervals until a total time of 15 minutes has elapsed. To help maintain a constant temperature in the ice-water bath, stir it occasionally.

8. After completing the temperature readings, remove the test tube-thermometer assembly from the ice bath, **keeping the thermometer in place.** Immerse the lower portion of the test tube in a beaker of warm water to melt the frozen acetic acid. **Do not discard this acid;** it will be used in Step 9.

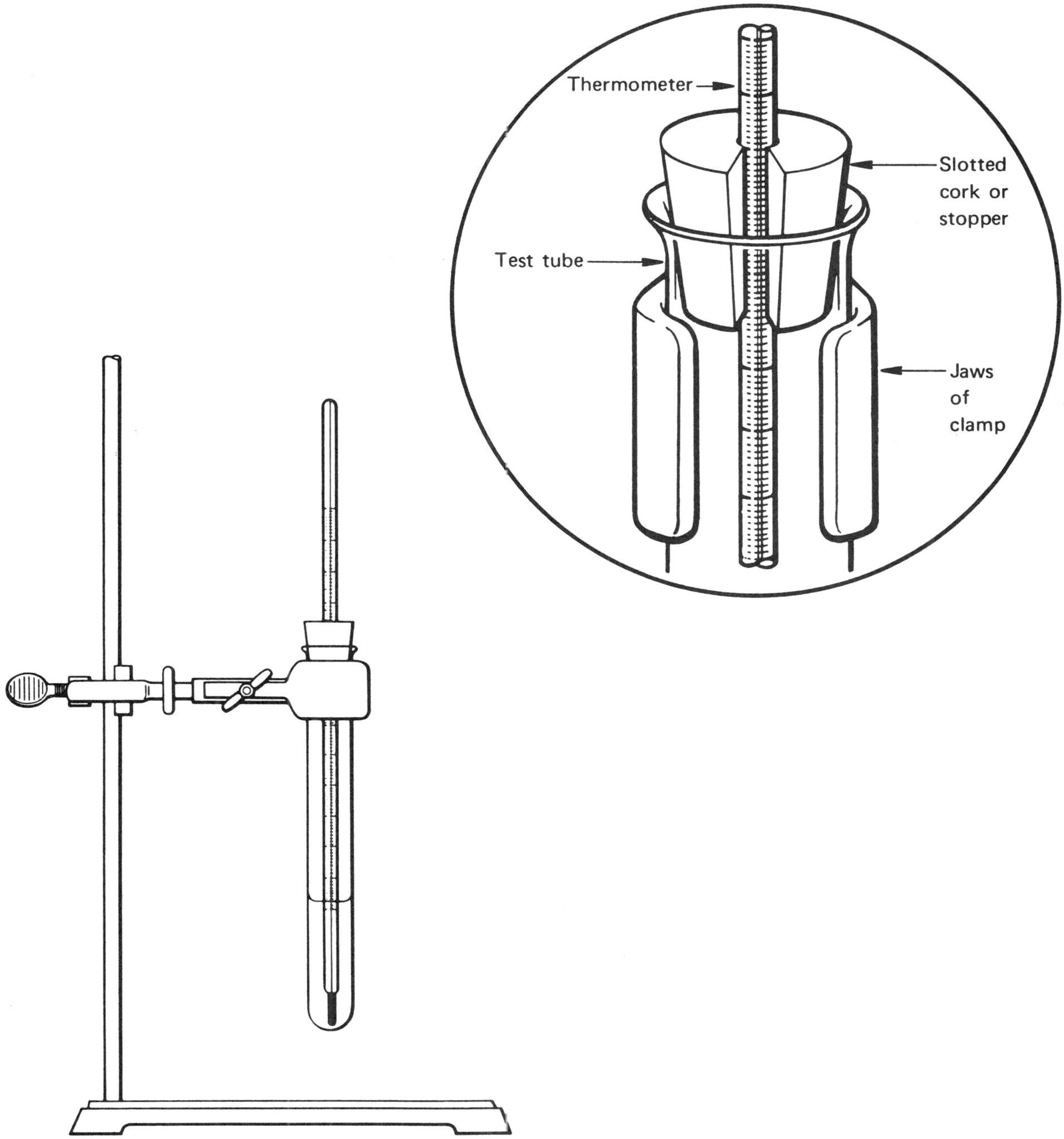


**Figure 6.1. Setup for freezing-point determination.**

9. Accurately weigh 0.450 g (450. mg) of benzoic acid crystals. Now remove the thermometer from the test tube of acetic acid and lay it on the table, being careful not to contaminate the thermometer or lose any acid. Carefull add **all** of the benzoic acid to the acetic acid. Stir gently with the thermometer until all of the crystals have dissolved. Stir for an additional minute or two to ensure a uniform solution. Adjust the temperature of the solution to approximately 25°C.

10. Repeat Steps 4 through 7 of this procedure to obtain freezing point data for the glacial acetic acid-benzoic acid solution. Use a beaker of fresh ice and water for these steps.

8.4cm

2.54cm = 1in

# EXPERIMENT 7

# Properties of Solutions

## MATERIALS AND EQUIPMENT

Solids: ammonium chloride ($NH_4Cl$), barium chloride ($BaCl_2$), barium sulfate ($BaSO_4$), fine and coarse crystals of sodium chloride ($NaCl$), and sodium sulfate ($Na_2SO_4$). Liquids: 1,1,1-trichloroethane ($CCl_3CH_3$), isopropyl alcohol ($C_3H_7OH$), and kerosene. Solutions: saturated iodine-water ($I_2$), and saturated potassium dichromate ($K_2Cr_2O_7$).

## DISCUSSION

The term **solution** is used in chemistry to describe a system in which one substance (or more) is dissolved in another substance.

A solution has two components, a solute and a solvent. The **solute** is the substance that is dissolved. The **solvent** is the dissolving medium; it is the substance present in greater quantity. The name of the solution is taken from the name of the solute. Thus when sodium chloride is dissolved in water, sodium chloride is the solute, water is the solvent, and the solution is called a sodium chloride solution.

Like other mixtures a solution has variable composition, since more or less solute may be dissolved in a given quantity of a solvent. But unlike other mixtures a solution is homogeneous, since the solute remains uniformly dispersed throughout the solution after mixing. The size of the dissolved particles (molecules or ions) is of the order of $10^{-8}$ to $10^{-7}$ cm (1-10Å).

For many substances whose reactions are those of their ions, it is necessary to use solutions in order to obtain satisfactory reactions. For example, when the two solids sodium chloride ($NaCl$) and silver nitrate ($AgNO_3$) are mixed, no detectable reaction is observed. However, when water solutions of these salts are mixed, a white precipitate of silver chloride ($AgCl$) is formed immediately.

The formation of a solution depends on the nature of the solute and the solvent. In general, water, which is polar, is a better solvent for inorganic than for organic substances. On the other hand, nonpolar solvents such as benzene, 1,1,1-trichloroethane, and ether are good solvents for many organic substances that are practically insoluble in water.

The **rate of dissolving** of a solute depends on:

1. The particle size of the solute.
2. Agitation or stirring of the solution.
3. The temperature of the solution.
4. The concentration of the solute in solution.

In addition, the **amount of solute** that will dissolve per unit mass or unit volume of solvent depends on the temperature of the solution.

**Solubility** expresses the amount of solute that will dissolve in a given amount of solvent. Terms used to describe the relative solubility of a solute are: **insoluble, slightly soluble, moderately soluble, soluble,** and **very soluble.** When two liquids are **immiscible** they do not form a solution or are generally insoluble in each other.

The **concentration** of a solution expresses the relative content of dissolved solute. Common ways to express concentration are as follows:

1. **Dilute Solution:** A solution that contains a relatively small amount of solute per unit volume of solution.

2. **Concentrated Solution:** A solution that contains a relatively large amount of solute per unit volume of solution.

3. **Saturated Solution:** A solution that contains as much solute as can dissolve at a given temperature and pressure; the dissolved solute is in equilibrium with undissolved solute:

$$\text{Solute (solid)} \rightleftarrows \text{Solute (dissolved)}$$

A saturated solution may be dilute or concentrated.

4. **Unsaturated Solution:** A solution containing less solute per unit volume than the corresponding saturated solution.

5. **Supersaturated Solution:** A solution containing more dissolved solute than is normally present in the corresponding saturated solution. However, a supersaturated solution is in a metastable condition and will form a saturated solution if disturbed. For example, when a small crystal of a dissolved salt is dropped into a supersaturated solution, crystallization begins at once, and salt precipitates until a saturated solution is formed.

6. **Mass-percent Solution:** The percent by mass of the solute in a solution. Thus a 10 percent sodium hydroxide solution contains 10 g of NaOH in 100 g of solution (10 g NaOH + 90 g $H_2O$; 2 g NaOH + 18 g $H_2O$; etc.). The formula for calculating mass percent is:

$$\text{Mass percent} = \frac{\text{g solute}}{\text{g solute} + \text{g solvent}} \times 100$$

7. **Molarity:** The number of moles (molar mass) of solute per liter of solution. Thus a solution containing 1 mole of NaOH (40.0 g) per liter is 1 molar (abbreviated 1 M). The concentration of a solution containing 0.5 mole in 500 mL of solution is also 1 molar, etc. The equation is:

$$\text{Molarity} = \frac{\text{moles of solute}}{\text{liter of solution}}$$

## PROCEDURE

**Wear protective glasses.**

### A. Concentration of a Saturated Solution

NOTE: Use the same balance for all weighings.

Weigh an evaporating dish to the highest precision possible with the balance available to you. Obtain 6 mL of saturated potassium dichromate solution from the reagent shelf and pour it into the dish. Weigh the dish and the solution. Record these masses on the report form.

Place the evaporating dish on a 400 mL beaker that is half full of boiling water (see Figure 1.6 for setup). Continue to boil the water until the dichromate solution has evaporated almost to dryness (about 25 to 30 minutes), **adding more water to the beaker as needed.** Remove the evaporating dish and beaker from the wire gauze and dry the bottom of the dish with a towel. Put the dish on the wire gauze and heat gently to evaporate the last traces of water. Do not heat too strongly because (1) there is danger of sample loss by spattering, and (2) potassium dichromate decomposes at temperatures above 500°C. Allow the dish to cool for 5 to 10 minutes and reweigh.

**While the evaporation is proceeding, continue with other parts of the experiment.**

### B. Relative Solubility of a Solute in Two Solvents

1. Add about 2 mL of trichloroethane and 5 mL of water to a test tube and note the relative position of each substance. Stopper the tube, shake for about 5 seconds, and allow the liquid layers to separate.

2. Now add 5 mL of saturated iodine-water to the test tube, note the color of each layer, insert stopper, and shake gently for about 20 seconds. Allow the liquids to separate and again note the color of each layer.

3. NOTE: Dispose of mixtures containing trichloroethane in the container provided.

### C. Miscibility of Liquids

Take three **dry** test tubes and into the first place 1 mL of kerosene and 1 mL of isopropyl alcohol; into the second place 1 mL of kerosene and 1 mL of water; and into the third, 1 mL of water and 1 mL of isopropyl alcohol. Mix by shaking for about 5 seconds and note which pairs are miscible.

### D. Rate of Dissolving versus Particle Size

Fill a **dry** test tube to a depth of about 0.5 cm with fine crystals of sodium chloride. Fill another **dry** tube to the same depth with coarse sodium chloride crystals. Add 10 mL of tap water to each tube and stopper. Shake both tubes at the same time, noting the number of seconds required to dissolve the salt in each tube. (Don't shake the tubes for more than two minutes.)

### E. Rate of Dissolving versus Temperature

1. Weigh out two 0.5 g samples of fine sodium chloride crystals.

2. Take a 100 mL and a 150 mL beaker and add 50 mL tap water to each. Heat the water in the 150 mL beaker to boiling and allow it to cool for about 1 minute.

3. Add the 0.5 g samples of salt to each beaker and observe the time necessary for the crystals to dissolve in the hot water.

4. As soon as the crystals are dissolved in the hot water, take the beaker containing the hot solution in your hand, slowly tilt it back and forth, and observe the layer of denser salt solution on the bottom. Repeat with the cold-water solution.

### F. Solubility versus Temperature; Saturated and Unsaturated Solutions

1. Label four pieces of weighing paper as follows: 1.0 g NaCl, 1.4 g NaCl, 1.0 g $NH_4Cl$, and 1.4 g $NH_4Cl$, respectively. Weigh the stated amount of salt onto each piece of paper. Place each of the 1.0 g samples into separate, labeled test tubes. Add 5 mL of water, stopper, and shake until each salt is dissolved.

2. Add the 1.4 g-sample of sodium chloride to the tube containing sodium chloride solution and the 1.4 g-sample of ammonium chloride to the tube containing ammonium chloride solution. Shake for about 3 minutes and note whether all of the crystals have dissolved in each tube.

3. Place both tubes (unstoppered) into a beaker of boiling water, shake occasionally, and note the results after about 5 minutes.

4. Remove the tubes and cool in running tap water for about 1 minute. Let stand for a few minutes and observe.

### G. Ionic Reactions in Solution

Into four test tubes place pea-sized quantities of barium chloride, sodium sulfate, sodium chloride, and barium sulfate, one salt in each tube. Add 5 mL of water to each tube and shake to dissolve. One of the four salts does not dissolve. Mix the barium chloride and the sodium sulfate solutions together and note the results. (Sodium chloride and barium sulfate are the products of this reaction.)

# EXPERIMENT 8

# Composition of Potassium Chlorate

## MATERIALS AND EQUIPMENT

Solids: Reagent Grade potassium chlorate ($KClO_3$) and potassium chloride (KCl). Solutions: dilute (6 M) nitric acid ($HNO_3$) and 0.1 M silver nitrate ($AgNO_3$). Two No. 0 crucibles with covers; Ceramfab pad.

## DISCUSSION

The **percentage composition** of a compound is the percentage by mass of each element in the compound. If the formula of a compound is known, the percentage composition can be calculated from the molar mass and the total mass of each element in the compound. The **molar mass** of a compound is determined by adding up the atomic masses of all the atoms making up the formula. The **total mass** of an element in a compound is determined by multiplying the atomic mass of that element by the number of atoms of that element in the formula. The percentage of each element is then calculated by dividing its total mass in the compound by the molar mass of the compound and multiplying by 100.

The percentage composition of many compounds may be directly determined or verified by experimental methods. In this experiment the percentage composition of potassium chlorate will be determined both experimentally and from the formula.

When potassium chlorate is heated to high temperatures (above 400°C) it decomposes to potassium chloride and elemental oxygen, according to the following equation:

$$2\ KClO_3 \longrightarrow 2\ KCl + 3\ O_2\uparrow$$

The relative amounts of oxygen and potassium chloride are measured by heating a weighed sample of potassium chlorate and determining the amount of residue (potassium chloride) remaining. The decrease in mass brought about by heating represents the amount of oxygen originally present in the sample.

From the experiment we obtain the following three values:

1. Mass of original sample ($KClO_3$).
2. Mass lost when sample was heated (Oxygen).
3. Mass of residue (KCl).

From these three experimental values (and a table of atomic masses) we may calculate the following:

4. Percentage oxygen in sample (Experimental value)

$$= \frac{\text{Mass lost by sample}}{\text{Original sample mass}} \times 100$$

5. **Percentage KCl in sample (Experimental value)**

$$= \frac{\text{Mass of residue}}{\text{Original sample mass}} \times 100$$

6. Percentage oxygen in $KClO_3$ from formula (Theoretical value)

$$= \frac{\text{3 at. masses of oxygen}}{\text{Molar mass of } KClO_3} \times 100 = \frac{3 \times 16.00\text{ g}}{122.6\text{ g}} \times 100$$

7. Percentage KCl in $KClO_3$ from formula (Theoretical value)

$$= \frac{\text{Molar mass of KCl}}{\text{Molar mass of } KClO_3} \times 100 = \frac{74.55\text{ g}}{122.6\text{ g}} \times 100$$

8. Percentage error in experimental oxygen determination

$$= \frac{\text{(Experimental value)} - \text{(Theoretical value)}}{\text{Theoretical value}} \times 100$$

## PROCEDURE

PRECAUTIONS: Since potassium chlorate is a strong oxidizing agent it may cause fires or explosions if mixed or heated with combustible (oxidizable) materials. Observe the following safety precautions when working with potassium chlorate:

1. **Wear protective glasses.**

2. Use clean crucibles which have been heated and cooled prior to adding potassium chlorate.

3. Use Reagent Grade potassium chlorate.

4. Dispose of any excess or spilled potassium chlorate as directed by your instructor. (Potassium chlorate may start fires if mixed with paper or other solid wastes.)

5. Heat samples slowly and carefully to avoid spattering molten material—and to avoid poor experimental results.

NOTES:

1. Make all weighings to the highest precision possible with the balance available to you. Use the same balance to make all weighings for a given sample. Record all data directly on the report sheet as they are obtained.

2. Duplicate samples of potassium chlorate are to be analyzed, if two crucibles are available.

3. For utmost precision, handle crucibles with tongs after the initial heating.

## A. Determining Percentage Composition

Place a clean, dry crucible (uncovered) on a clay triangle and heat for 2 or 3 minutes at the maximum flame temperature. The tip of the sharply defined inner-blue cone of the flame should almost touch and heat the crucible bottom to redness. Allow the crucible to cool. If two crucibles are being used, carefully transfer the first to a Ceramfab pad and heat the second while the first crucible is cooling.

Weigh the cooled crucible and its cover (see Note 1); add between 1 and 1.5 g of potassium chlorate; weigh again.

NOTE: The crucible must be covered when potassium chlorate is being heated in it.

Place the covered crucible on the clay triangle and heat gently for 8 minutes with the tip of the inner-blue cone of the flame 6 to 8 cm (about 2.5 to 3 in.) below the crucible bottom. Then **carefully** lower the crucible or raise the burner until the tip of the sharply defined inner-blue cone just touches the bottom of the crucible, and heat for an additional 10 minutes. **The bottom of the crucible should be heated to a dull red color during this period.**

Grasp the crucible just below the cover with the **concave part** of the tongs and **very carefully** transfer it to a ceramfab pad. Allow to cool (about 10 minutes) and weigh. Begin analysis of a second sample while the first is cooling.

After weighing, heat the first sample for an additional 6 minutes at the maximum flame temperature (bottom of the crucible heated to a dull red color); cool and reweigh. The last two weighings should be in agreement. If the mass decreased more than 0.05 g between these two weighings, repeat the heating and weighing until two successive weighings agree within 0.05 g.

Complete the analysis of the second sample, following the same procedure used for the first.

## B. Qualitative Examination of Residue

This part of the experiment should be started as soon as the final heating and weighing of the first sample is completed and while the second sample is in progress.

Number and place three clean test tubes in a rack. Put a pea-sized quantity of potassium chloride into tube No. 1 and a like amount of potassium chlorate into tube No. 2. Add 10 mL of **distilled water** to each of these two tubes and shake to dissolve the salts. Now add 10 mL of **distilled water** to the crucible containing the residue from the first sample. Heat the crucible gently for about 1 minute; transfer 1 to 2 mL of the resulting solution from the crucible to tube No. 3; add about 10 mL of **distilled water** and mix.

Test the solution in each tube as follows: Add 5 drops of dilute (6M) nitric acid and 5 drops of 0.1 M silver nitrate solution. Mix thoroughly. Record your observations. This procedure using nitric acid and silver nitrate is a general test for chloride ions. The formation of a white precipitate is a positive test and indicates the presence of chloride ions. A positive test is obtained with any substance that produces chloride ions in solution.

# EXPERIMENT 9

# Double Displacement Reactions

## MATERIALS AND EQUIPMENT

Solid: sodium sulfite ($Na_2SO_3$). Solutions: dilute (6 M) ammonium hydroxide ($NH_4OH$), 0.1 M ammonium chloride ($NH_4Cl$), 0.1 M barium chloride ($BaCl_2$), 0.1 M calcium chloride ($CaCl_2$), 0.1 M copper(II) sulfate ($CuSO_4$), dilute (6 M) hydrochloric acid (HCl), concentrated (12 M) hydrochloric acid (HCl), 0.1 M iron(III) chloride ($FeCl_3$), dilute (6 M) nitric acid ($HNO_3$), 0.1 M potassium nitrate ($KNO_3$), 0.1 M silver nitrate ($AgNO_3$), 0.1 M sodium carbonate ($Na_2CO_3$), 0.1 M sodium chloride (NaCl), 10 percent sodium hydroxide (NaOH), dilute (3 M) sulfuric acid ($H_2SO_4$), and 0.1 M zinc nitrate [$Zn(NO_3)_2$].

## DISCUSSION

Double displacement reactions are among the most common of the simple chemical reactions and are comparatively easy to study.

In each part of this experiment two water solutions, each containing positive and negative ions, will be mixed in a test tube. Consider the hypothetical reaction

$$AB + CD \longrightarrow AD + CB$$

where AB exists as $A^+$ and $B^-$ ions in solution and CD exists as $C^+$ and $D^-$ ions in solution. As the ions come in contact with each other, there are six possible combinations that might conceivably cause chemical reaction. Two of these combinations are the meeting of ions of like charge; that is, $A^+ + C^+$ and $B^- + D^-$. But since like charges repel, no reaction will occur. Two other possible combinations are those of the original two compounds; that is, $A^+ + B^-$ and $C^+ + D^-$. Since we originally had a solution containing each of these pairs of ions, they can mutually exist in the same solution; therefore they do not recombine. Thus the two possibilities for chemical reaction are the combination of each of the positive ions with the negative ion of the other compound; that is, $A^+ + D^-$ and $C^+ + B^-$. Let us look at some examples.

**Example 1.** When solutions of sodium chloride and potassium nitrate are mixed, the equation for the double displacement reaction (hypothetical) is

$$NaCl + KNO_3 \longrightarrow KCl + NaNO_3$$

We get the hypothetical products by simply combining each positive ion with the other negative ion. But has there been a reaction? When we do the experiment we see no evidence of reaction. There is no precipitate formed, no gas evolved, and no obvious temperature change. Thus we must conclude that no reaction occurred. Both hypothetical products are soluble salts, so the ions are still present in solution. We can say that we simply have a solution of the four kinds of ions, $Na^+$, $Cl^-$, $K^+$, and $NO_3^-$.

The situation is best expressed by changing the equation to

$$NaCl + KNO_3 \longrightarrow \text{No reaction}$$

**Example 2.** When solutions of sodium chloride and silver nitrate are mixed, the equation for the double displacement reaction (hypothetical) is

$$NaCl + AgNO_3 \longrightarrow NaNO_3 + AgCl$$

A white precipitate is produced when these solutions are mixed. This precipitate is definite evidence of a chemical reaction. One of the two products, sodium nitrate ($NaNO_3$) or silver chloride ($AgCl$), is insoluble. Although the precipitate can be identified by further chemical testing, we can instead look at the **Solubility Table in Appendix 5** to find that sodium nitrate is soluble but silver chloride is insoluble. We may then conclude that the precipitate is silver chloride and indicate this in the equation with a downward arrow, ↓. Thus

$$NaCl + AgNO_3 \longrightarrow NaNO_3 + AgCl\downarrow$$

**Example 3.** When solutions of sodium carbonate and hydrochloric acid are mixed, the equation for the double displacement reaction (hypothetical) is

$$Na_2CO_3 + 2\,HCl \longrightarrow 2\,NaCl + H_2CO_3$$

Bubbles of a colorless gas are evolved when these solutions are mixed. Although this gas is evidence of a chemical reaction, neither of the indicated products is a gas. But carbonic acid, $H_2CO_3$, is an unstable compound and readily decomposes into carbon dioxide and water.

$$H_2CO_3 \longrightarrow H_2O + CO_2$$

Therefore, $CO_2$ and $H_2O$ are the products that should be written in the equation. The original equation then becomes

$$Na_2CO_3 + 2\,HCl \longrightarrow 2\,NaCl + H_2O + CO_2\uparrow$$

The evolution of a gas is indicated by an upward arrow, ↑.

Examples of other substances that decompose to form gases are sulfurous acid ($H_2SO_3$) and ammonium hydroxide ($NH_4OH$):

$$H_2SO_3 \longrightarrow H_2O + SO_2\uparrow$$

$$NH_4OH \longrightarrow H_2O + NH_3\uparrow$$

**Example 4.** When solutions of sodium hydroxide and hydrochloric acid are mixed, the equation for the double displacement reaction (hypothetical) is

$$NaOH + HCl \longrightarrow NaCl + H_2O$$

The mixture of these solutions produces no visible evidence of reaction, but on touching the test tube we notice that it feels warm. The evolution of heat is evidence of a chemical reaction. This example and Example 1 appear similar because there is no visible evidence of reaction. However, the difference is very important. In Example 1 all four ions are still uncombined. In the present example the hydrogen ions ($H^+$) and hydroxide ions ($OH^-$) are no longer free in solution but have combined to form water. The reaction of $H^+$ (an acid) and $OH^-$ (a base) is called **neutralization.** The formation of the slightly ionized compound (water) caused the reaction to occur and was the source of the heat liberated.

Water is the most common slightly ionized substance formed in double displacement reactions; other examples are acetic acid ($HC_2H_3O_2$), oxalic acid ($H_2C_2O_4$), and phosphoric acid ($H_3PO_4$).

From the four examples cited we see that a double displacement reaction will occur if at least one of the following classes of substances is formed by the reaction:

1. A precipitate.
2. A gas.
3. A slightly ionized compound.

## PROCEDURE

**Wear protective glasses.**

Each part of the experiment (except No. 12) consists of mixing equal volumes of two solutions in a test tube. Use about a **3 mL sample** of each solution (about 1.5 cm of liquid in a standard test tube). It is not necessary to measure each volume accurately. Record your observations at the time of mixing. Where there is no visible evidence of reaction, feel each tube, or check with a thermometer, to determine if heat is evolved (exothermic reaction). In each case where a reaction has occurred, complete and balance the equation, properly indicating precipitates and gases. When there is no evidence of reaction, write the words "No reaction" as the right-hand side of the equation.

1. Mix 0.1 M sodium chloride and 0.1 M potassium nitrate solutions. No Reaction
2. Mix 0.1 M sodium chloride and 0.1 M silver nitrate solutions.
3. Mix 0.1 M sodium carbonate and conc. (12 M) hydrochloric acid solutions.
4. Mix 10 percent sodium hydroxide and dil. (6 M) hydrochloric acid solutions.
5. Mix 0.1 M barium chloride and dil. (3 M) sulfuric acid solutions.
6. Mix **dilute** (6 M) ammonium hydroxide and **dilute** (3 M) sulfuric acid solutions.
7. Mix 0.1 M copper(II) sulfate and 0.1 M zinc nitrate solutions.
8. Mix 0.1 M sodium carbonate and 0.1 M calcium chloride solutions.
9. Mix 0.1 M copper(II) sulfate and 0.1 M ammonium chloride solutions.
10. Mix 10 percent sodium hydroxide and dil. (6 M) nitric acid solutions.
11. Mix 0.1 M iron(III) chloride and dil. (6 M) ammonium hydroxide solutions.
12. **Do this part in the hood.** Add 1 g of solid sodium sulfite to 3 mL of water and shake to dissolve. Add about 1 mL of conc. (12 M) hydrochloric acid solution, dropwise, using a medicine dropper.

# EXPERIMENT 10

# Single Displacement Reactions

## MATERIALS AND EQUIPMENT

Solids: strips of sheet copper, lead, and zinc measuring about 1 X 2 cm; and sandpaper or emery cloth. Solutions: 0.1 M copper(II) nitrate [$Cu(NO_3)_2$], 0.1 M lead(II) nitrate [$Pb(NO_3)_2$], 0.1 M magnesium sulfate ($MgSO_4$), 0.1 M silver nitrate ($AgNO_3$), and dilute (3 M) sulfuric acid ($H_2SO_4$).

## DISCUSSION

The chemical reactivity of elements varies over an immense range. Some, like sodium and fluorine, are so reactive that they are never found in the free or uncombined state in nature. Others, like xenon and platinum, are nearly inert and can be made to react with other elements only under special conditions.

The **reactivity** of an element is related to its tendency to lose or gain electrons; that is, to be oxidized or reduced. In principle it is possible to arrange nearly all the elements into a single series in order of their reactivities. A series of this kind indicates which free elements are capable of displacing other elements from their compounds. Such a list is known as an **activity** or **electromotive series.** To illustrate the preparation of an activity series we will experiment with a small group of selected elements and their compounds.

A generalized single displacement reaction is represented in the form

$$A + BC \longrightarrow B + AC$$

Element A is the more active element and replaces element B from the compound BC. But if element B is more active than element A, no reaction will occur.

Let us consider two specific examples, using copper and mercury.

**Example 1.** A few drops of mercury metal are added to a solution of copper(II) chloride ($CuCl_2$).

**Example 2.** A strip of metallic copper is immersed in a solution of mercury(II) chloride ($HgCl_2$).

In Example 1 no change is observed even after the solution has been standing for a prolonged time, and we conclude that there is no reaction. In Example 2 the copper strip is soon coated with metallic mercury and the solution becomes pale green. From this evidence we conclude that mercury will not displace copper in copper compounds but copper will displace mercury in mercury compounds. Therefore copper is a more reactive metal than mercury and is above mercury in the activity series. In terms of chemical equations these facts may be represented as

**Example 1.**

$$Hg + CuCl_2 \longrightarrow \text{No reaction}$$

**Example 2.**

$$Cu + HgCl_2 \longrightarrow Hg + CuCl_2$$

The second equation shows that, in terms of oxidation numbers, the chloride ion remained unchanged, mercury changed from +2 to 0, and copper changed from 0 to +2. The +2 oxidation state of copper is the one normally formed in solution.

Expressed another way, the actual reaction that occurred was the displacement of a mercury ion by a copper atom. This can be expressed more simply in equation form:

$$Cu^0 + Hg^{2+} \longrightarrow Cu^{2+} + Hg^0$$

In contrast to double displacement reactions, single displacement reactions involve changes in oxidation number and therefore are also classified as **oxidation-reduction reactions.**

## PROCEDURE

**Wear protective glasses.**

NOTES:

1. With some of the combinations used in these experiments the reactions may be slow or difficult to detect. If you see no immediate evidence of reaction, set the tube aside and allow it to stand for about 10 minutes, then reexamine it.

2. Evidence of reaction will be either evolution of a gas or appearance of a metallic deposit on the surface of the metal strip. Metals deposited from a solution are often black or gray (in the case of copper, very dark reddish brown) and bear little resemblance to commercially prepared metals.

3. Dispose of unreacted metal strips in the special container provided.

Obtain three pieces of sheet zinc, two of copper, and one of lead. Clean the metal pieces with fine sandpaper or emery cloth to expose fresh metal surfaces. Place six clean test tubes in a rack and number or position-code them, then add the following reagents:

Tube 1: Copper strip and about 4 mL silver nitrate.

Tube 2: Lead strip and about 4 mL copper(II) nitrate.

Tube 3: Zinc strip and about 4 mL lead(II) nitrate.

Tube 4: Zinc strip and about 4 mL magnesium sulfate.

Tube 5: Copper strip and about 4 mL dilute sulfuric acid.

Tube 6: Zinc strip and about 4 mL dilute sulfuric acid.

Observe the contents of each tube carefully and record any evidence of chemical reaction.

# EXPERIMENT 11

# Ionization—Acids, Bases, and Salts

## MATERIALS AND EQUIPMENT

*Demonstration.* Solids: sodium chloride ($NaCl$) and sugar ($C_{12}H_{22}O_{11}$). Liquid: glacial acetic acid ($HC_2H_3O_2$). Solutions: 0.1 M ammonium chloride ($NH_4Cl$), 1 M ammonium hydroxide ($NH_4OH$), 1 M acetic acid ($HC_2H_3O_2$), saturated barium hydroxide [$Ba(OH)_2$], 0.1 M copper(II) sulfate ($CuSO_4$), 1 M hydrochloric acid ($HCl$), 0.1 M potassium chromate ($K_2CrO_4$), 0.1 M sodium bromide ($NaBr$), 1 M sodium hydroxide ($NaOH$), 0.1 M sodium nitrate ($NaNO_3$), and dilute (3 M) sulfuric acid ($H_2SO_4$). Conductivity apparatus; magnetic stirrer and stirring bar.

Solids: calcium hydroxide [$Ca(OH)_2$], calcium oxide ($CaO$), iron wire (paper clips), magnesium ribbon ($Mg$), magnesium oxide ($MgO$), marble chips ($CaCO_3$), sodium bicarbonate ($NaHCO_3$), sulfur ($S$), and wood splints. Solutions: dilute (6 M) acetic acid ($HC_2H_3O_2$), concentrated (15 M) ammonium hydroxide ($NH_4OH$), dilute (6 M) hydrochloric acid ($HCl$), dilute (6 M) nitric acid ($HNO_3$), phenolphthalein, 10 percent sodium hydroxide ($NaOH$), and dilute (3 M) sulfuric acid ($H_2SO_4$).

## DISCUSSION

The Brønsted-Lowry acid-base theory defines an **acid** as a substance that will liberate or give up a proton (a hydrogen ion, $H^+$) and a **base** as a substance that will combine with or accept a proton. In short, an **acid** is a **proton donor** and a **base** is a **proton acceptor**.

Many compounds may be recognized as acids from their written formulas. The ionizable hydrogen atoms, which are responsible for the acidity, are written first, followed by the symbols of the other elements in the formula. Examples are

| | | | |
|---|---|---|---|
| $HCl$ | Hydrochloric acid | $H_2CO_3$ | Carbonic acid |
| $HNO_3$ | Nitric acid | $HNO_2$ | Nitrous acid |
| $H_2SO_4$ | Sulfuric acid | $H_2SO_3$ | Sulfurous acid |
| $HC_2H_3O_2$ | Acetic acid | $H_2C_2O_4$ | Oxalic acid |
| $H_3PO_4$ | Phosphoric acid | | |

In a like manner common bases may be recognized by their formulas as a hydroxide ion ($OH^-$) combined with a metal or other positive ion. Examples are

| | |
|---|---|
| $NaOH$ | Sodium hydroxide |
| $KOH$ | Potassium hydroxide |
| $Ca(OH)_2$ | Calcium hydroxide |
| $Mg(OH)_2$ | Magnesium hydroxide |
| $NH_4OH$ | Ammonium hydroxide |

An aqueous solution will be either acidic, basic, or neutral, depending on the nature of the dissolved solute. The acidic properties of an acid are due to **hydronium ions** ($H_3O^+$) in solution. A **hydronium** ion is a hydrated hydrogen ion ($H^+ \cdot H_2O$). To simplify writing equations, the formula of the hydronium ion is often abbreviated $H^+$. However, free hydrogen ions do not actually exist in solution. The basic properties of a base are due to hydroxide ions in solution.

Water behaves as both an acid and a base, as illustrated by the equation

$$H_2O + H_2O \rightleftarrows H_3O^+ + OH^-$$

One water molecule has donated a proton (acted as an acid) and another water molecule has accepted a proton (acted as a base). Since hydronium and hydroxide ions are produced in equal concentrations, water is neutral.

Aqueous solutions contain both hydronium and hydroxide ions. In acidic solutions the concentration of the hydronium ions is greater than that of the hydroxide ions. In basic solutions the concentration of the hydroxide ions is greater than that of the hydronium ions. The terms **alkali** and **alkaline solutions** are used synonymously with base and basic solutions.

Substances used to determine whether a solution is acidic or basic are known as **indicators.** These indicators are usually organic compounds that change color at a particular hydronium or hydroxide ion concentration. For example, litmus, a vegetable dye, shows a pink color in acidic solutions and a blue color in basic solutions. Another common indicator is phenolphthalein; it is colorless in acidic solutions and pink in basic solutions.

The reaction of an acid and a base to form water and a salt is known as **neutralization.** For example,

$$HCl + NaOH \longrightarrow H_2O + NaCl$$

An indicator can be used to determine when the acid in a solution has been neutralized. For instance, when sodium hydroxide solution is added to a hydrochloric acid solution containing phenolphthalein, the solution turns faintly pink when the acid is neutralized and one more drop of the base is added.

**Salts** consist of a positively charged ion ($H^+$ excluded) and a negatively charged ion ($O^{2-}$ and $OH^-$ excluded). Salts may be formed by the reaction of acids and bases, or by replacing the hydrogen atoms in an acid with a metal, or by the interaction of two other salts. There are many more salts than acids and bases. For example, from a single acid such as HCl we may produce many chloride salts (e.g., NaCl, KCl, RbCl, $CaCl_2$, $NH_4Cl$, $NiCl_2$, $FeCl_3$, etc.).

Pure water will not conduct an electric current. However, certain classes of compounds such as acids, bases, and salts, when dissolved in water cause the resulting solution to become a conductor of electricity. These substances whose water solutions are conductors of electricity are called **electrolytes.** Compounds, such as hydrogen chloride and soluble oxides, that react with water to form acids or bases are also electrolytes. Metal oxides that react with water form bases and are called **basic anhydrides.** Nonmetal oxides that react with water form acids and are called **acid anhydrides.** Substances whose water solutions are nonconductors are called **nonelectrolytes.** Examples of nonelectrolytes are sugar and alcohol.

Solutions conduct electricity because they contain ions (electrically charged atoms or groups of atoms), which are free to move. For instance, when sodium chloride dissolves in water, the sodium and chloride ions in the salt crystals disperse throughout the solution. The electrical current through the solution is the movement of these ions to the positive and negative electrodes.

When hydrogen chloride gas, abbreviated HCl(g), dissolves in water it reacts with some of the water to form hydronium and chloride ions. The resulting solution is a conductor and is known as hydrochloric acid. The process of forming ions in this manner is known as **ionization.** The equation for the reaction is

$$HCl(g) + H_2O \longrightarrow H_3O^+ + Cl^-$$

The necessity for water in this ionization process is illustrated by the fact that, when hydrogen chloride is dissolved in benzene, no ions are formed and the solution is a nonconductor.

The formula HCl is often used to represent both hydrogen chloride and hydrochloric acid.

Electrolytes are classified as strong or weak depending on the extent to which they exist as ions in solution. **Strong electrolytes** are essentially 100 percent ionized; that is, they exist totally as ions in solution. **Weak electrolytes** are considerably less ionized; only a small amount of the dissolved substance exists as ions, the remainder being in the un-ionized or molecular form. Most salts are strong electrolytes; acids and bases occur as both strong and weak electrolytes. Examples are as follows:

| Strong Electrolytes | Weak Electrolytes |
|---|---|
| Most salts | $HC_2H_3O_2$ |
| HCl | $H_2SO_3$ |
| $H_2SO_4$ | $HNO_2$ |
| $HNO_3$ | $H_2CO_3$ |
| NaOH | $H_2S$ |
| KOH | $H_2C_2O_4$ |
| $Ba(OH)_2$ | $H_3PO_4$ |
| $Ca(OH)_2$ | $NH_4OH$ |

## PROCEDURE

**Wear protective glasses.**

### A. Ionization—Instructor Demonstration

All of the following tests (except number 8) are performed in 18 X 150 mm test tubes, using the conductivity apparatus shown in Figure 11.1 or other suitable conductivity apparatus. The electrodes should be rinsed thoroughly with distilled water between the testing of different solutions.

Each test is performed by filling a test tube about half full of the liquid to be tested, then raising the test tube up around a pair of electrodes. When a measurable number of ions are in solution, the solution will conduct the electric current and the light will glow. A dimly glowing

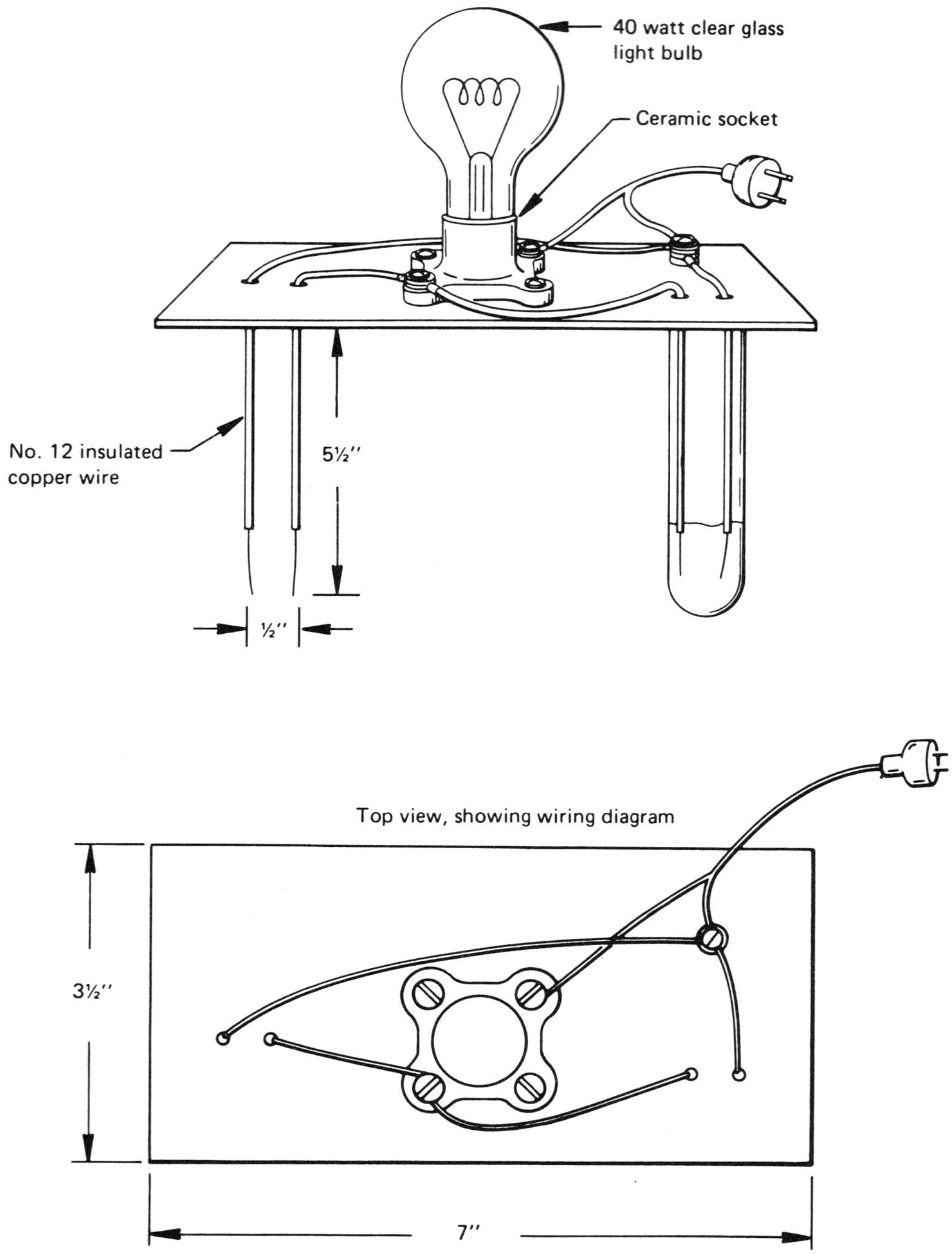


**Figure 11.1. Conductivity apparatus.**

light indicates a relatively small number of ions in solution; a brightly glowing light indicates a relatively large number.

> NOTE: The student should complete the data table in the report sheet at the time the demonstration is performed.

1. Test the conductivity of distilled water.

2. Test the conductivity of tap water.

3. Add a small amount of sugar to a test tube that is half full of distilled water. Dissolve the sugar and test the solution for conductivity.

4. Add a small amount of sodium chloride to a test tube that is half full of distilled water. Dissolve the salt and test the solution for conductivity.

5. Remove the plug from the electrical outlet, clean and dry the electrodes, and reconnect the plug.

(a) Test the conductivity of glacial acetic acid.

(b) Pour out half of the acid, replace with distilled water, mix, and test the solution for conductivity.

(c) Pour out half of the solution in 5(b), replace with distilled water, mix, and test the solution for conductivity.

6. Strong and weak acids and bases. Test the following 1 molar solutions for conductivity: acetic acid, hydrochloric acid, ammonium hydroxide, sodium hydroxide. If the conductivity apparatus has two sets of electrodes, as shown in Figure 11.1, the relative conductivity of the strong and weak acids or bases may be compared by alternately raising a tube of each solution around the electrodes.

7. Test the following 0.1 M salt solutions for conductivity: sodium nitrate, sodium bromide, potassium chromate, copper(II) sulfate, and ammonium chloride.

8. Clean the electrodes well. Place about 25 mL of distilled water and 1 drop of dil. (3 M) sulfuric acid in a 150 mL beaker. Place the beaker on a magnetic stirrer and dip one set of electrodes into the solution. With the stirrer turning, add saturated barium hydroxide solution dropwise until the light goes out completely. Add a few more milliliters of barium hydroxide solution.

## B. Properties of Acids

### 1. Reaction with a Metal

(a) Into four consecutive test tubes place about 5 mL of dil. (6 M) hydrochloric, (3 M) sulfuric, (6 M) nitric, and (6 M) acetic acids.

(b) Place a small strip of magnesium ribbon into each tube, one at a time, and test the gas evolved for hydrogen by bringing a burning splint to the mouth of the tube. If the liberation of gas is slow, stopper the test tube loosely for a minute or two before testing for hydrogen.

2. **Effect on Indicators**

(a) Test dilute solutions of hydrochloric acid, acetic acid, and sulfuric acid by placing a drop of each acid from a stirring rod onto a strip of red and onto a strip of blue litmus paper. Note any color changes.

(b) Add 2 drops of phenolphthalein solution to about 5 mL of distilled water. Add several drops of dilute hydrochloric acid, mix, and note any color change.

3. **Reaction with Carbonates and Bicarbonates**

(a) Cover the bottom of a 150 mL beaker with a small quantity of sodium bicarbonate powder. Now add about 4 to 5 mL of dil. (6 M) hydrochloric acid to the beaker and cover with a glass plate. After about 30 seconds lower a burning splint into the beaker and observe the results.

(b) Repeat the above experiment, using a few granules of marble chips (calcium carbonate) instead of sodium bicarbonate. Allow the reaction to proceed for 2 minutes before testing with the burning splint.

4. **Reaction with Bases—Neutralization.** To about 25 mL of water in a beaker, add 3 drops of phenolphthalein solution and 5 drops of dil. (6 M) hydrochloric acid. Using a medicine dropper, add 10 percent sodium hydroxide solution dropwise, stirring after each drop, until the indicator in the solution changes color. Then add dilute hydrochloric acid, drop by drop. stirring after each drop, until the indicator becomes colorless again. Repeat the additions of base and acid one or two more times.

NOTE: Dispose of unreacted marble chips in the wastebasket, not the sink.

5. **Nonmetal Oxide plus Water**

(a) **Do this part in the hood.** Place a small lump of sulfur in a deflagrating spoon and start it burning by heating in the burner flame. Lower the burning sulfur into a wide-mouth bottle containing 15 mL of distilled water and let the sulfur burn for 2 minutes. Remove the deflagration spoon and quench the excess burning sulfur in a beaker of water. Cover the bottle with a glass plate and shake the bottle back and forth to dissolve the sulfur dioxide gas. Test the solution with blue litmus paper.

(b) Fit a test tube with a one-hole stopper containing a glass delivery tube long enough to extend to the bottom of another test tube. Place several pieces of marble chips and a few milliliters of dil. (6 M) hydrochloric acid into the tube and insert the stopper. Bubble the liberated carbon dioxide into another test tube containing 10 mL water, 2 drops of 10 percent sodium hydroxide solution, and 2 drops of phenolphthalein solution (see Figure 11.2). Record the results.

### C. Properties of Bases

1. **"Feel" Test.** Make very dilute base solutions by adding 3 drops of concentrated ammonium hydroxide to 10 mL of water in a test tube and 3 drops of 10 percent sodium hydroxide solution to 10 mL of water in another test tube. Rub a small amount of each very dilute solution between your fingers to obtain the characteristic "feel" of a hydroxide (base) solution. Wash your hands thoroughly immediately after making the "feel" test. Save the very dilute base solutions for Part C. 2.

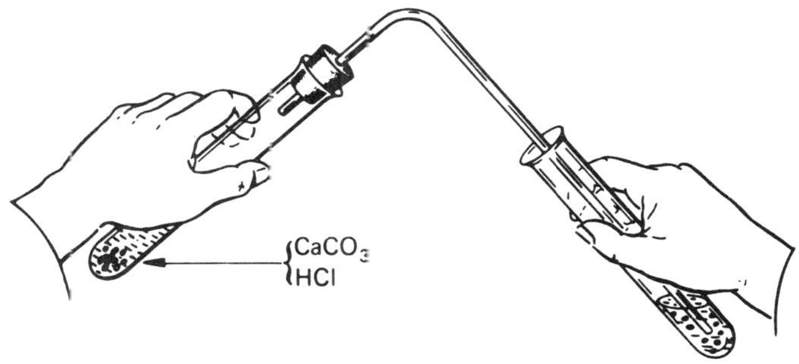


**Figure 11.2. Generator for carbon dioxide.**

2. **Effect on Indicators**

(a) Test the two base (alkaline) solutions prepared in C.1 with both red and blue litmus paper. Note any color changes.

(b) Add 2 drops of phenolphthalein solution to each of the two alkaline solutions. Note any color changes.

3. **Metal Oxides plus Water**

(a) Place 10 mL of water and 2 drops of phenolphthalein solution in each of 3 test tubes. Add a pinch of calcium oxide to the first, magnesium oxide to second, and calcium hydroxide to the third tube. Note and record the results.

(b) Wind the end of a 5 cm piece of iron wire (or paper clip) around a small marble chip. Grasp the wire with tongs and heat the marble chip in the hottest part of the burner flame (see Figure 1.2) for about 2 minutes–the edges of the chip should become white hot while being heated. Allow the chip to cool; then drop it into a beaker containing 15 mL of water and 2 drops of phenolphthalein solution. For comparison, repeat this part of the experiment with a marble chip which has not been heated. Note the results. Return the iron wire to the reagent shelf.

4. **Reaction with Acids–Neutralization**

Review Part B.4.

NAME ____________________

SECTION ________ DATE ____________

## *REPORT FOR EXPERIMENT 11*

INSTRUCTOR ________________

# Ionization—Acids, Bases, and Salts

### A. Ionization—Instructor Demonstration

1. Complete the table for each of the substances tested in the ionization demonstration. Place an "X" in the column where the property of the substance tested fits the column description.

| | Nonelectrolyte | Strong Electrolyte | Weak Electrolyte |
|---|---|---|---|
| Distilled water | | | |
| Tap water | | | |
| Sugar | | | |
| NaCl | | | |
| $HC_2H_3O_2$ (glacial) | | | |
| 1st dilution | | | |
| 2nd dilution | | | |
| 1 M $HC_2H_3O_2$ | | | |
| 1M HCl | | | |
| 1 M $NH_4OH$ | | | |
| 1 M NaOH | | | |
| $NaNO_3$ | | | |
| NaBr | | | |
| $K_2CrO_4$ | | | |
| $CuSO_4$ | | | |
| $NH_4Cl$ | | | |

2. In the conductivity tests, what controlled the brightness of the light?

3. Write an equation to show how acetic acid reacts with water to produce ions in solution.

4. What classes of compounds tested are electrolytes?

5. (a) Write an equation for the chemical reaction that occurred between sulfuric acid and barium hydroxide.

   (b) Explain in terms of the properties of the products formed why the light went out when barium hydroxide was added to sulfuric acid solution, even though both of these reactants are electrolytes.

   (c) Explain why the light came on again when additional barium hydroxide was added.

## B. Properties of Acids

### 1. Reaction with a Metal

(a) Write the formulas of the acids which liberated hydrogen gas when reacting with magnesium metal.

____________________

(b) Write equations to represent the reactions in which hydrogen gas was formed.

2. **Effect on Indicators**

(a) What is the effect of acids on the color of red litmus?

(b) What is the effect of acids on the color of blue litmus?

(c) What color is phenolphthalein in an acid solution? ____________

3. **Reaction with Carbonates and Bicarbonates**

(a) What gas is formed in these reactions?

Name ____________ Formula ____________

(b) What happened to the burning splint when it was thrust into the beaker?

(c) What do you conclude about one of the properties of the gas in the beaker, based on the behavior of the burning splint?

(d) Complete and balance the equations representing the reactions:

$NaHCO_3 + HCl \longrightarrow$

$CaCO_3 + HCl \longrightarrow$

4. **Reaction with Bases—Neutralization**

(a) Write an equation for the neutralization reaction of HCl and NaOH.

(b) How did you know when all the acid was neutralized?

5. **Nonmetal Oxide plus Water**

(a) Write an equation for the combustion of sulfur in air.

(b) What acid is formed when the product of the sulfur combustion reacts with water?

Name ______________________________ Formula ____________________

Write the equation for its formation.

(c) What evidence in this experiment leads you to believe that carbon dioxide in water has acidic properties?

(d) What acid is formed when carbon dioxide reacts with water?

Name ______________________________ Formula ____________________

## C. Properties of Bases

1. **"Feel" Test.** What is the characteristic feel of basic solutions?

2. **Effect on Indicators**

(a) What is the effect of bases on the color of red litmus?

(b) What is the effect of bases on the color of blue litmus?

(c) What color is phenolphthalein in a basic solution? ____________________

3. **Metal Oxides plus Water**

(a.1) Color (if any) produced by phenolphthalein.

Color with $CaO$ ____________________

Color with $MgO$ ____________________

Color with $Ca(OH)_2$ ____________________

(a.2) Complete and balance these equations:

$CaO + H_2O \longrightarrow$

$MgO + H_2O \longrightarrow$

(b.1) The formula for marble is $CaCO_3$. What compounds are formed when it is heated strongly?

(b.2) Write the equation representing this decomposition:

$CaCO_3 \xrightarrow{\Delta}$

(b.3) What evidence led you to formulate the composition of the solid residue after heating the marble chip?

## ADDITIONAL QUESTIONS AND PROBLEMS

1. State whether each of the formulas below represents an acid, a base, a salt, or none of these types of compound:

| | | | |
|---|---|---|---|
| $CuF_2$ | ____________ | $CaSO_4$ | ____________ |
| $Ba(OH)_2$ | ____________ | $C_2H_4$ | ____________ |
| $LiOH$ | ____________ | $ICl$ | ____________ |
| $HBrO_3$ | ____________ | $HI$ | ____________ |
| $RaCO_3$ | ____________ | $SiO_2$ | ____________ |
| $KNO_2$ | ____________ | $HCN$ | ____________ |
| $H_2C_2O_4$ | ____________ | $ZnS$ | ____________ |

2. Complete and balance the following equations and name the product formed. (Only one product is formed in each case.)

(a) $K_2O + H_2O \longrightarrow$ ____________

(b) $SrO + H_2O \longrightarrow$ ____________

(c) $SO_3 + H_2O \longrightarrow$ ____________

(d) $N_2O_5 + H_2O \longrightarrow$ ____________

# EXPERIMENT 12

# Identification of Selected Anions

## MATERIALS AND EQUIPMENT

Liquid: 1,1,1-trichlorethane ($CCl_3CH_3$). Solutions: 0.1 M barium chloride ($BaCl_2$), **freshly prepared chlorine water** ($Cl_2$), dilute (6 M) hydrochloric acid (HCl), dilute (6 M) nitric acid ($HNO_3$), 0.1 M silver nitrate ($AgNO_3$), 0.1 M sodium arsenate ($Na_3AsO_4$), 0.1 M sodium bromide (NaBr), 0.1 M sodium carbonate ($Na_2CO_3$), 0.1 M sodium chloride (NaCl), 0.1 M sodium iodide (NaI), 0.1 M sodium phosphate ($Na_3PO_4$), 0.1 M sodium sulfate ($Na_2SO_4$), and unknown solutions.

## DISCUSSION

The examination of a sample of inorganic material to identify the ions that are present is called **qualitative analysis.** To introduce qualitative analysis, we will analyze for seven anions (negatively charged ions). The ions selected for identification are chloride ($Cl^-$), bromide ($Br^-$), iodide ($I^-$), sulfate ($SO_4^{2-}$), phosphate ($PO_4^{3-}$), carbonate ($CO_3^{2-}$), and arsenate ($AsO_4^{3-}$).

Qualitative analysis is based on the fact that no two ions behave identically in all of their chemical reactions. Identification depends on appropriate chemical tests coupled with careful observation of such characteristics as solution color, formation and color of precipitates, evolution of gases, etc. Test reactions are selected to identify the ions in the fewest steps possible. In this experiment only one anion is assumed to be present in each sample. If two or more anions must be detected in a single solution, the scheme of analysis can be considerably more complex.

### Silver Nitrate Test

When solutions of the sodium salts of the seven anions are reacted with silver nitrate solution, the following precipitates are formed: AgCl, AgBr, AgI, $Ag_3PO_4$, $Ag_2CO_3$, and $Ag_3AsO_4$. $Ag_2SO_4$ is moderately soluble and does not precipitate at the concentrations used in these solutions. When dilute nitric acid is added, the precipitates $Ag_3PO_4$, $Ag_2CO_3$, and $Ag_3AsO_4$ dissolve; AgCl, AgBr, and AgI remain undissolved.

In some cases a tentative identification of an anion may be made from the silver nitrate test. This identification is based on the color of the precipitate and on whether or not the precipitate is soluble in nitric acid. However, since two or more anions may give similar results, second or third confirmatory tests are necessary for positive identification.

### Barium Chloride Test

When barium chloride solution is added to solutions of the sodium salts of the seven anions, precipitates of $BaSO_4$, $Ba_3(PO_4)_2$, $BaCO_3$, and $Ba_3(AsO_4)_2$ are obtained. No precipitate is obtained with $Cl^-$, $Br^-$ or $I^-$.

When dilute hydrochloric acid is added, the precipitates $Ba_3(PO_4)_2$, $BaCO_3$, and $Ba_3(AsO_4)_2$ dissolve; $BaSO_4$ does not dissolve.

**Trichloroethane Test**

The silver nitrate test can prove the presence of a halide ion ($Cl^-$, $Br^-$, or $I^-$) because the silver precipitates of the other four anions dissolve in nitric acid. But the colors of the three silver halides do not differ sufficiently to establish which halide ion is present.

Adding chlorine water ($Cl_2$ dissolved in water) to halide salts in solution will oxidize bromide ion to free bromine ($Br_2$) and iodide ion to free iodine ($I_2$). The free halogen may be extracted from the water solution by adding trichloroethane and shaking vigorously. The colors of the three halogens in trichloroethane are quite different; $Cl_2$ is pale yellow, $Br_2$ is yellow-orange to reddish-brown, and $I_2$ is pink to violet. After adding chlorine water and shaking, a yellow-orange to reddish-brown color in the trichloroethane layer indicates that $Br^-$ was present in the original solution; a pink to violet color in the trichloroethane layer indicates that $I^-$ was present. However, a pale yellow color does not indicate $Cl^-$, since $Cl_2$ was added as a reagent. But if the silver nitrate test gives a white precipitate that is insoluble in nitric acid, and the trichloroethane test shows no $Br^-$ or $I^-$, then you can conclude that $Cl^-$ was present.

Though we have described many of the expected results, it is necessary to test known solutions to actually see the results of the tests and to develop satisfactory experimental techniques.

In this experiment two "unknown" solutions, each containing one of the seven anions, will be analyzed. When an unknown is analyzed the results should agree in all respects with one of the known anions. If the results do not fully agree with one of the seven known ions, either the testing has been poorly done or the unknown does not contain any of the specified ions.

Three different kinds of equations may be used to express the behavior of ions in solution. For example, the reaction of the chloride ion (from sodium chloride) may be written

(1) $NaCl + AgNO_3 \longrightarrow AgCl\downarrow + NaNO_3$

(2) $Na^+ + Cl^- + Ag^+ + NO_3^- \longrightarrow AgCl\downarrow + Na^+ + NO_3^-$

(3) $Cl^- + Ag^+ \longrightarrow AgCl\downarrow$

Equation (1) is the **un-ionized equation**; it shows the formulas of the substances in the equation as they are normally written. Equation (2) is the **total ionic equation**; it shows the substances as they occur in solution. Strong electrolytes are written as ions; weak electrolytes, precipitates, and gases are written in their un-ionized or molecular form. Equation (3) is the **net ionic equation**; it includes only those substances or ions in Equation (2) that have undergone a chemical change. Thus $Na^+$ and $NO_3^-$ (sometimes called the "spectator" ions) have not changed and do not appear in the net ionic equation. In either a total ionic equation or a net ionic equation, the atoms and charges must be balanced.

## PROCEDURE

**Wear protective glasses.**

Clean nine test tubes and rinse each twice with distilled water (5 mL of distilled water is sufficient to rinse a test tube). Identify these tubes, either with gummed paper labels or with a marking pencil. The order for the known solutions will be NaCl, NaBr, NaI, $Na_2SO_4$, $Na_3PO_4$, $Na_2CO_3$, and $Na_3AsO_4$, with the last two tubes reserved for the unknown solutions.

Clean, rinse (with distilled water), and put blank labels on two additional test tubes; take these tubes to your instructor for the unknown solutions. To avoid possible confusion set the two numbered test tubes—received from your instructor and containing your unknowns—aside in a beaker. Record the numbers of these unknowns in the top right-hand columns of your report sheet. Next pour about 2 mL (no more) of each of the seven known solutions—one solution per tube—into the first seven tubes in the rack. Now number the last two empty tubes to match the unknown solutions, and pour 2 mL of the corresponding unknown into each. Save the remaining portions of the unknown solutions for tests B and C.

NOTES:

1. You can save considerable time by measuring out 2 mL into the first test tube and approximating this volume, on the basis of the height of the liquid, for other 2 mL samples.

2. Read the instructions for recording observations before you record the results of your tests.

## A. Silver Nitrate Test

⚠ CAUTION: If silver nitrate solution gets on your hands, wash it off immediately to avoid stains.

Add about 1 mL of 0.1 M silver nitrate solution to each test tube. Record the results. Now add about 3 mL of dil. (6 M) nitric acid to each test tube; stopper and shake well. Record the results.

## B. Barium Chloride Test

Wash all nine test tubes and rinse each tube twice with distilled water. Again put about 2 mL of the specified solution into each of the nine test tubes. Add about 2 mL of 0.1 M barium chloride solution to each test tube and mix. Record the results. Now add 3 mL of dilute hydrochloric acid to each test tube; stopper and shake well. Record the results.

## C. Trichloroethane Test

Again wash and rinse all nine test tubes. Again put about 2 mL of the specified solution into each of the nine test tubes. Now add about 2 mL of trichloroethane and about 2 mL of chlorine water to each test tube; stopper and shake well. Record the results.
Dispose of mixtures containing trichloroethane in the container provided.

---

## Recording Observations

The following observations should be recorded for each of the solutions tested.

1. Parts A and B: Record whether or not a precipitate (ppt.) is formed. If a precipitate is formed, record "ppt. formed" and include its color. If no precipitate is formed, record "no ppt."

2. If a precipitate is formed, observe whether or not it dissolves when treated with dil. $HNO_3$ in Part A or with dilute HCl in Part B. Record: "ppt. dissolved" or "ppt. didn't dissolve."

3. Part C: Record the color of the $CCl_3CH_3$ layer.

4. Part D in Report Form: Record the formula of the anion present (either known from the formula of the salt tested, or proved by the analysis of the unknown solution).

# EXPERIMENT 13

# Identification of Halide Ions in Mixtures

## MATERIALS AND EQUIPMENT

Liquid: 1,1,1-trichloroethane ($CCl_3CH_3$). Solutions: dilute (6 M) ammonium hydroxide ($NH_4OH$), freshly prepared chlorine water ($Cl_2$), dilute (6 M) nitric acid ($HNO_3$), 0.1 M sodium nitrite ($NaNO_2$), and 0.1 M silver nitrate ($AgNO_3$). Mixed halide solutions: (1) chloride and iodide; (2) bromide and iodide; (3) chloride and bromide; and unknown solutions:

## DISCUSSION

The test procedures given in Experiment 12 are satisfactory for solutions that contain only one kind of anion. More involved procedures are needed for the detection of more than one anion in a single solution. This experiment deals with three halide ions—chloride ($Cl^-$), bromide ($Br^-$), and iodide ($I^-$). The procedures given identify each of these anions, either alone or in the presence of one or both of the others.

With the tests used in this experiment, chloride ions or iodide ions can be detected in the presence of either or both of the other two ions. However, bromide ions cannot be detected in the presence of iodide ions because the intense colors derived from iodide and iodine mask the colors derived from bromide and bromine. Fortunately, of the three halide ions, iodide is the strongest reducing agent. Therefore, iodide ions can be oxidized selectively to iodine and removed from the solution, allowing the detection of bromide ions.

### Iodide Test

This test proves the presence of iodide and removes it from the solution so that the bromide ion can be detected in the next test. In the presence of dilute nitric acid, potassium nitrite will oxidize iodide ions to iodine ($I_2$). When trichloroethane is added, the iodine will preferentially dissolve in the trichloroethane layer, turning it to a pink or red tinged with violet color. This color proves that iodide was present in the original sample. To test for bromide, the remaining water solution must be free of iodine and iodide ions.

### Bromide Test

The water solution from the previous test can now be tested for bromide ion, since the iodide has been removed. Chlorine water is added to the solution, oxidizing bromide ions to bromine ($Br_2$). Bromine will preferentially dissolve in trichloroethane, turning it yellow-orange to reddish-brown color, depending on concentration. This color proves that bromide was present in the original sample. Any unreacted chlorine will also dissolve in the trichloroethane, but it is almost colorless.

**Chloride Test**

In Experiment 12 we found that silver chloride, silver bromide, and silver iodide were all insoluble in water and dilute nitric acid. The test for chloride ion is based on the fact that silver chloride is many times more soluble in dilute ammonium hydroxide than either silver bromide or silver iodide. In the procedure the halide ions in the sample are first precipitated with silver nitrate. The precipitate is then treated with dilute ammonium hydroxide which dissolves the silver chloride. The resulting solution is then acidified with dilute nitric acid. The presence of a cloudy white precipitate (silver chloride) in the acidified solution proves that chloride was present in the original sample.

## PROCEDURE

**Wear protective glasses.**

1. NOTE: To prevent contamination, rinse all glassware (including medicine droppers) with distilled water before each use. For this purpose fill a flask or wash bottle with distilled water and use at your laboratory bench.
2. Dispose of mixtures containing trichloroethane in the container provided.

Five solutions containing halide ions are to be analyzed in this experiment. Three of these solutions are available in the laboratory and contain these anion mixtures: (1) $Cl^-$ and $I^-$; (2) $Br^-$ and $I^-$; (3) $Cl^-$ and $Br^-$. The other two solutions are obtained from your instructor as unknowns, and each of these unknowns may contain **one, two,** or **three** halide ions. To obtain your two unknown solutions, put blank labels on two clean, distilled-water rinsed, test tubes and take them to your instructor. After your instructor has given you the unknowns, set the tubes aside in a beaker and record their code numbers on your report sheet. Samples for testing the unknowns are taken from the tubes in the beaker as needed.

The following instructions are given for testing an **individual** solution sample; however, to save time and make comparisons more readily, each test is run **concurrently on samples of all five solutions.** Place five test tubes in the test tube rack and add the specified samples of the solutions to be tested. Record your observations in the data table as they are made. As each test for each sample is completed, draw a conclusion as to whether the anion being tested for is present or absent. On the last line of the data table write the formulas of all anions proven to be present in each solution tested.

### A. Iodide Test

Use about 2 mL (no more of solution for this test. Add 6 drops of dil. (6 M) nitric acid. Now add 15 drops of 0.1 M sodium nitrite solution and 4 mL of trichloroethane. Stopper and shake the tube vigorously for at least 30 seconds. If the trichloroethane layer (the bottom layer) is not pink, iodide was not present in the sample, and the test tube should be labeled and set aside for Test B.

If the trichloroethane layer is pink, iodide is present in the sample. In this case all of the iodide must be removed by the following procedure in order to test the sample for bromide. Pour the contents of the test tube into your smallest beaker. Hold the beaker up to the light and tilt it at a 45 degree angle. With a clean medicine dropper remove the water layer (top layer), collecting it in a clean test tube. It is not necessary to get every drop of the water layer; in fact, it is more important not to pick up any of the trichloroethane layer. Discard the material remaining in the beaker. To the solution in the test tube add two more drops of sodium nitrite solution.

If darkening occurs, add additional drops of sodium nitrite solution until no further darkening is apparent. Now add 2 mL of trichloroethane; stopper and shake the test tube vigorously for at least 20 seconds. Again pour the contents of the test tube into a small clean beaker and remove the water layer, placing it in a clean test tube. Discard the material remaining in the beaker. To the solution in the test tube again add 2 mL of trichloroethane and shake. If the trichloroethane layer has an appreciable pink color, separate the water layer again and repeat the extraction with 2 mL of trichloroethane. If the trichloroethane layer is not pink, the test tube should be labeled and set aside for Test B.

## B. Bromide Test

You should now have five iodide-free solutions set aside in test tubes, one from each of your samples. Each of these tubes should contain a layer of trichloroethane and the solution to be tested for bromide. To each tube add 20 drops of chlorine water; stopper and shake the tube. If the trichloroethane layer turns orange to reddish-brown, bromide was present in the sample. If the trichloroethane layer remains almost colorless, bromide was not present in the sample.

If you have difficulty in judging whether the color observed in the trichloroethane layer of your unknown is due to bromine, compare it with tube 1 (no $Br_2$ present) and with tubes 2 and 3 ($Br_2$ present).

## C. Chloride Test

Since chlorine was added to the solution to test for bromide ion, a new sample of the original solution must be used for the chloride test.

Place 3 mL (no more) of the solution to be tested in a test tube and add 4 mL of 0.1 M silver nitrate solution. Heat the tube in boiling water for about 1 minute to coagulate the precipitate. When the precipitate is fairly well settled to the bottom of the tube, carefully pour off (decant) and discard the liquid, leaving the precipitate in the tube. (If a small amount of the precipitate is lost when the liquid is decanted, it will not affect the final results.)

To wash the precipitate, add 10 mL of distilled water, 1 drop of dil. (6 M) nitric acid, and shake. When the precipitate has settled, decant and discard the liquid. To the precipitate remaining in the test tube add 4 mL of distilled water, 8 drops (no more) of dil. (6 M) ammonium hydroxide, and 10 drops of 0.1 M silver nitrate solution. Stopper and shake the test tube. When the precipitate has settled to the bottom of the tube and the solution above the precipitate is reasonably clear, decant most of the clear liquid into a clean test tube. Now add 12 drops of dil. (6 M) nitric acid to this liquid. If the solution has a definite milky appearance (not just slightly cloudy), chloride was present in the sample. In making the judgment of the chloride test, compare the appearance of the unknown solutions with those of the known solutions.

NAME____________________

SECTION________ DATE________

INSTRUCTOR________________

## *REPORT FOR EXPERIMENT 13*

# Identification of Halide Ions in Mixtures

**Data Table**

| | $Cl^- + I^-$ | $Br^- + I^-$ | $Cl^- + Br^-$ | Unknown No.____ | Unknown No.____ |
|---|---|---|---|---|---|
| A. **Iodide Test** Observations | | | | | |
| Conclusions | | | | | |
| B. **Bromide Test** Observations | | | | | |
| Conclusions | | | | | |
| C. **Chloride Test** Observations | | | | | |
| Conclusions | | | | | |
| Anions present in each solution tested | | | | | |

# EXPERIMENT 14

# Properties of Lead(II), Silver, and Mercury(I) Ions

## MATERIALS AND EQUIPMENT

Solutions: concentrated (15 M) ammonium hydroxide ($NH_4OH$), dilute (6 M) hydrochloric acid (HCl), 0.1 M lead(II) nitrate [$Pb(NO_3)_2$], 0.1 M mercury(I) nitrate [$Hg_2(NO_3)_2$], dilute (6 M) nitric acid ($HNO_3$), 0.1 M potassium chromate ($K_2CrO_4$), and 0.1 M silver nitrate ($AgNO_3$); "known" solution containing $Pb^{2+}$, $Ag^+$, and $Hg_2^{2+}$ ions; and "unknown" solutions to be analyzed.

## DISCUSSION

The object of this experiment is to investigate some of the chemical properties of lead(II), silver, and mercury(I) cations and to identify these ions in solution. In the broader scheme of qualitative analysis these three cations are known as the silver group. They are grouped together because of the common property of forming water insoluble chlorides, a property which enables them to be separated from most other cations.

First we will run some selected chemical reactions for each of these ions. Then a scheme of analysis based on these reactions will be used to separate and identify these cations in a "known" solution containing all three ions and in an "unknown" solution containing one or more of the ions.

If you have not done Experiment 12, see page 94 for an explanation of un-ionized, total ionic, and net ionic equations.

The net ionic equations representing the reactions to be observed are:

Lead(II) [$Pb(NO_3)_2$ solution]

$Pb^{2+} + 2\ Cl^- \longrightarrow PbCl_2 \downarrow$ (white ppt forms)

$PbCl_2 \xrightarrow[H_2O]{hot} Pb^{2+} + 2\ Cl^-$ (ppt dissolves)

$Pb^{2+} + CrO_4^{\ 2-} \longrightarrow PbCrO_4 \downarrow$ (bright yellow ppt forms)

Silver [$AgNO_3$ solution]

$Ag^+ + Cl^- \longrightarrow AgCl \downarrow$ (white ppt forms)

$AgCl + 2\ NH_4OH \longrightarrow Ag(NH_3)_2^{\ +} + Cl^- + 2\ H_2O$ (ppt dissolves)

$Ag(NH_3)_2^{\ +} + Cl^- + 2\ H^+ \longrightarrow AgCl \downarrow + 2\ NH_4^{\ +}$ (white ppt forms)

Mercury(I) [$Hg_2(NO_3)_2$ solution]

$Hg_2^{2+} + 2\ Cl^- \longrightarrow Hg_2Cl_2 \downarrow$ (white ppt forms)

$Hg_2Cl_2 + 2\ NH_4OH \longrightarrow Hg \downarrow + HgNH_2Cl \downarrow + 2\ H_2O + NH_4^{\ +} + Cl^-$

(black ppt forms)

## PROCEDURE

**Wear protective glasses.**

NOTES:

1. Use distilled water throughout this experiment.

2. Since hot water is frequently needed in the procedure, fill a 400 mL beaker half full of water and start heating it to boiling before beginning to work with your cation solutions.

3. There are numerous places in the procedure where 1 mL of reagent is used. You can save considerable time by determining how many drops are needed to deliver 1 mL from your medicine dropper and using this number of drops whenever 1 mL of reagent is required.

4. Submit a clean, labeled test tube to your instructor for your unknown solution.

5. Precipitates are washed to remove soluble ions. Washing is accomplished by adding the specified amount of solvent (usually water), agitating the mixture by gently shaking back and forth, and pouring off (decanting) the washing solvent. Shaking back and forth, rather than up and down, prevents the accumulation of large amounts of precipitate on the walls of the tube.

6. Record your observations on the report form, as directed, immediately after you make them.

7. Dispose of solutions and precipitates of these heavy metals as directed by your instructor.

### A. Tests for Lead(II) Ion

1. To 2 mL (no more) of 0.1 M lead(II) nitrate solution in a test tube add 1 mL of dil. (6 M) hydrochloric acid.

2. Allow the precipitate to settle and then separate it from the liquid by decanting the liquid. (Decanting means to pour the liquid off carefully, leaving the solid behind.) A loss of a small amount of precipitate at this point has no effect on the experimental results.

3. Wash the precipitate with 2 mL of cold water, again decanting the liquid after the precipitate settles.

4. Add 5 mL of water to the precipitate and place the tube in the beaker of boiling water. Heat the mixture for about 2 minutes, **shaking frequently.** All the precipitate should dissolve.

5. Remove the tube from the beaker and add a few drops of 0.1 M potassium chromate solution. A yellow precipitate of lead(II) chromate confirms the presence of lead(II) ions.

### B. Tests for Silver Ion

1. To 2 mL of 0.1 M silver nitrate solution in a test tube add 1 mL of dilute hydrochloric acid.

2. Allow the precipitate to settle and then decant the liquid.

3. Wash the precipitate with 2 mL of cold water, again decanting the liquid after the precipitate settles.

4. Add concentrated ammonium hydroxide dropwise to the precipitate until it all dissolves. (This should take less than 1 mL.)

5. Now add dil. (6 M) nitric acid dropwise **until the solution is acidic** (test with blue litmus paper). A white precipitate of silver chloride confirms the presence of silver ions.

### C. Tests for Mercury(I) Ion

1. To 2 mL of 0.1 M mercury(I) nitrate solution in a test tube add 1 mL of dilute hydrochloric acid.

2. Allow the precipitate to settle and then decant the liquid.

3. Wash the precipitate with 2 mL of cold water, again decanting the liquid after the precipitate settles.

4. Add about 1 mL of concentrated ammonium hydroxide to the precipitate. The formation of a black precipitate confirms the presence of mercury(I) ions. This precipitate is composed of finely divided black mercury (Hg) and white mercury(II) amido chloride ($HgNH_2Cl$) and appears black overall.

Since the known solution contains all three cations present, you should see evidence of all the reactions indicated for each of these ions. The unknown may not contain all three cations. Therefore, you will not see any evidence of reaction for a cation that is not present.

### D. Analysis of a Known and an Unknown Solution

The sequence of steps that follows is to be performed on a known solution containing all three silver group cations and on an unknown solution obtained from your instructor. With the known solution you should see evidence of all reactions described for each cation. The unknown may contain one, two, or three cations; you will not see any evidence of reaction for a cation that is not present. If you have two funnels, run the known and unknown simultaneously; if you have only one funnel, run the known first.

1. To 2 mL of the sample in a test tube add 1 mL of dil. (6 M) hydrochloric acid.

2. Mix by shaking, allow the precipitate to settle, and decant the liquid.

3. Wash the precipitate with 2 mL of cold water and decant the liquid. Wash the precipitate again with 2 mL of cold water and decant the liquid.

4. Add 5 mL of water to the precipitate and place the tube in the beaker of boiling water for about 3 minutes, shaking frequently. Put about 10 mL of water into each of two additional tubes and heat them in the boiling water. While the liquids are heating prepare a filter cone, place it in a funnel, wet the paper, and place the funnel in a test tube in the rack.

5. Mix and quickly pour the hot solution and the precipitate from D.4 into the funnel, catching the filtrate in the test tube. (Save this filtrate for part D.7.)

6. Transfer the funnel to another test tube. Wash the precipitate **twice** by pouring 5 mL portions of hot water over the precipitate in the funnel. Use the hot water from one of the tubes that you have been heating in the boiling water. Discard this filtrate.

7. Test the filtrate from D.5 for lead(II) ions by adding a few drops of 0.1 M potassium chromate solution to it. A yellow precipitate confirms the presence of lead(II) ion.

8. Transfer the funnel to another test tube and add 1 mL of concentrated ammonium hydroxide to the precipitate in the funnel. The formation of a black residue confirms the presence of mercury(I) ion.

9. Add dil. (6 M) nitric acid to the filtrate from D.8 **until the solution is acidic to litmus** or a permanent white precipitate is formed. The formation of this precipitate confirms the presence of silver ion. (A slightly cloudy filtrate from D.8 is due to incomplete removal of the lead(II) chloride but does not interfere with this test.)

**Diagram of the Silver Group Analysis.**

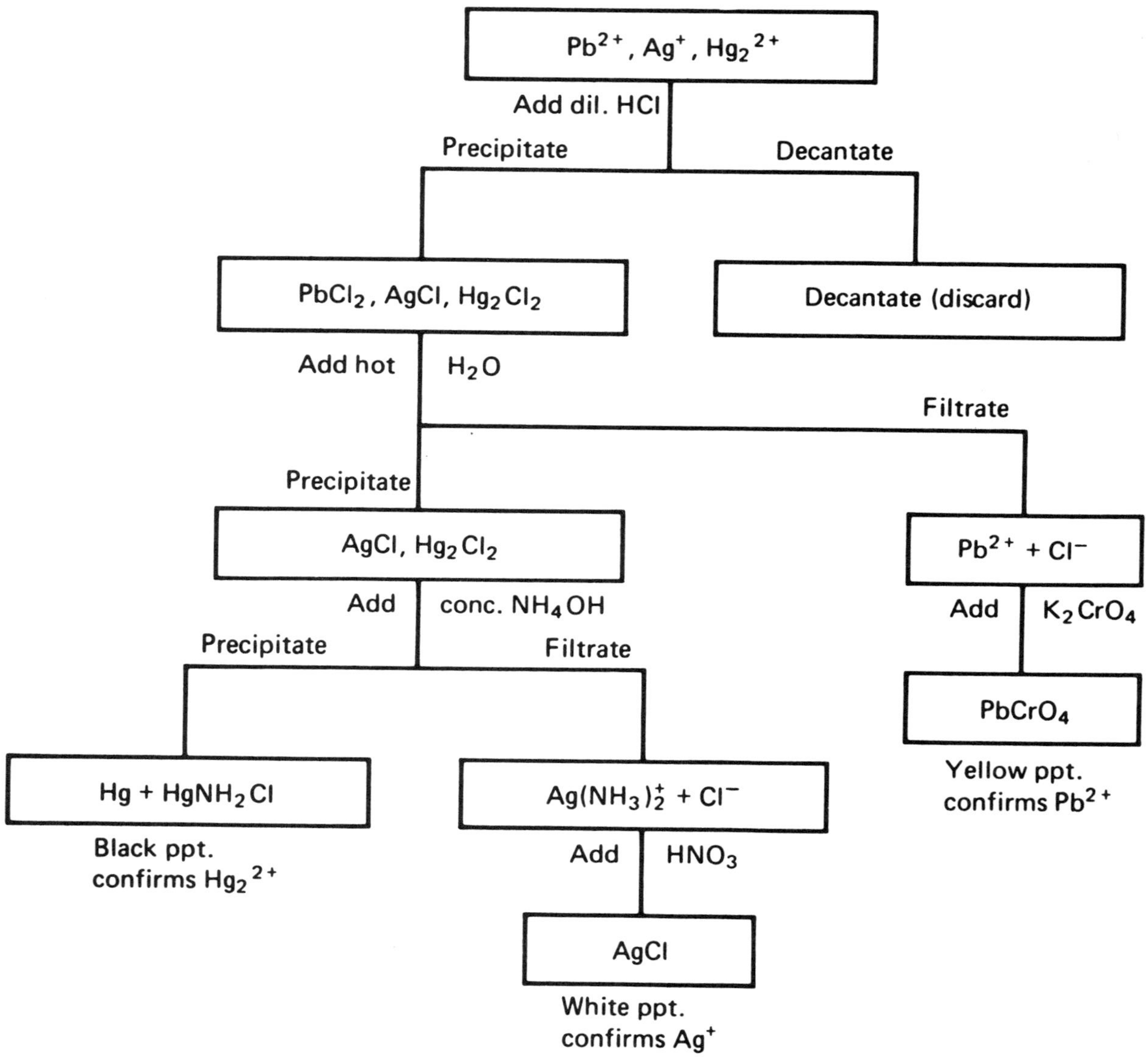

# *EXPERIMENT 15*

# Quantitative Precipitation of Chromate Ion

## MATERIALS AND EQUIPMENT

Solid: potassium chromate ($K_2CrO_4$). Solution: 0.50 M lead(II) nitrate [$Pb(NO_3)_2$].

## DISCUSSION

In this experiment you will examine and verify the mole and mass relationships involved in the quantitative precipitation of chromate ion as lead(II) chromate. Potassium chromate is the source of the chromate ion, and lead(II) ion from lead(II) nitrate is the precipitating agent. The chemistry is expressed in these equations.

$$K_2CrO_4 + Pb(NO_3)_2 \longrightarrow PbCrO_4\downarrow + 2\ KNO_3 \qquad \text{(Un-ionized equation)}$$

$$CrO_4^{2-} + Pb^{2+} \longrightarrow PbCrO_4\downarrow \qquad \text{(Net ionic equation)}$$

The first equation above shows that potassium chromate and lead(II) nitrate react with each other in a 1-to-1 mole ratio. Furthermore, for every mole of chromate ions present, 1 mole of lead(II) chromate is formed. From these molar relationships we can calculate the amount of lead(II) chromate that is theoretically obtainable from any specified amount of potassium chromate in the reaction. This theoretical value can then be compared to experimental value.

To conduct the experiment quantitatively, we need to precipitate all the chromate ion from a known amount of potassium chromate in solution and to isolate the precipitate in as pure a form as feasible. To ensure complete precipitation of the chromate, an excess of lead(II) nitrate is used. Lead(II) chromate is only slightly soluble ($2 \times 10^{-7}$ mole/liter), and the excess lead(II) nitrate reduces the solubility even further.

Use the following relationships in your calculations:

1. Moles $K_2CrO_4$ reacted = moles $Pb(NO_3)_2$ reacted = moles $PbCrO_4$ produced

2. $$\text{Moles} = \frac{\text{g of solute}}{\text{molar mass of solute}}$$

3. $$\text{Molarity} = \frac{\text{moles}}{\text{liter}} = \frac{\text{g of solute}}{\text{molar mass of solute} \times \text{liters of solution}}$$

   Note that molarity is an expression of concentration, the units of which are moles of solute per liter of solution. Thus a 1.00 molar (1.00 M) solution contains 1.00 mole of solute in 1 liter of solution.

4. $$\text{Percentage error} = \frac{(\text{experimental value}) - (\text{theoretical value})}{\text{theoretical value}} \times 100$$

## PROCEDURE

**Wear protective glasses.**

NOTES:

1. Make all weighings to the highest precision possible with the balance available to you.

2. Use the same balance for all weighings.

3. Record all the data directly on the report sheet as they are obtained.

The sequence of major experimental steps in this experiment is as follows:

1. Weigh beaker
2. Weigh potassium chromate into beaker.
3. Dissolve potassium chromate in distilled water.
4. Precipitate lead(II) chromate by adding lead(II) nitrate solution.
5. Weigh filter paper.
6. Filter and wash precipitated lead(II) chromate.
7. Heat and dry lead(II) chromate.
8. Determine the mass of lead(II) chromate precipitated.

Weigh a clean, **dry** 150 mL beaker. Add between 0.7 and 0.8 gram of potassium chromate to the beaker; weigh again. Add 40 mL distilled water and dissolve the potassium chromate. Heat this solution to almost boiling. Obtain 10.0 mL of 0.50 M $Pb(NO_3)_2$ solution in a graduated cylinder and slowly add, while stirring, to the hot potassium chromate solution. (To avoid losses of solution and precipitate, keep the stirring rod in the beaker.) Again heat the mixture almost to the boiling point while stirring.

Accurately weigh a piece of filter paper. Fold it into a cone, place it into a funnel, wet the paper with distilled water and press the top of the paper cone against the funnel. Separate the liquid from the precipitate by carefully decanting the liquid into the funnel, **keeping as much of the precipitate as possible in the beaker.** The decantation is best accomplished without losses by pouring the liquid down a stirring rod (see Figure 1.9 in Experiment 1).

The lead(II) chromate must now be washed free of excess lead(II) nitrate. Wash the precipitate by adding 25 mL of distilled water to the beaker and stirring. Allow the precipitate to settle for about a minute. Decant the wash water into the funnel, again keeping as much of the precipitate as possible in the beaker. Repeat the washing once more with another 25 mL of distilled water. After the last washing, spread the precipitate evenly on the bottom of the beaker with the stirring rod.

Remove the filter paper from the funnel and use it to wipe off any lead(II) chromate adhering to the stirring rod. Place the filter cone down in the beaker as shown in Figure 15.1.

**The following method of drying must be strictly adhered to in order to avoid spattering and scorching the precipitate and the filter paper.** Adjust the burner so you have a nonluminous, 10 to 15 cm flame without a distinct inner cone. Place the beaker on a wire gauze on a ring stand so that the wire gauze is 10 to 15 cm (4 to 6 in.) above the tip of the flame. Heat the beaker and contents for 60 minutes. If spattering of the precipitate occurs remove the burner and either lower the flame or raise the beaker before continuing heating. Make a similar adjustment if you see evidence that the paper is scorching. **Do not leave the drying setup unattended.**

**While the precipitate is drying start working on the report form.**

Cool the beaker, weigh, and reheat for an additional 10 minutes. Cool again and reweigh. If the second weighing is within 0.05 g of the first, the contents of the beaker may be considered dry. If the second weighing has decreased more than 0.05 g, a third heating and weighing is necessary. The experiment is complete after obtaining constant weight (within 0.05 g).
Dispose the $PbCrO_4$ and filter paper in the special container provided.

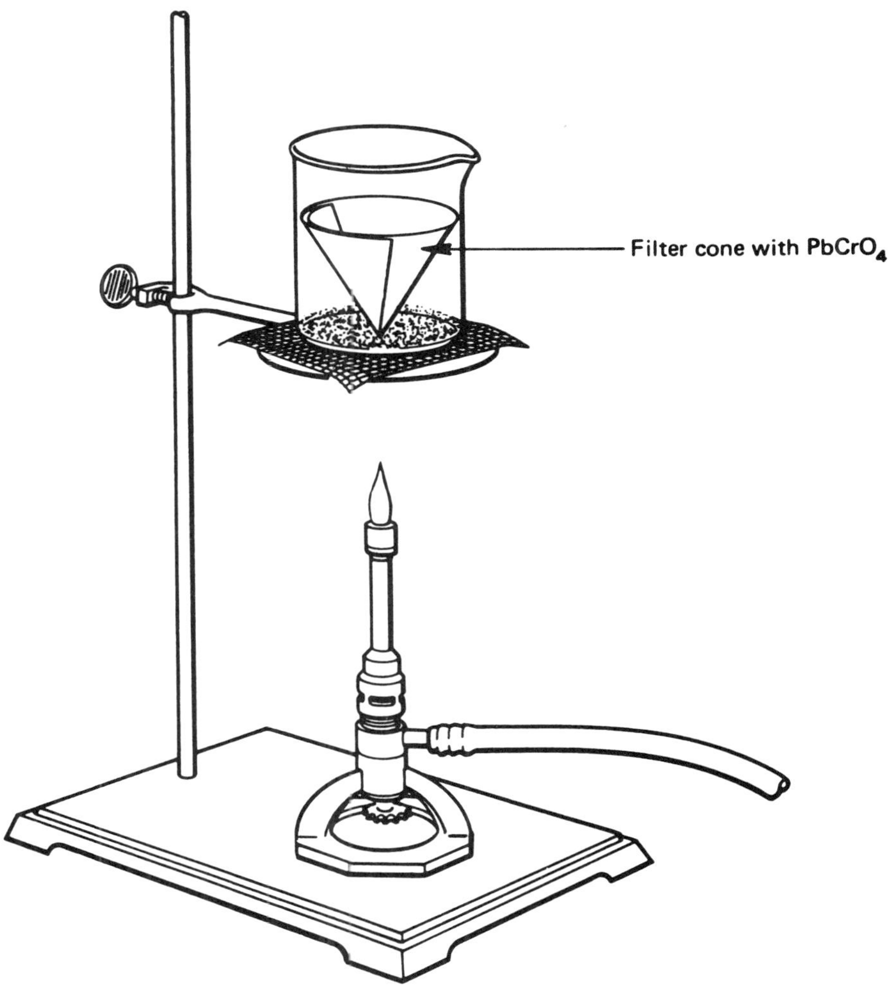


**Figure 15.1. Setup for drying lead(II) chromate.**

*EXPERIMENT 16*

# Boyle's Law

## MATERIALS AND EQUIPMENT

Boyle's Law Apparatus. (The apparatus described in this experiment is available from Central Scientific Company.)

## DISCUSSION

In this experiment you will examine and verify the quantitative relationship between the volume and the pressure in a sample of gas. This relationship was first recognized by the British scientist Robert Boyle in 1662. It is summarized as Boyle's law: at constant temperature, the volume of a sample of gas varies inversely with the pressure exerted by the gas. This statement may be symbolized as follows:

$$V \propto \frac{1}{P} \qquad \text{(constant T)} \qquad (1)$$

$$V = k \times \frac{1}{P} \qquad \text{(constant T)} \qquad (2)$$

$$PV = k \qquad \text{(constant T)} \qquad (3)$$

where k is a constant that depends on the mass and the temperature of the gas.

Equation (2) emphasizes the inverse relationship between pressure and volume. As the pressure is increased, the volume is decreased, and vice versa.

Equation (3) obtained by simply rearranging equation (2) shows that: at constant temperature the product of the pressure and volume of a given mass of gas is constant. From equation (3) it follows that

$$P_1V_1 = P_2V_2 \qquad \text{(constant T)} \qquad (4)$$

where $P_1V_1$ is the pressure-volume product at one set of conditions and $P_2V_2$ is the product at a second set of conditions. Solving equation (4) for $V_2$ gives

$$V_2 = V_1 \times \frac{P_1}{P_2} \qquad \text{(constant T)} \qquad (5)$$

This equation is commonly used in Boyle's law calculations. To calculate a new volume ($V_2$), the initial volume is multiplied by the ratio of the initial pressure over the final pressure ($P_1/P_2$).

In this experiment pressure and volume data are taken on a sample of air at various pressures. These data are plotted as volume vs. pressure to obtain a curve typical of an inverse relationship. The pressure-volume product (PV) is calculated for each set of data. The constancy of these PV products proves the validity of equation (3) and thus the validity of Boyle's law.

The procedure given in this experiment is based on using the Boyle's law apparatus shown diagramatically in Figure 16.1. Other types of apparatus may be used, one of which is shown in Figure 16.2. The apparatus in Figure 16.1 consists of three vertical tubes A, B, and C, connected at their lower ends. The tubes A and B are open at the top while tube C is closed at the top. The large tube A serves as a reservoir for mercury, and the mercury levels in tubes B and C are controlled by means of a wooden plunger D inserted in the reservoir. A millimeter scale is located between tubes B and C for reading the mercury levels in these tubes.

The pressure on the surface of the mercury in the open tube B is atmospheric and is determined by reading a barometer. The pressure on the air confined in the closed tube C differs from atmospheric pressure by an amount equal to the pressure exerted by the column of mercury of height $h$ ($h$ is the difference between the mercury levels in tubes B and C). The pressure in tube C is greater or less than atmospheric pressure depending on whether the mercury level in B is above or below that in tube C. If the mercury is higher in tube B, the height $h$ should be added to atmospheric pressure (both in mm Hg) to give the total pressure on the gas sample. If the mercury is lower in tube B, the height $h$ should be subtracted from atmospheric pressure. The length the air volume, $V$, is measured in millimeters and is directly proportional to the volume of the confined air in tube C.

## PROCEDURE

**Wear protective glasses.**

> ⚠ PRECAUTIONS: Move the wooden plunger down and up in the mercury reservoir slowly and carefully. Do not remove the plunger from the reservoir. The apparatus is fragile and expensive, and spilled mercury is very hazardous!
>
> NOTE: Record all readings of mercury levels to the nearest millimeter.

Read the barometer and record the atmospheric pressure. Record the scale reading of the top of tube C.

**Slowly** insert the plunger D in tube A and note how the mercury levels vary in tubes B and C as the plunger is depressed and withdrawn. With the plunger held just above the mercury, take initial readings of the mercury levels in tubes B and C. Push the plunger into the reservoir until the mercury rises forty to fifty millimeters in the open tube B. Carefully hold the plunger in a fixed position and again read the mercury levels in B and C. Continuing in this manner, take a series of eight readings up to the maximum attainable pressure (plunger all the way down). Record all the data in the Data Table.

For each set of data calculate the values for the remaining four columns in the Data Table. V is obtained by subtracting the mercury level reading in C from the scale reading at the top of tube C; h is the difference between the B and C mercury level readings; and P is the barometric pressure plus h (or minus h if the mercury level in B is lower than in C). No units are used for PV because V is given in units of length rather than volume.

GRAPHING THE DATA: Choose scale values that will spread your data over more than half the length of each coordinate (axis) on the graph paper provided. Select suitable starting values and number major scale divisions along each axis. Now plot the P vs. V data and draw the smooth line that best fits the eight plotted points. This line should be slightly curved and need not pass through each point (see Study Aid 3).

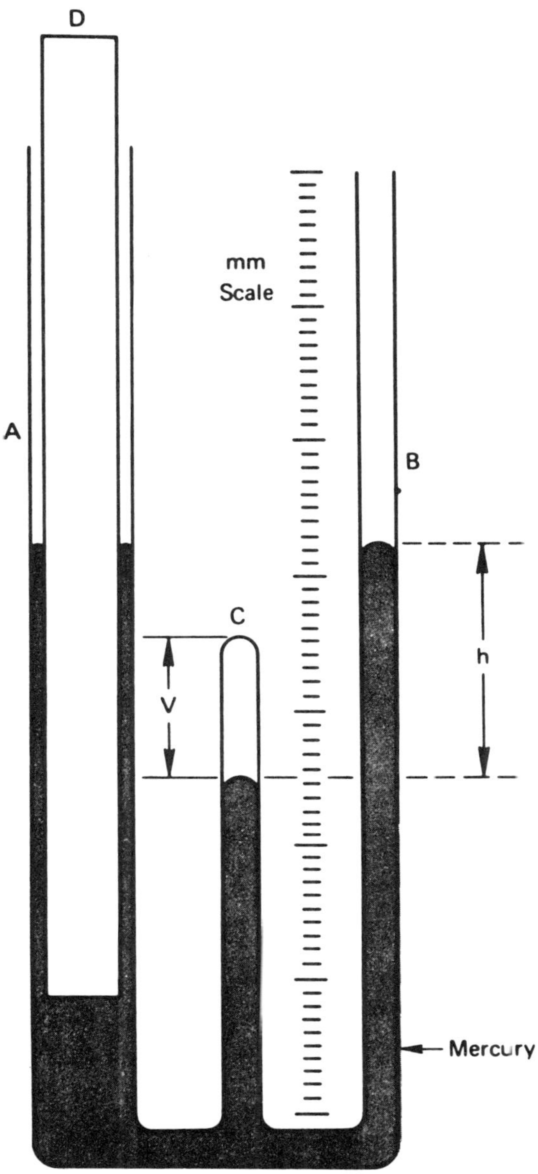


**Figure 16.1.** Diagram of Boyle's Law Apparatus as described in the experiment.

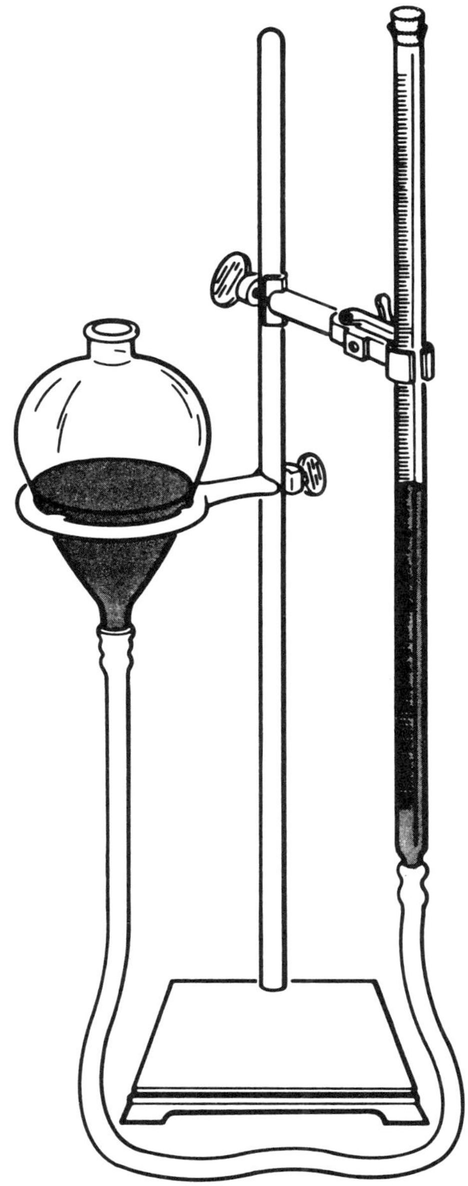

**Figure 16.2.** An alternate Boyle's Law Apparatus

NAME ______________________

SECTION ________ DATE __________

INSTRUCTOR ________________

# *REPORT FOR EXPERIMENT 16*

## Boyle's Law

Atmospheric pressure ____________ mm Hg

Scale reading of the top of tube C ____________ mm

**Data Table**

| Mercury Level in tube B (mm) | Mercury Level in Tube C (mm) | Volume* V (mm) | Pressure Difference h (mm Hg) | Pressure† P (mm Hg) | Pressure-Volume Product PV |
|---|---|---|---|---|---|
| | | | | | |
| | | | | | |
| | | | | | |
| | | | | | |
| | | | | | |
| | | | | | |
| | | | | | |
| | | | | | |

*This is not actually a volume, but is the length of the air column which is directly proportional to the volume of confined air.

†See Discussion for the method of determining the Pressure.

## QUESTIONS AND PROBLEMS

1. What part of the tabulations in the data table proves Boyle's law? How?

2. How would a *P* vs. *V* curve for data taken at a lower temperature compare with the curve you obtained?

3. Tube C has an air space 25.0 cm long. The mercury in tube B stands 95 mm lower than in tube C. If more mercury is poured into the open arm A until the level in tube B is 170. mm higher than in tube C, how long will the air space be? The barometric pressure is 740. mm Hg.

____________

4. The pressure of a 500. mL sample of oxygen is 750. torr. What will be the pressure if the oxygen is compressed to 380. mL, assuming temperature is constant?

____________

Pressure-Volume Relationship of a Sample of Gas (Temperature Constant)

Pressure, mm Hg

Volume (length of air column in mm)

# EXPERIMENT 17

# Charles' Law

## MATERIALS AND EQUIPMENT

125 mL Erlenmeyer flask, one-hole rubber stopper, glass and rubber tubing, pneumatic trough, thermometer, screw clamp.

## DISCUSSION

The quantitative relationship between the volume and the absolute temperature of a gas is summarized in Charles' law. This law states: at constant pressure, the volume of a particular sample of gas is directly proportional to the absolute temperature.

Charles' law may be expressed mathematically:

$$V \propto T \text{ (constant pressure)} \tag{1}$$

$$V = kT \quad \text{or} \quad \frac{V}{T} = k \text{ (constant pressure)} \tag{2}$$

where V is volume, T is Kelvin temperature, and k is a proportionality constant dependent on the number of moles and the pressure of the gas.

If the volume of the same sample of gas is measured at two temperatures, $V_1/T_1 = k$ and $V_2/T_2 = k$, and we may say that:

$$\frac{V_1}{T_1} = \frac{V_2}{T_2} \quad \text{or} \quad V_2 = V_1 \times \frac{T_2}{T_1} \text{ (constant pressure)} \tag{3}$$

where $V_1$ and $T_1$ represent one set of conditions and $V_2$ and $T_2$ a different set of conditions.

This experiment checks the validity of Charles' law by two methods–A and B. By method A the validity is verified by

1. Measuring the volume of a sample of air at two temperatures,
2. Using Charles' law to predict the volume at the lower temperature based on the volume of the air at the higher temperature,
3. Comparing the predicted volume at the lower temperature with the experimental volume found at the lower temperature.

By method B the validity of Charles' law is checked by comparing the V/T ratios for the two different temperatures.

In the experiment the air volume at the higher temperature is measured at atmospheric pressure in a dry Erlenmeyer flask. To simplify the calculations this air is assumed to be dry, although it actually contains a small amount of moisture due to atmospheric humidity.

The air volume at the lower temperature is measured over water. Because this volume of air is saturated with water vapor that contributes to the total pressure in the flask, the volume must be corrected to that of dry air at atmospheric pressure. This correction is made by (1) subtracting the partial pressure (vapor pressure) of the water from the total pressure (atmospheric pressure) in the flask to obtain the partial pressure of dry air alone and (2) calculating the volume that this dry air would occupy at atmospheric pressure using this equation:

$$V_{DA} = V_{WA} \times \frac{P_{DA}}{P_{Atm}} \tag{4}$$

where

$V_{DA}$ = Volume of dry air at atmospheric pressure

$V_{WA}$ = Volume of wet air at atmospheric pressure

$P_{DA}$ = Pressure of dry air $\left(P_{Atm} - P_{H_2O}\right)$

$P_{Atm}$ = Atmospheric pressure

## PROCEDURE

**Wear protective glasses.**

NOTE: It is essential that the Erlenmeyer flask and rubber stopper assembly be as dry as possible in order to obtain reproducible results.

Dry a 125 mL Erlenmeyer flask by gently heating the entire outer surface with a burner flame. Care must be used in heating to avoid breaking the flask. If the flask is wet, first wipe the inner and outer surfaces with a towel to remove nearly all the water. Then, holding the flask with a test tube holder, gently heat the entire flask. Avoid placing the flask directly in the flame. Allow to cool.

While the flask is cooling select a 1-hole rubber stopper to fit the flask and insert a 5 cm piece of glass tubing into the stopper so that the end of the tubing is flush with the bottom of the stopper. Attach a 3 cm piece of rubber tubing to the glass tubing (see Figure 17.1). Insert the stopper into the flask and mark (wax pencil) the distance that it is inserted. Clamp the flask so that it is submerged as far as possible in water contained in a 400 mL beaker (without the flask touching the bottom of the beaker) (see Figure 17.2).

Heat the water to boiling. Keep the flask in the gently boiling water for at least 8 minutes to allow the air in the flask to attain the temperature of the boiling water. Add water as needed to maintain the water level in the beaker. Read and record the temperature of the boiling water.

While the flask is still in the boiling water, seal it by clamping the rubber tubing tightly with a screw clamp. Remove the flask from the hot water and submerge it in a pan of cold water **keeping the top down at all times** to avoid losing air. Remove the screw clamp letting the cold water flow into the flask. Keep the flask totally submerged for about 6 minutes to allow the flask and contents to attain the temperature of the water. Read and record the temperature of the water in the pan.

In order to equalize the pressure inside the flask with that of the atmosphere, bring the water level in the flask to the same level as the water in the pan by raising or lowering the flask (see Figure 17.3). With the water levels equal, pinch the rubber tubing to close the flask. Remove the flask from the water and set it down on the laboratory bench.

Using a graduated cylinder carefully measure and record the volume of liquid in the flask.

Repeat the entire experiment. Use the same flask and flame dry again; make sure that the rubber stopper assembly is thoroughly dried inside and outside.

After the second run fill the flask to the brim with water and insert the stopper assembly to the mark, letting the glass and rubber tubing fill to the top and overflow. Measure the volume of water in the flask. Since this volume is the total volume of the flask, record it as the volume of air at the higher temperature. Because the same flask is used in both runs, it is necessary to make this measurement only once.

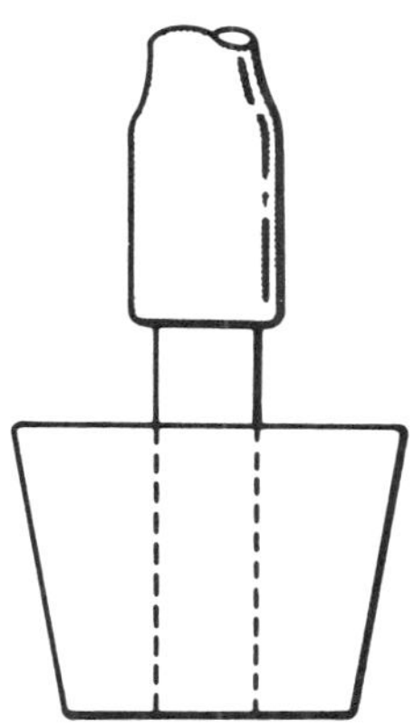

**Figure 17.1. Rubber stopper assembly.**

**Figure 17.2. Heating the flask (and air) in boiling water.**

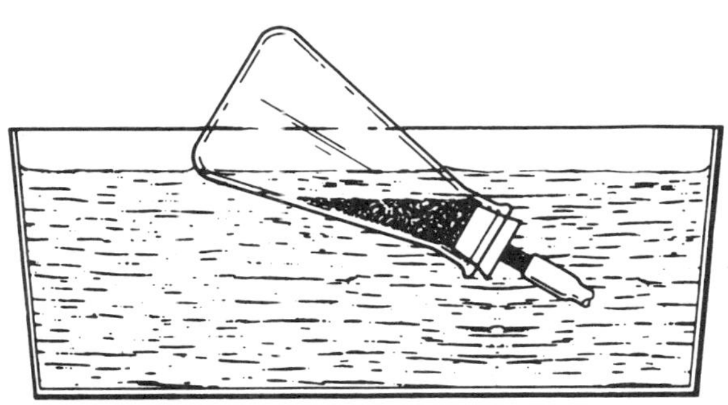

**Figure 17.3. Equalizing the pressure in the flask. The water level inside the flask is adjusted to the level of the water in the pan by raising or lowering the flask.**

# EXPERIMENT 18

# Gaseous Diffusion

## MATERIALS AND EQUIPMENT

Solutions: concentrated ammonium hydroxide (in dropping bottle) and concentrated hydrochloric acid (in dropping bottle). Liquids: acetone and isopropyl alcohol. Glass diffusion tube, 65-80 cm long × 8 mm diameter (firepolished ends); two medicine dropper bulbs; utility clamp; rubber stopper, No. 4 or larger, one-hole, cut open on one side; metric ruler or meter stick; clock or watch with second hand (or stopwatch).

## DISCUSSION

The Kinetic Molecular Theory assumes that (1) the molecules of gases are in rapid random motion and (2) their average velocities (speeds) are proportional to the absolute (Kelvin) temperature. It also assumes that (3) at the same temperature, the average kinetic energies of the molecules of different gases are equal.

Graham's Law of Diffusion is based on the foregoing three basic assumptions of the Kinetic Molecular Theory and can be stated in either of two alternate forms: (1) The rates of diffusion of different gases are inversely proportional to the square roots of their densities, or (2) the rates of diffusion of different gases are inversely proportional to the square roots of their molar masses.

It is difficult to make experimental measurements of the velocities of individual gas molecules. But the **relative molecular velocities** of certain gases can be compared with the aid of some simple laboratory equipment. In this experiment you will determine the relative molecular velocities of ammonia ($NH_3$) gas and of hydrogen chloride (HCl) gas. This information will be used to calculate experimental values for (1) the molar mass of hydrogen chloride and (2) the ratio of the average molecular velocity of ammonia gas to that of hydrogen chloride gas.

Ammonia and hydrogen chloride gases react, upon contact, to form a white cloud (or smoke) of microscopic particles of solid ammonium chloride. This fact makes it fairly easy to **experimentally measure** the relative rates of diffusion of these gases. This is accomplished by simultaneously introducing ammonia and hydrogen chloride gases into the opposite ends of a glass tube and noting the time needed for the appearance of the faint white cloud of ammonium chloride. Experimental values for the relative rates of diffusion of $NH_3$ and HCl through the air in the tube can then be obtained by measuring the distance traveled by each gas and dividing by the time required for the appearance of the faint white cloud.

The kinetic molecular theory holds that, at the same temperature, the average kinetic energies of the molecules of different gases are equal. The kinetic energy (KE) of any moving body—regardless of whether it is a molecule or an automobile—is given by the equation, $KE = \frac{1}{2}mv^2$

(where $m$ is the mass of the body and $v$ is its velocity). Hence $KE_{(NE_3)} = KE_{(HCl)}$, and the masses and average velocities of $NH_3$ and HCl molecules must be related in this fashion:

$$\frac{m_{NH_3} v^2_{NH_3}}{2} = \frac{m_{HCl} v^2_{HCl}}{2} \qquad (1)$$

By multiplying Equation (1) by 2 and rearranging, we obtain Equation (2)

$$\frac{v^2_{NH_3}}{v^2_{HCl}} = \frac{m_{HCl}}{m_{NH_3}} \qquad (2)$$

By taking the square root of Equation (2) we obtain

$$\frac{v_{NH_3}}{v_{HCl}} = \frac{\sqrt{m_{HCl}}}{\sqrt{m_{NH_3}}} \qquad (3)$$

Since the ratio of the molecular masses is proportional to the ratio of the molar masses, Equation (3) may be rewritten as

$$\frac{v_{NH_3}}{v_{HCl}} = \frac{\sqrt{M_{HCl}}}{\sqrt{M_{NH_3}}} \qquad \text{(where } M \text{ represents molar mass)} \qquad (4)$$

Equation (4) is, of course, the mathematical statement of Graham's Law of Diffusion in this form: The rates of diffusion of different gases are inversely proportional to the square roots of their molar masses.

From Equation (4) the ratio of the average molecular velocity of ammonia to that of hydrogen chloride can be calculated.

$$\frac{v_{NH_3}}{v_{HCl}} = \frac{\sqrt{M_{HCl}}}{\sqrt{M_{NH_3}}} = \frac{\sqrt{36.5}}{\sqrt{17.0}} = 1.47$$

This calculation tells us that, at a given temperature, the average velocity of ammonia molecules is theoretically 1.47 times greater than that of hydrogen chloride molecules.

## PROCEDURE

**Wear protective glasses.**

NOTES:

1. Use concentrated ammonium hydroxide and concentrated hydrochloric acid in dropping bottles or other containers designated by your instructor. Keep these containers tightly closed when not in use.

2. In performing Step 1, make sure that each liquid wets the entire inner wall of the tube. This is accomplished by rotating the tube while holding it at about a 45° angle while pouring the liquids through it.

3. Steps 4 and 5 are most efficiently performed by two persons working together. Therefore, unless otherwise directed by your instructor, arrange to cooperate with another student in the performance of these steps.

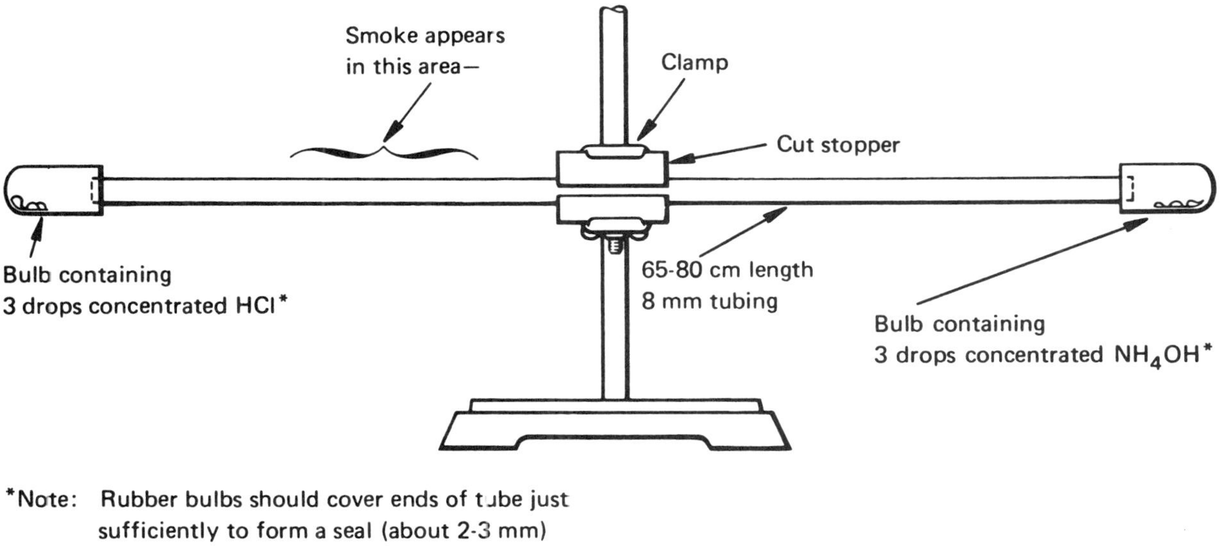


**Figure 18.1. Setup for Gaseous Diffusion.**

⚠ PRECAUTION: Isopropyl alcohol and acetone are volatile and highly flammable liquids.

1. Clean the glass diffusion tube by:

   (a) pouring warm water through it and rinsing with 5 mL of distilled water;

   (b) pouring about 3 mL of isopropyl alcohol through the tube;

   (c) pouring about 3 mL of acetone through the tube and wiping the outside clean with a towel or tissue;

   (d) drawing air through the tube using an aspirator pump or vacuum line for 5 minutes. **The tube must be perfectly dry for satisfactory results.**

2. Clean two medicine dropper bulbs by rinsing with water, then with 1 mL isopropyl alcohol, and finally with 1 mL acetone. Dry the bulbs thoroughly.

3. Slip a cut stopper in place at the center of the glass tube. Clamp the tube by means of this stopper and a utility clamp so that it is in a horizontal position about four inches above the desk top as shown in Figure 18.1.

4. Put **three drops** of conc. hydrochloric acid solution into one of the dropper bulbs, and at the same time have your partner put **three drops** of conc. ammonium hydroxide into the other bulb. Keep the bulbs turned open-end up so these liquids can't escape.

5. Without delay note the position of the second hand on your watch or clock; and as it crosses the next half minute or minute mark, working with your partner, simultaneously place the dropper bulb containing the ammonium hydroxide on the right end of the tube and the bulb containing hydrochloric acid on the left end. The bulbs should be **gripped lightly by the upper or rim portions,** and put on the ends of the glass tubing **only far enough to form a seal** (2 or 3 mm). **Do not squeeze the bulbs or push them very far onto the glass tube** because either of these actions will force air/gas mixtures into the glass tube and have an adverse effect on the experiment. Be sure to note and record the exact time when the bulbs were put on the tube.

6. Watch the tube for the very first visible trace of white smoke—it will usually appear within 5 to 10 minutes as a very faint ring resembling a wisp of cigarette smoke. Note and record the time of the first appearance of smoke to the nearest 0.1 minute (6 seconds = 0.1 minute). Mark the tube with a ballpoint pen at the point where the smoke first appears.

7. Measure and record the distances traveled by both HCl and $NH_3$ to the nearest 0.1 cm

8. Make a duplicate run by repeating the foregoing procedure starting with Step 1.

# EXPERIMENT 19

# Liquids—Vapor Pressure and Boiling Points

## MATERIALS AND EQUIPMENT

Liquids: flasks containing acetone [$(CH_3)_2C{=}O$], ethanol ($C_2H_5OH$), methanol ($CH_3OH$), and water. Boiling chips, fine copper wire (24 gauge), screw clamp, one-gallon metal can with rubber stopper (one can per class).

## DISCUSSION

The molecules that escape (vaporize) from the surface of a liquid in an open container diffuse into the surroundings. The probability of these escaped molecules returning to condense to a liquid is very small. The volume of the liquid therefore decreases. But the molecules that vaporize from a liquid in a closed container cannot permanently escape. The volume of the liquid therefore does not decrease. An equilibrium is established in which the rate of return to the liquid (condensation) is equal to the rate of escape (evaporation). The pressure exerted by the vapor in this equilibrium is known as the **vapor pressure** of the liquid. In this experiment you will observe some of the phenomena associated with the vapor pressures of liquids.

Vapor pressure is dependent on two factors: the temperature of the liquid, and the nature of the liquid.

As the temperature of the liquid is raised, the vapor pressure increases because the average kinetic energy of the molecules is increased causing more of them to escape from the liquid. When the vapor pressure has increased until it equals the pressure of the atmosphere above it, the liquid boils. The temperature at which the vapor pressure equals exactly one atmosphere (760 torr) is called the **normal boiling point.**

Each liquid has a unique pattern of vapor pressure behavior. Liquids which have a relatively high vapor pressure evaporate easily and are said to be **volatile.** In comparing two liquids at a given temperature, the more volatile of the two will have a higher vapor pressure. As a consequence of its higher vapor pressure the more volatile liquid will boil at a lower temperature.

A volatile liquid in an open beaker becomes colder as a result of evaporation. One way of explaining this cooling effect is to assume that the faster moving, "hotter," molecules escape from the liquid leaving the slower moving, "colder," molecules behind.

Part A of this experiment demonstrates the cooling effect of evaporation. The bulb of a thermometer is wrapped with filter paper and wetted with a liquid. The evaporation of the liquid causes cooling and a drop in temperature on the thermometer. The more volatile the liquid, the faster it will evaporate, and thus the more the temperature will be lowered.

$$\text{Liquid} + \text{Heat} \xrightarrow{\text{Evaporation}} \text{Vapor}$$

The equation above indicates that evaporation consumes heat; thus, if no external heat is supplied the temperature of the system will drop as evaporation occurs.

Part B of this experiment relates boiling point to vapor pressure. Since the boiling point is the temperature at which the vapor pressure equals the pressure of the atmosphere above a liquid, it follows that the boiling point will be reduced if the pressure above the liquid is reduced.

In the experiment, water is heated to boiling in both the open and stoppered tubes shown in Figure 19.1. The screw clamp attached to the stoppered tube is left open. The pressure above the water in both tubes is equal to the atmospheric pressure. As the water boils, steam is produced. From the open tube the steam escapes into the atmosphere. From the stoppered tube a mixture of air and steam is expelled through the delivery tube. After a few minutes boiling, the air has been almost totally expelled and the vapor in the system is almost pure steam (at atmospheric pressure).

Now heating of both tubes is stopped, and the screw clamp on the stoppered tube system is closed. As the system begins to cool, steam begins to condense on the walls of the stoppered tube and in the delivery tube. The condensing steam reduces the pressure within the system to less than atmospheric pressure. Consequently, the water in the stoppered tube continues to boil for some time because its vapor pressure exceeds the pressure of the atmosphere within the closed system. Because it is boiling, the water in the stoppered tube cools faster than the water in the open tube.

A vapor gives off heat when condensing to a liquid.

$$\text{Vapor} \xrightarrow{\text{Condensation}} \text{Liquid} + \text{Heat}$$

Because steam (vapor) is condensing on the walls of the stoppered tube in the closed system, the walls remain hot for a relatively long time. Since the water in the open tube is not boiling, there is little vapor condensing on the walls of this tube. Therefore, the walls cool faster than do the walls of the stoppered tube in the closed system.

In Part C, the condensation of steam reduces the pressure in the closed can. This is the same phenomenon that occurs in Part B. But a metal can does not have the relative strength of a test tube, so the walls of the can are crushed because of the difference between the atmospheric pressure outside and the reduced pressure inside the can.

## PROCEDURE

**Wear protective glasses.**

### A. Cooling Effect of Evaporation

⚠ PRECAUTIONS:

1. Avoid breathing methanol vapor since it is poisonous.
2. Acetone, methanol, and ethanol are flammable liquids.

Wrap the bulb of a thermometer with a half of a circle of filter paper and fasten the paper securely with fine copper wire. Record the temperature reading of the thermometer (room temperature).

The liquids for this experiment are contained in 125 mL Erlenmeyer flasks. Dip the covered bulb of the thermometer into the acetone. Remove the thermometer from the liquid and suspend it from a ring stand. Note that the temperature decreases, and record the lowest temperature attained.

Repeat the experiment, using first methanol, then ethanol, and then water. Attach a fresh half-circle of filter paper each time. **Do the experiment with methanol in the hood.**

## B. Relationship of Vapor Pressure and Boiling Point

NOTE: Work in pairs in this part of the experiment.

Assemble the apparatus as shown in Figure 19.1. The 25 X 200 mm test tubes must be scrubbed with detergent and rinsed with distilled water. Use the usual precautions when inserting the thermometer and the glass tubing through the rubber stopper (see page 5). Place two boiling chips and 25 mL of distilled water into each test tube. The stopper must be inserted very securely to avoid leaks. The screw clamp must be open.

Heat the water in both test tubes to the boiling point by moving a burner flame from one to the other. It is desirable to have the water in both tubes reach the boiling point at nearly the same time. Read the temperature in each tube while the water is boiling. Record these temperatures as the first entries in the data table (Part B). Continue boiling the water in both tubes until there are essentially no bubbles of gas rising to the surface through the water in the trough. When this condition has been reached, heat for an additional minute.

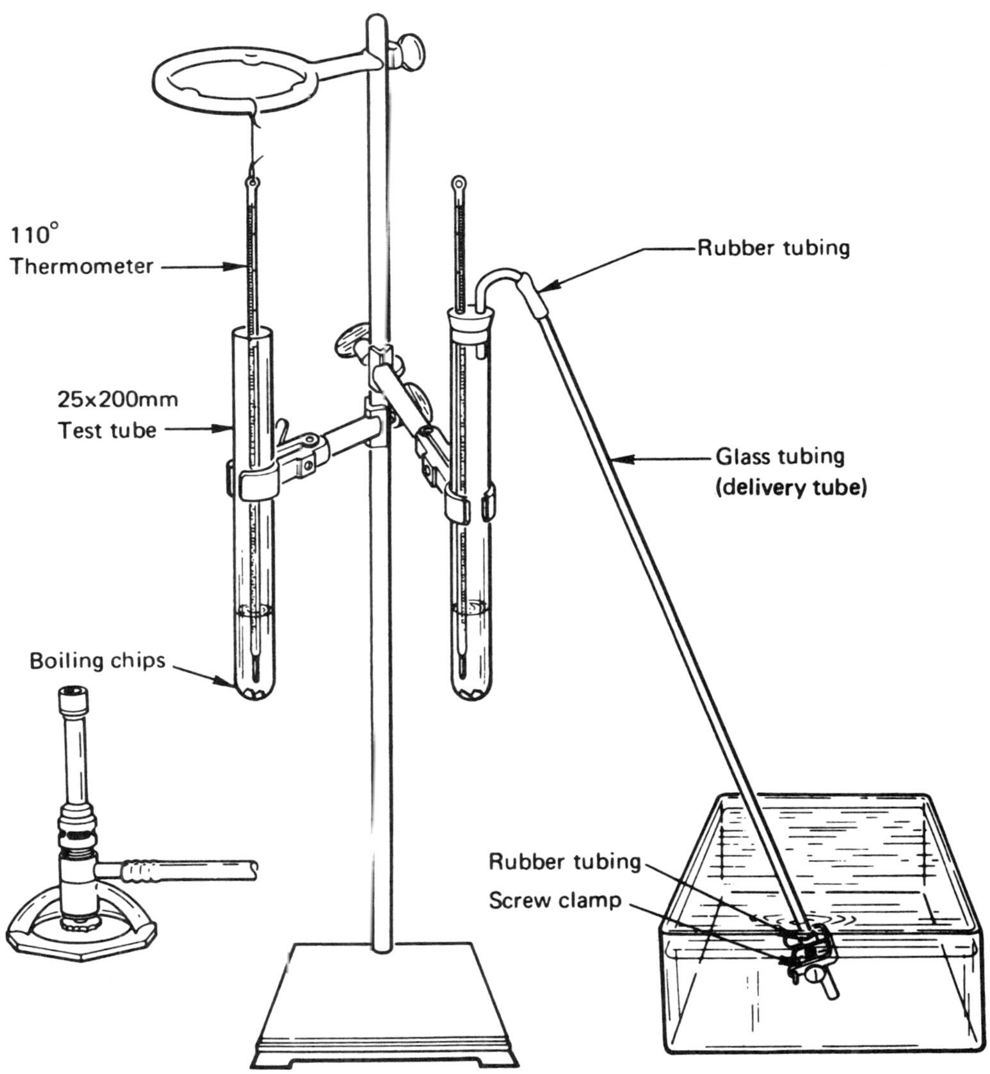


**Figure 19.1. Setup for Part B.**

Now in **rapid succession** (a) pinch the rubber tubing shut just below the screw clamp, (b) stop heating, (c) while keeping the tubing pinched shut, close the screw clamp **tightly.** When the clamp is tightly closed, note the amount of water in the delivery tube. If leaks occur, reassemble the apparatus, checking all the rubber connections, and repeat the experiment. (It may be necessary to replace the rubber tubing.)

Take a series of temperature readings, noting at what temperature boiling stops in each tube. Each time the thermometer in the closed tube cools 5°C, read and record the temperature in both tubes. It is not critical that you have a reading for every 5° temperature drop, but it is important that both temperatures be read at the same time for each recorded reading.

When the water in the open tube has cooled to about 85°C, touch the upper parts of both tubes and note which one is hotter. Also note the condition of the rubber tubing at the top of the closed tube. When the mercury column is no longer visible because of the rubber stopper, do not try to change the position of the thermometer. Simply wait for the mercury to reappear below the stopper and then continue to take temperature readings. Discontinue taking readings when the temperature in the closed tube falls below 40°C. If the water in either tube is still boiling, record the temperature at which boiling stopped as "below 40°C."

When boiling has stopped in the closed tube, observe the amount of water in the delivery tube. Keeping the rubber tubing under water in the trough, open the screw clamp and note what happens.

### C. Effect of Vapor Pressure Change and Atmospheric Pressure (Instructor Demonstration)

Obtain a metal gallon can and a rubber stopper which will fit the mouth of the can.

Pour about 50 mL of water into the can, place it on a ring stand, and heat with a burner.

NOTE: At this point the can should not be stoppered.

After the water has begun to boil and condensing steam has appeared at the mouth of the can, heat for about five more minutes. Stop heating and immediately stopper the can very tightly. **Careful:** the can is very hot at this point. Set the can on the desk, allow it to cool, and observe what happens.

NAME ______________________

SECTION ______ DATE __________

*REPORT FOR EXPERIMENT 19* INSTRUCTOR ______________

# Liquids—Vapor Pressure and Boiling Points

## A. Cooling Effect of Evaporation

Room temperature __________

| Liquid wetting thermometer bulb | Lowest temperature attained °C | Normal boiling point* °C |
|---|---|---|
| Acetone | | |
| Methanol | | |
| Ethanol | | |
| Water | | |

*See Appendix 7.

1. Why does the temperature drcp when the thermometer covered with wet filter paper is suspended in the air?

2. On the basis of the observed behavior of these four liquids, what is the relationship between the normal boiling points and effectiveness of liquids in cooling by evaporation?

## B. Relationship of Boiling Point and Pressure

**Data Table**

| Temperature readings | |
|---|---|
| Closed tube | Open tube |
| | |
| | |
| | |
| | |
| | |
| | |
| | |
| | |
| | |
| | |
| | |

1. Temperature at which boiling stopped:
   (a) in the open test tube ______________
   (b) in the closed test tube ______________

2. How much water was in the delivery tube? (In terms of approximate length of water column above screw clamp):
   (a) at the time of clamping ______________ cm
   (b) after boiling has stopped ______________ cm

3. When the water had cooled to 85°C in the open tube and you touched the upper part of both test tubes, which was hotter?
   ______________

   Why?

4. During the cooling, what happened to the rubber tubing at the top of the closed tube? Why?

5. When you opened the screw clamp under water, what happened?

6. Why did the water in the closed test tube continue to boil at lower temperatures than the water in the open test tube?

7. Why did the water in the closed test tube cool faster than the water in the open test tube?

8. Explain clearly what caused the reduced pressure inside the closed test tube during cooling.

9. Why was there more water in the delivery tube after the apparatus cooled (before the clamp was opened) than at the time it was clamped shut?

**C. Effect of Vapor Pressure Change and Atmospheric Pressure**

1. (a) What did you observe at the mouth of the can just before it was stoppered?

   (b) What does this indicate about the composition of the gas in the can when the can was stoppered?

2. What happened to the can after the heating was stopped and it was stoppered tightly?

3. If the can had been stoppered and heating stopped when the water first began to boil, how might the results have been different? Why?

4. If the can had been allowed to cool about three minutes before stoppering, how might the results have been different? Why?

5. If the can had been stoppered about three minutes before heating was stopped, how might the results have been different? Why?

# EXPERIMENT 20

# Neutralization—Titration I

## MATERIALS AND EQUIPMENT

Solid: potassium acid phthalate, abbreviated KHP ($KHC_8H_4O_4$). Phenolphthalein indicator, unknown base solution (NaOH), one buret (25 mL or 50 mL), buret clamp.

## DISCUSSION

The reaction of an acid and a base to form a salt and water is known as **neutralization.** In this experiment potassium acid phthalate (abbreviated KHP) is used as the acid. Potassium acid phthalate is an organic substance having the formula $HKC_8H_4O_4$, and like HCl, has only one acid hydrogen atom per molecule. Because of its complex formula, potassium acid phthalate is commonly called KHP. Despite its complex formula we see that the reaction of KHP with sodium hydroxide is similar to that of HCl. One mole of KHP reacts with one mole of NaOH.

$$HKC_8H_4O_4 + NaOH \longrightarrow NaKC_8H_4O_4 + H_2O$$

$$HCl + NaOH \longrightarrow NaCl + H_2O$$

**Titration** is the process of measuring the volume of one reagent required to react with a measured volume or mass of another reagent. In this experiment we will determine the molarity of a base (NaOH) solution from data obtained by titrating KHP with the base solution. The base solution is added from a buret to a flask containing a weighed sample of KHP dissolved in water. From the mass of KHP used we calculate the moles of KHP. Exactly the same number of moles of base is needed to neutralize this number of moles of KHP since one mole of NaOH reacts with one mole of KHP. We then calculate the molarity of the base solution from the titration volume and the number of moles of NaOH in that volume.

In the titration, the point of neutralization, called the **end-point**, is observed when an indicator, placed in the solution being titrated, changes color. The indicator selected is one that changes color when the stoichiometric quantity of base (according to the chemical equation) has been added to the acid. A solution of phenolphthalein, an organic compound, is used as the indicator in this experiment. Phenolphthalein is colorless in acid solution but changes to pink when the solution becomes slightly alkaline. When the number of moles of sodium hydroxide added is equal to the number of moles of KHP originally present, the reaction is complete. The next drop of sodium hydroxide added changes the indicator from colorless to pink.

Use the following relationships in your calculations:

1. According to the equation for the reaction,

   Moles of KHP reacted = Moles of NaOH reacted

2. $\text{Moles} = \dfrac{\text{g of solute}}{\text{molar mass of solute}}$

3. Molarity is an expression of concentration, the units of which are moles of solute per liter of solution.

$$\text{Molarity} = \frac{\text{moles}}{\text{liter}}$$

Thus a 1.00 molar (1.00 M) solution contains 1.00 mole of solute in 1 liter of solution. A 0.100 M solution, then, contains 0.100 mole of solute in 1 liter of solution.

4. The number of moles of solute present in a known volume of solution of known concentration can be calculated by multiplying the volume of the solution (in liters) by the molarity of the solution:

$$\text{Moles} = \text{liters} \times \text{molarity} = \text{liters} \times \frac{\text{moles}}{\text{liter}}$$

## PROCEDURE

**Wear protective glasses.**

Obtain some solid KHP in a test tube or vial. Weigh two samples of KHP into 125 mL Erlenmeyer flasks, numbered for identification. (The flasks should be rinsed with distilled water, but need not be dry on the inside.) First weigh the flask to the highest precision of the balance. Add KHP to the flask by tapping the test tube or vial until 1.000 to 1.200 g has been added (see Figure 20.1). Determine the mass of the flask and the KHP. In a similar manner weigh another sample of KHP into the second flask. To each flask add approximately 30 mL of distilled water. If some KHP is sticking to the walls of the flask, rinse it down with water from a wash bottle. Warm the flasks slightly and swirl them until all the KHP is dissolved.

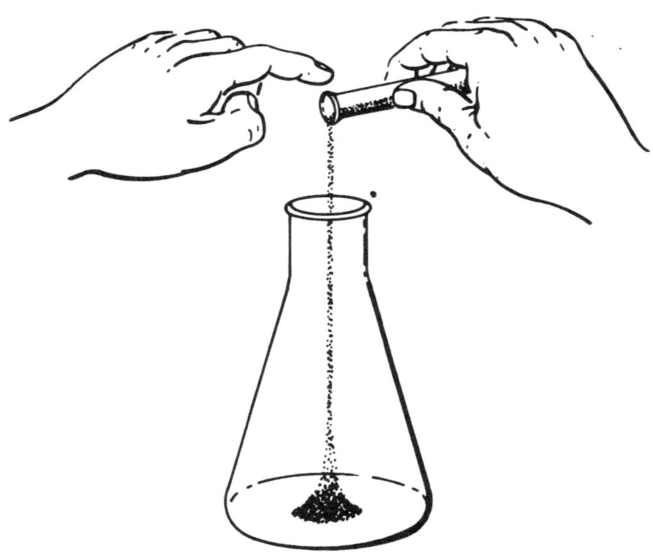

**Figure 20.1. Method of adding KHP from a vial to a weighed Erlenmeyer flask.**

Obtain one buret and clean it. See "Use of the Buret," p. 158, for instructions on cleaning and using the buret. Read and record all buret volumes to the nearest 0.01 mL.

Obtain about 250 mL of a base (NaOH) of unknown molarity in a clean, dry 250 mL Erlenmeyer flask as directed by your instructor. Record the number of this unknown.

NOTES:

1. Keep your base solution stoppered when not in use.

2. The 250 mL sample of base is intended to be used in both this experiment and Experiment 21. **Be sure to label and save it.**

Rinse the buret with two **5 to 10 mL** portions of the base, running the second rinsing through the buret tip. Discard the rinsings. Fill the buret with the base, making sure that the tip is completely filled and contains no air bubbles. Adjust the level of the liquid in the buret so that the bottom of the meniscus is at exactly 0.00 mL. Record the initial buret reading (0.00 mL) in the space provided on the report sheet.

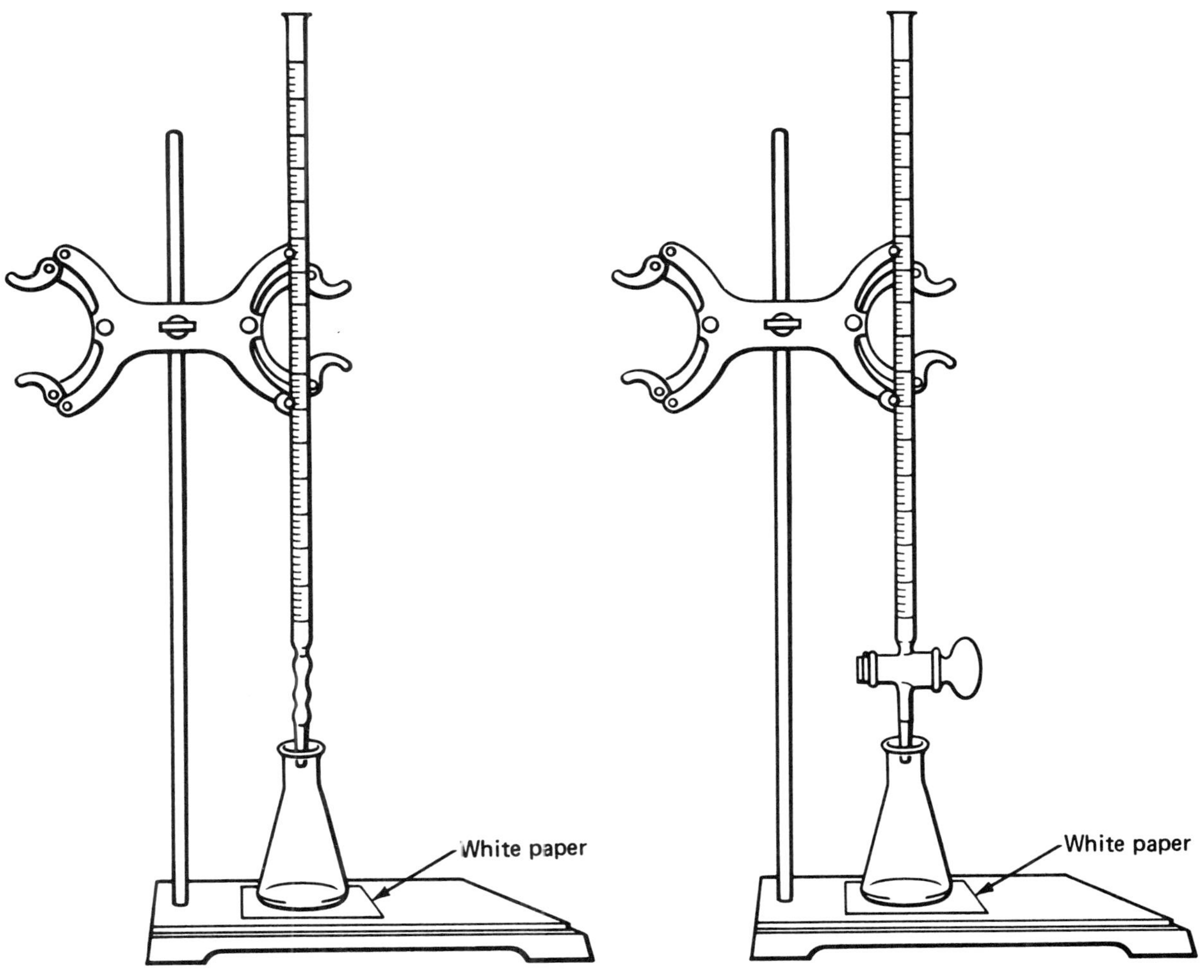


a. Setup with Mohr buret

b. Setup with stopcock buret

**Figure 20.2. Titration Setup.**

Add 3 drops of phenolphthalein solution to each 125 mL flask containing KHP and water. Place the first flask (Sample 1) on a piece of white paper under the buret extending the tip of the buret into the flask (see Figure 20.2).

Titrate the KHP by adding base until the end-point is reached. The titration is conducted by swirling the solution in the flask with the right hand (if you are right handed) while manipulating the buret with the left. As base is added you will observe a pink color caused by localized high base concentration. Toward the end-point the color flashes throughout the solution, remaining for a longer time. When this occurs, add the base drop by drop until the end-point is reached, as indicated by the first drop of base which causes a faint pink color to remain in the entire solution for at least 30 seconds. Read and record the final buret reading (see Figure 20.5). Refill the buret to the zero mark and repeat the titration with Sample 2. Then calculate the molarity of the base in each sample. If these molarities differ by more than 0.004, titrate a third sample.

When you are finished with the titrations, empty and rinse the buret at least twice (including the tip) with tap water and once with distilled water. Return the vial and the unused KHP.

### Use of the Buret

A buret is a volumetric instrument that is calibrated to deliver a measured volume of solution. The 50 mL buret is calibrated from 0 to 50 mL in 0.1 mL increments and is read to the nearest 0.01 mL. All volumes delivered from the buret should be between the calibration marks. (Do not estimate above the 0 mL mark or below the 50 mL mark.)

1. **Cleaning the Buret.** The buret must be clean in order to deliver the calibrated volume. Drops of liquid clinging to the sides as the buret is drained are evidence of a dirty buret.

To clean the buret first rinse it a couple of times with tap water, pouring the water from a beaker. Then scrub it with a detergent solution, using a long-handled buret brush. Rinse the buret several times with tap water and finally with distilled water. Check for cleanliness by draining the distilled water through the tip and observe whether droplets of water remain on the inner walls of the buret.

2. **Using the Buret.** After draining the distilled water, rinse the buret with two 5 to 10 mL portions of the solution to be used in it. This rinsing is done by holding the buret in a horizontal position and rolling the solution around to wet the entire inner surface. Allow the final rinsing to drain through the tip.

Fill the buret with the solution to slightly above the 0 mL mark and adjust it to 0.00 mL, or some other volume below this mark, by draining the solution through the tip. The buret tip must be completely filled to deliver the volume measured.

To deliver the solution from the buret, turn the stopcock with the forefinger and the thumb of your left hand (if you are right handed) to allow the solution to enter the flask. (See Figure 20.3). This procedure leaves your right hand free to swirl the solution in the flask during the titration. With a little practice you can control the flow so that increments as small as 1 drop of solution can be delivered.

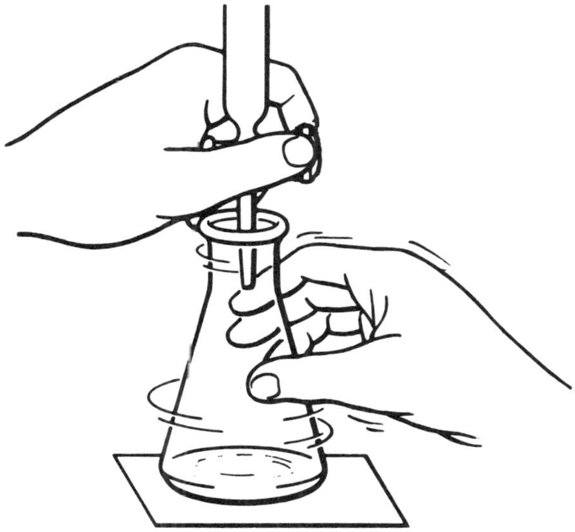

Figure 20.3. Titration technique.

3. **Reading the Buret**. The smallest calibration mark of a 50 mL buret is 0.1 mL. However, the buret is read to the nearest 0.01 mL by estimating between the calibration marks. When reading the buret be sure your line of sight is level with the bottom of the meniscus in order to avoid parallax errors (see Figure 20.4). The exact bottom of the meniscus may be made more prominent and easier to read by allowing the meniscus to pick up the reflection from a heavy dark line on a piece of paper (see Figure 20.5).

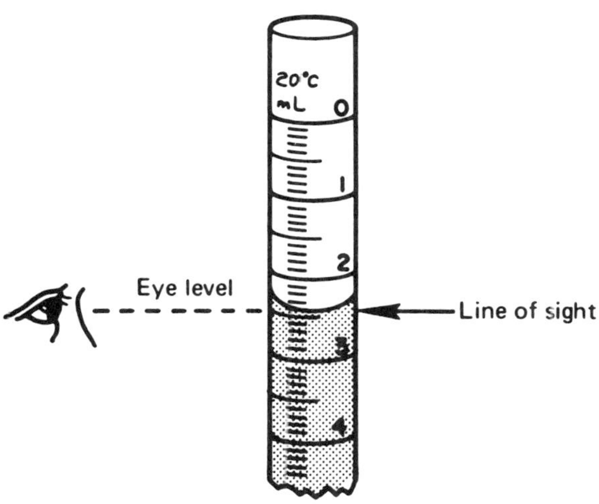


Figure 20.4. Reading the buret. The line of sight must be level with the bottom of the meniscus to avoid parallax.

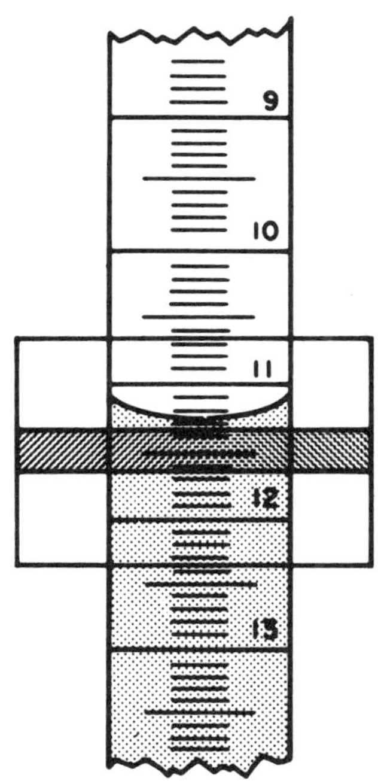


Figure 20.5. Reading the meniscus. A heavy dark line brought to within one division of the meniscus will make the meniscus more prominent and easier to read. The volume reading is 11.28 mL.

NAME ____________________

SECTION ______ DATE __________

INSTRUCTOR ________________

## *REPORT FOR EXPERIMENT 20*

# Neutralization—Titration I

**Data Table**

| | Sample 1 | Sample 2 | Sample 3 (if needed) |
|---|---|---|---|
| Mass of flask and KHP | | | |
| Mass of empty flask | | | |
| Mass of KHP | | | |
| **Final buret reading** | | | |
| **Initial buret reading** | | | |
| **Volume of base used** | | | |

**CALCULATIONS:** In the spaces below show calculation setups for Sample 1 only. Show answers for both samples in the box. Remember to use the proper numbers of significant figures in all calculations. (The number 0.005 has only one significant figure.)

| | Sample 1 | Sample 2 | Sample 3 (if needed) |
|---|---|---|---|
| 1. Moles of acid (KHP, Form. wt = 204.2) | ______ | ______ | ______ |
| 2. Moles of base used to neutralize (react with) the above number of moles of acid. | ______ | ______ | ______ |
| 3. Molarity of base (NaOH) | ______ | ______ | ______ |

4. Average molarity of base ______________

5. Unknown base number ______________

## QUESTIONS AND PROBLEMS

1. If you had added 50 mL of water to a sample of KHP instead of 30 mL, would the titration of that sample then have required more, less, or the same amount of base? Explain.

2. A student weighed out 1.066 g of KHP. How many moles was that?

_______________ moles

3. A titration required 18.08 mL of 0.1574 M NaOH solution. How many moles of NaOH were in this volume?

_______________ moles

4. A student weighed a sample of KHP and found it weighed 1.176 g. Titration of this KHP required 19.84 mL of base (NaOH). Calculate the molarity of the base.

_______________ M

5. Forgetful Freddy weighed his KHP sample, but forgot to bring his report sheet along, so he recorded his weights on a paper towel. During his titration, which required 18.46 mL of base, he spilled some base on his hands. He remembered to wash his hands, but forgot about the data on the towel, and used it to dry his hands. When he went to calculate the molarity of his base, Freddy discovered that he didn't have the mass of his KHP. His kindhearted instructor told Freddy that his base was 0.2987 M. Calculate the mass of Freddy's KHP sample.

_______________ g

6. What mass of solid NaOH would be needed to make 545 mL of Freddy's NaOH solution?

_______________ g

# EXPERIMENT 21

# Neutralization—Titration II

## MATERIALS AND EQUIPMENT

Acid of unknown molarity, standard base solution (NaOH), vinegar, phenolphthalein indicator, suction bulb, buret, buret clamp, 10 mL volumetric pipet.

## DISCUSSION

This experiment may follow Experiment 20 or it may be completed independently of Experiment 20. In either case the discussion section of Experiment 20 supplements the following discussion.

The reaction of an acid and a base to form water and a salt is known as **neutralization.** Hydrochloric acid and sodium hydroxide, for example, react to form water and sodium chloride.

$$HCl + NaOH \longrightarrow H_2O + NaCl$$

The ionic reaction in neutralizations of this type is that of hydrogen (or hydronium) ion reacting with hydroxide ion to form water.

$$H^+ + OH^- \longrightarrow H_2O$$

or

$$H_3O^+ + OH^- \longrightarrow 2\ H_2O$$

A monoprotic acid–i.e., an acid having one ionizable hydrogen atom per molecule–reacts with sodium hydroxide (or any other monohydroxy base) on a 1:1 mole basis. This fact is often utilized in determining the concentrations of solutions of acids by titration.

**Titration** is the process of measuring the volume of one reagent required to react with a measured volume or mass of another reagent. In this experiment an acid solution of unknown concentration is titrated with a base solution of known concentration. Phenolphthalein is used as an **indicator**. This substance is colorless in acid solution, but changes to pink when the solution becomes slightly basic or alkaline. The change of color, caused by a single drop of the base solution in excess over that required to neutralize the acid, marks the **end-point** of the titration.

Molarity (M) is the concentration of a solution expressed in terms of moles of solute per liter of solution.

$$\text{Molarity} = \frac{\text{moles}}{\text{liter}}$$

Thus a solution containing 1.00 mole of solute in 1.00 liter of solution is 1.00 molar (1.00 M). If only 0.155 mole is present in 1.00 liter of solution, it is 0.155 M, etc. To determine the molarity of any quantity of solution it is only necessary to divide the total number of moles of solute present in the solution by the total volume (in liters).

To determine the number of moles of solute present in a known volume of solution, multiply the volume in liters by the molarity.

$$\text{Moles} = \text{liters} \times \text{molarity} = \text{liters} \times \frac{\text{moles}}{\text{liter}}$$

For titrations involving monoprotic acids and monohydroxy bases (one hydroxide ion per formula unit), the number of moles of acid is identical to the number of moles of base required to neutralize the acid. In this experiment we measure the volume of base of known molarity required to neutralize a measured volume of acid of unknown molarity. The molarity of the acid can then be calculated.

$$\text{Moles base} = \text{liters} \times \text{molarity} = \text{liters base} \times \frac{\text{moles base}}{\text{liters base}}$$

$$\text{Moles acid} = \text{moles base} \times \frac{1 \text{ mole acid}}{1 \text{ mole base}}$$

$$\text{Molarity of acid} = \frac{\text{moles acid}}{\text{liters acid}}$$

In order to determine the molarity of an acid solution, it is not actually necessary to know what the acid is—only whether it is monoprotic, diprotic, or triprotic. The calculations in this experiment are based on the assumption that the acid in the unknown is monoprotic.

If the molarity and the formula of the solute are known, the concentration in grams of solute per liter of the solution may be calculated by multiplying the molar mass.

$$\text{Molarity} \times \text{molar mass} = \frac{\text{moles}}{\text{liter}} \times \frac{\text{grams}}{\text{mole}} = \frac{\text{grams}}{\text{liter}}$$

In determining the acid content of commercial vinegar, it is customary to treat the vinegar as a dilute solution of acetic acid, $HC_2H_3O_2$. The acetic acid concentration of the vinegar may be calculated as grams of acetic acid per liter or as percent acetic acid by mass. If the acetic acid content is to be expressed on a mass percent basis, the density of the vinegar must also be known.

## PROCEDURE

**Wear protective glasses. Do not pipet by mouth.**

### A. Molarity of an Unknown Acid

Obtain a sample of acid of unknown molarity in a clean, dry 125 mL Erlenmeyer flask as directed by your instructor.

With a volumetric pipet transfer a 10.00 mL sample of the acid to a clean, but not necessarily dry, Erlenmeyer flask. See "Use of the Pipet," p. 165, for instructions on cleaning and using the pipet. Pipet a duplicate 10.00 mL sample into a second flask. (If pipets are not available, a buret which has been carefully cleaned and rinsed may be used to measure the acid samples.)

You will need about 150 mL of base of known molarity (standard solution). Your instructor will give you the exact molarity of the base solution that you used in Experiment 20 or he may supply you with another solution of sodium hydroxide of known molarity. Record the exact molarity of this solution. Keep the flask containing the base stoppered when not in use.

Clean and set up a buret. See "Use of the Buret," p.158, for instructions on cleaning and using the buret.

Rinse the buret with two 5 to 10 mL portions of the base, running the second rinsing through the buret tip. Discard the rinsings. Fill the buret with the base, making sure that the tip is completely filled and contains no air bubbles. Adjust the level of the liquid in the buret so that the bottom of the meniscus is at exactly 0.00 mL. Record the initial buret reading (0.00 mL) in the space provided on the report sheet.

Add three drops of phenolphthalein solution and about 25 mL of distilled water to the flask containing the 10.00 mL of acid. Place this flask on a piece of white paper under the buret and lower the buret tip into the flask (see Figure 20.2).

Titrate the acid by adding base until the end-point is reached. During the titration swirl the solution in the flask with the right hand (if you are right handed) while manipulating the buret with the left. As the base is added you will observe a pink color caused by localized high base concentration. Near the end-point this color flashes throughout the solution, remaining for increasingly longer periods of time. When this occurs, add the base drop by drop until the end-point is reached, as indicated by the first drop of base which causes the entire solution to retain a faint pink color for at least 30 seconds. Record the final buret reading.

Refill the buret with base and adjust the volume to the zero mark. Titrate the duplicate sample of acid. If the volumes of base used differ by more than 0.20 mL, titrate a third sample. In the calculations, assume that the unknown acid reacts like HCl (one mole of acid reacts with one mole of base).

### B. Acetic Acid Content of Vinegar

Obtain about 40 mL of vinegar in a clean, dry 50 mL beaker. Record the sample number, if any, of this vinegar.

Titrate duplicate 10.00 mL samples of vinegar using exactly the same procedure outlined in Part A. Remember to rinse the pipet with vinegar before pipeting the vinegar samples.

When you are finished with the titrations, empty the buret and rinse it and the pipet at least twice with tap water and once with distilled water.

### Use of the Pipet

A volumetric (transfer) pipet (Figure 21.1) is calibrated to deliver a specified volume of liquid to a precision of about ±0.02 mL in a 10 mL pipet. To achieve this precision, the pipet must be clean and used in a specified manner.

Liquids are drawn into a pipet by means of a rubber suction bulb (Figure 21.2), or by a rubber tube connected to a water aspirator pump. Suction by mouth has also been used to draw liquids into a pipet, but this is a dangerous practice and is not recommended.

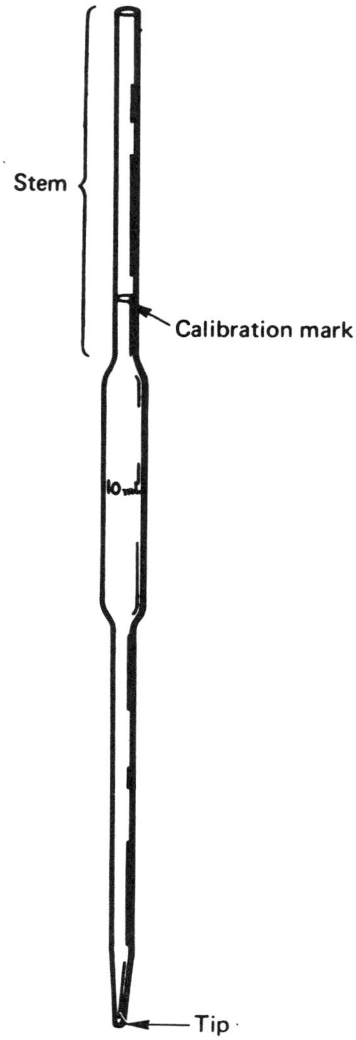


**Figure 21.1. A volumetric (transfer) pipet.**

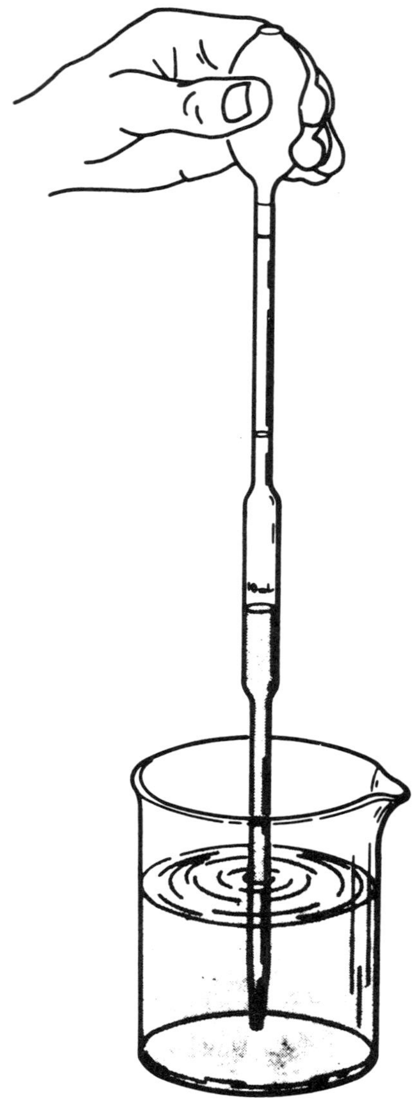

**Figure 21.2. Liquid is drawn into the pipet with a rubber suction bulb. Keep the tip of the pipet below the liquid level during suction.**

1. **Cleaning the Pipet.** Use a rubber suction bulb to draw up enough detergent solution to fill about two-thirds of the body or bulb of the pipet. Retain this solution in the pipet by pressing the forefinger tightly against the top of the pipet stem (Figure 21.3). Turn the pipet to a nearly horizontal position and gently shake and rotate it until the entire inside surface is wetted. Allow the pipet to drain and rinse it at least three times with tap water and once with distilled water.

2. **Using the Pipet.** Unless the pipet is known to be clean and absolutely dry on the inside, it must be rinsed twice with small portions of the liquid that is to be pipeted. This is done as in the washing procedure described above. These rinses are discarded in order to avoid contamination of the liquid being pipeted. A pipet does not need to be rinsed between successive pipettings of the same solution.

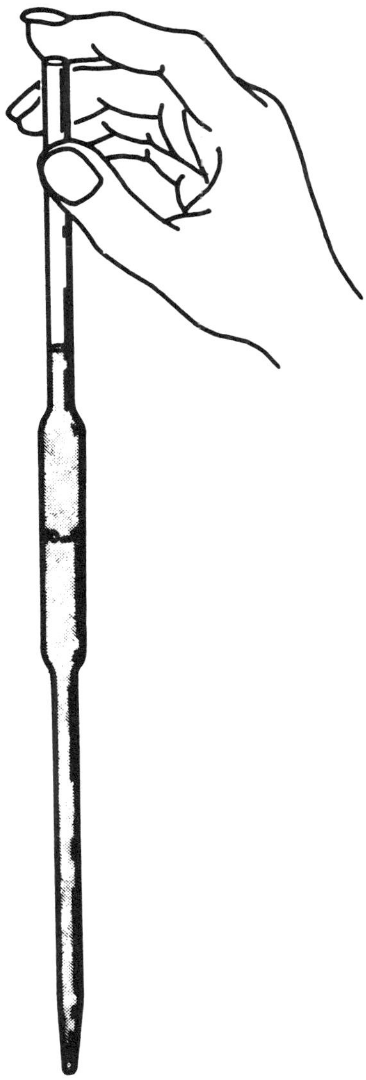

Figure 21.3. **Liquid is retained in the pipet by applying pressure with the forefinger to the top of the stem.**

Figure 21.4. **The pipet is calibrated to deliver the specified volume, leaving a small amount of liquid in the tip.**

To transfer a measured volume of a liquid, collapse a suction bulb by squeezing and place it tightly against the top of a pipet (do not try to push the bulb on to the pipet). Draw the liquid into the pipet until it is filled to about 5 cm above the calibration mark by allowing the bulb to slowly expand. Be careful–do not allow the liquid to get into the bulb. Remove the bulb and quickly place your forefinger over the top of the pipet stem. The liquid will be retained in the pipet if the finger is pressed tightly against the top of the stem. Keeping the pipet in a vertical position, decrease the finger pressure very slightly and allow the liquid level to drop slowly toward the calibration mark. When the liquid level has almost reached the calibration mark, again increase the finger pressure and stop the liquid when the bottom of the meniscus is exactly on the calibration mark. Touch the tip to the wall of the flask to remove the adhering drop of liquid.

Move the pipet to the flask which is to receive the sample and allow the liquid to drain while holding the pipet in a vertical position. About 10 seconds after the liquid has stopped running from the pipet, touch the tip to the inner wall of the sample flask to remove the drop of liquid adhering to the tip. A small amount of liquid will remain in the tip (Figure 21.4). Do not blow or shake this liquid into the sample; the pipet is calibrated to deliver the volume specified without this small residual.

If you have never used a volumetric pipet, it is advisable to practice by pipetting some samples of distilled water until you have mastered the technique.

NAME ______________________

SECTION ______ DATE ______

INSTRUCTOR ______________________

## REPORT FOR EXPERIMENT 21

# Neutralization—Titration II

### A. Molarity of an Unknown Acid

**Data Table**

| | Sample 1 | | Sample 2 | | Sample 3 (if needed) | |
|---|---|---|---|---|---|---|
| | Acid* | Base | Acid* | Base | Acid* | Base |
| Final buret reading | | | | | | |
| Initial buret reading | | | | | | |
| Volume used | | | | | | |

*If a pipet is used to measure the volume of acid, record only in the space for volume used.

Molarity of base (NaOH) ______________________

**CALCULATIONS:** In the spaces below show calculation setups for Sample 1 only. Show answers for both samples in the box.

| | Sample 1 | Sample 2 | Sample 3 (if needed) |
|---|---|---|---|
| 1. Moles of base (NaOH) | ______ | ______ | ______ |
| 2. Moles of acid used to neutralize (react with) the above number of moles of base | ______ | ______ | ______ |
| 3. Molarity of acid | ______ | ______ | ______ |

4. Average molarity of acid ______________

5. Unknown acid number ______________

## B. Acetic Acid Content of Vinegar

**Data Table**

| | Sample 1 | | Sample 2 | | Sample 3 (if needed) | |
|---|---|---|---|---|---|---|
| | Vinegar* | Base | Vinegar* | Base | Vinegar* | Base |
| Final buret reading | | | | | | |
| Initial buret reading | | | | | | |
| Volume used | | | | | | |

*If a pipet is used to measure the volume of vinegar, record only in the space for volume used.

Molarity of base (NaOH) ____________ Vinegar Number____________

**CALCULATIONS:** In the spaces below show calculation setups for Sample 1 only. Show answers for both samples in the box.

| | Sample 1 | Sample 2 | Sample 3 (if needed) |
|---|---|---|---|
| 1. Moles of base (NaOH) | ______ | ______ | ______ |
| 2. Moles of acid ($HC_2H_3O_2$) used to neutralize (react with) the above number of moles of base | ______ | ______ | ______ |
| 3. Molarity of acetic acid in the vinegar | ______ | ______ | ______ |

4. Average molarity of acetic acid in the vinegar ____________

5. Grams of acetic acid per liter (from average molarity) ____________

6. Mass percent acetic acid in vinegar sample (density of vinegar = 1.005 g/mL) ____________

# EXPERIMENT 22

# Chloride Content of Salts

## MATERIALS AND EQUIPMENT

Solutions: 0.10 M aluminum chloride ($AlCl_3$), 0.10 M barium chloride ($BaCl_2$), 0.10 M potassium chloride (KCl), 0.10 M silver nitrate ($AgNO_3$), and 0.10 M sodium chloride (NaCl). Dichlorofluorescein indicator, and "unknown" solutions.

## DISCUSSION

This experiment verifies the ratios of chloride ions in several metal chlorides. It also furnishes us with data for studying the relationship between the molar concentrations of a salt and that of its ions in solution.

We determine the amount of chloride ion in solution by reacting it with silver nitrate solution. The net ionic equation for this reaction is

$$Cl^- + Ag^+ \longrightarrow AgCl\downarrow$$

If we titrate different chloride salt solutions having the same molarity (moles/liter), the ratio of the amounts of $AgNO_3$ solution required should be the same as the ratio of the chloride contents of the respective salts. Consider the chloride available in 1 liter of each 1 molar NaCl and $BaCl_2$ solutions.

$$1 \text{ mole NaCl} \longrightarrow 1 \text{ mole } Na^+ + 1 \text{ mole } Cl^-$$

$$1 \text{ mole } BaCl_2 \longrightarrow 1 \text{ mole } Ba^{2+} + 2 \text{ moles } Cl^-$$

Thus it is evident that if equal volumes of the same molar concentration of NaCl and $BaCl_2$ solutions are reacted, the $BaCl_2$ should require twice as much $Ag^+$ (silver nitrate solution).

In the experiment, silver nitrate solution is added to the chloride solutions containing dichlorofluorescein indicator. A sudden change from white to pink indicates that all the chloride in solution is reacted and marks the end point of the titration.

An unknown solution can be analyzed for its chloride content by the above method. If the data include the volume of the unknown solution and the molarity and volume of the $AgNO_3$ solution, the chloride molarity of the unknown may be calculated.

The following relationships are useful for calculations in this experiment:

$$\text{Moles} = \frac{\text{g of solute}}{\text{molar mass of solute}}$$

$$\text{Molarity} = \frac{\text{moles}}{\text{liter}} = \frac{\text{g of solute}}{\text{molar mass of solute} \times \text{liters of solution}}$$

$$\text{Moles} = \text{liters} \times \text{molarity} = \frac{\text{liters} \times \text{moles}}{\text{liter}}$$

In a chemical reaction **where the reactants react in equal molar amounts** such as in $Cl^- + Ag^+ \longrightarrow AgCl$ (these ions react in a 1:1 mole ratio) we may say that

$$\text{moles } Cl^- \text{ reacted} = \text{moles } Ag^+ \text{ reacted}$$

Since

$$\text{moles} = \text{liters} \times \text{molarity}$$

$$\text{liters } Cl^- \times \text{molarity } Cl^- = \text{liters } Ag^+ \times \text{molarity } Ag^+$$

For the purpose of calculations, when the volumes of the two solutions are in the same units, whether they be liters, mL, or drops, we can write a more general statement equating the two reactants.

$$\text{Volume } Cl^- \times \text{molarity } Cl^- = \text{volume } Ag^+ \times \text{molarity } Ag^+$$

Using this relationship we may calculate the molarity of the chloride ion in solution when data for the other three values are known.

## PROCEDURE

**Wear protective glasses.**

NOTES:

1. Use the **same** medicine dropper to dispense the chloride solutions and the silver nitrate solution. Rinse the dropper with distilled water when changing from one solution to another.

2. Due to the different sizes of drops (even with the same medicine dropper), the volume of silver nitrate may vary about one drop in every ten drops in duplicate titrations. **It is very important to hold the medicine dropper in a vertical position in order to deliver drops of as uniform a volume as possible.**

3. Since less than 2 mL of most reagents are required, obtain only small volumes of reagents in clean, dry beakers.

Submit a small, clean, dry beaker (or test tube) to your instructor for an unknown chloride solution. Record the number of the unknown in the data table.

Place 10 clean test tubes in the rack. From a medicine dropper, add exactly 10 drops of 0.10 M NaCl to tubes one and two; 10 drops of 0.10 M KCl to tubes three and four; 10 drops 0.10 M $BaCl_2$ to tubes five and six; 10 drops of 0.10 M $AlCl_3$ to tubes seven and eight; and 10 drops of the unknown solution to tubes nine and ten. Add 1 drop of dichlorofluorescein indicator to each tube.

Obtain about 10 mL of 0.10 M $AgNO_3$ solution from the reagent bottle. Titrate the chloride samples by adding silver nitrate solution dropwise to each solution. Shake the solution during the titration. The end-point is reached when the color of the precipitate changes from white to pink and remains pink throughout the entire solution for at least ten seconds while shaking. **Count and record the number of drops of $AgNO_3$ solution required to reach the end-point in each tube.** If results for duplicate samples vary by more than one drop for every 10 drops of silver nitrate added, run another sample.

Complete the data table and answer the questions on the report form.

NAME ____________

SECTION ________ DATE ________

INSTRUCTOR ____________

## REPORT FOR EXPERIMENT 22

# Chloride Content of Salts

**Data Table**

| Solution | Drops of Silver Nitrate | | | |
|---|---|---|---|---|
| | Trial 1 | Trial 2 | Trial 3 (If needed) | Average |
| NaCl | | | | |
| KCl | | | | |
| $BaCl_2$ | | | | |
| $AlCl_3$ | | | | |
| Unknown No. ______ | | | | |

### QUESTIONS AND PROBLEMS

1. Tabulate the experimental ratios of the chloride ion contents of the salts tested. Obtain these ratios by dividing the average number of drops of $AgNO_3$ required for each solution by the average number of drops required for the NaCl solution. Express the ratios to the nearest tenth.

| Salt | NaCl | KCl | $BaCl_2$ | $AlCl_3$ |
|---|---|---|---|---|
| Ratio | 1.0 | | | |

2. Assuming that sodium chloride has one chloride ion per simplest formula unit

   (a) Do the data of this experiment prove that potassium chloride has one chloride ion per simplest formula unit? Explain.

   (b) Do the data of this experiment prove that barium chloride has two chloride ions per simplest formula unit? Explain.

(c) Do the data of this experiment prove that aluminum chloride has three chloride ions per formula unit? Explain.

(d) Do the data of this experiment prove that the simplest formula of aluminum chloride is $AlCl_3$? Explain.

3. Write and balance: (a) the un-ionized equation; (b) the total ionic equation; and (c) the net ionic equation for each of the following reactions:

$NaCl + AgNO_3 \longrightarrow$

$KCl + AgNO_3 \longrightarrow$

$BaCl_2 + AgNO_3 \longrightarrow$

$AlCl_3 + AgNO_3 \longrightarrow$

4. It was seen in the experiment that about 10 drops of $AgNO_3$ solution were required to react with 10 drops of NaCl solution; and about 20 drops of $AgNO_3$ solution were required to react with 10 drops of $BaCl_2$ solution.

   (a) What two factors are responsible for the fact that equal volumes of NaCl and $AgNO_3$ solutions react with each other?

   (b) What change(s) would you suggest to make equal volumes of $BaCl_2$ and $AgNO_3$ solutions react with each other?

5. What is the molarity of $Cl^-$ ion in each of the solutions tested?

   0.10 M NaCl ____________

   0.10 M KCl ____________

   0.10 M $BaCl_2$ ____________

   0.10 M $AlCl_3$ ____________

6. Calculate the chloride molarity of the unknown solution. Show setups.

   Unknown No. ____________

   Molarity ____________

7. How many drops of 0.10 M $AgNO_3$ would be required to react with each of the following solutions? Assume each individual solution is 0.10 M.

   (a) 5 drops NaCl + 10 drops KCl ____________

   (b) 10 drops NaCl + 5 drops $BaCl_2$ ____________

   (c) 5 drops $BaCl_2$ + 5 drops $AlCl_3$ ____________

   (d) 5 drops $SnCl_4$ ____________

   (e) 15 drops $NH_4Cl$ ____________

 NAME ______________________

8. (a) How many moles of sodium chloride are in 5 drops of 0.10 M NaCl solution? Assume 1 drop is 0.050 mL.

______________________

(b) Calculate the number of milligrams of potassium chloride in 1 drop of 0.10 M KCl solution.

______________________

9. How many grams of sodium chloride are required to prepare 300 mL of 0.15 M NaCl solution? Show calculations.

______________________

# EXPERIMENT 23

# Oxidation-Reduction—Sulfur Dioxide

## MATERIALS AND EQUIPMENT

Solids: iron(II) sulfide (FeS) and sodium bisulfite (also called sodium hydrogen sulfite) ($NaHSO_3$). Solutions: 0.1 M barium chloride ($BaCl_2$), concentrated (12 M) hydrochloric acid (HCl), dilute (6 M) hydrochloric acid, 3 percent hydrogen peroxide ($H_2O_2$), 0.1 M iodine in potassium iodide ($I_2$), 0.1 M potassium chlorate ($KClO_3$), 0.1 M potassium chromate ($K_2CrO_4$), and 0.1 M potassium permanganate ($KMnO_4$). Two 25 cm lengths of 3/16 in. (4.8 mm) I.D. rubber tubing; 25 × 200 mm test tube.

## DISCUSSION

Oxidation-reduction reactions, often called **redox** reactions, involve changes in oxidation states. In such reactions **oxidation** is recognized by an increase in the oxidation state of an element. **Reduction** is recognized by a decrease in the oxidation state of an element. Oxidation results from a loss of electrons; reduction, from a gain of electrons.

The oxidation state of an element, either free or in chemical combination, is indicated by an **oxidation number.** Oxidation numbers are assigned according to an arbitrary system. (Consult your text for details of this system and of the use of oxidation numbers in balancing redox equations.)

In this experiment you will use the reactions of some sulfur compounds to study the phenomenon of oxidation-reduction. Sulfur is a common nonmetallic element that is essential to life and of great industrial importance. The element exists in several oxidation states and unites with other elements to form many different kinds of compounds. The two most important oxides are sulfur dioxide ($SO_2$) and sulfur trioxide ($SO_3$); both are acid anhydrides.

Sulfur dioxide reacts with water to form sulfurous acid.

$$H_2O + SO_2 \rightleftarrows H_2SO_3$$

The equation is written reversibly (in equilibrium form) because sulfurous acid is unstable and decomposes to sulfur dioxide and water. Sulfurous acid is a weak acid which forms hydronium ions ($H_3O^+$), bisulfite ions ($HSO_3^-$), and sulfite ions ($SO_3^{2-}$) in water solutions. Two names are commonly used for the $HSO_3^-$ ion; they are bisulfite ion and hydrogen sulfite ion; we will refer to it as the bisulfite ion in this experiment.

The common oxidation states of sulfur are −2, 0, +4, and +6. In sulfur dioxide, bisulfites, and sulfites, the oxidation number of sulfur is +4. Sulfur in this oxidation state may be further oxidized to +6 or reduced to 0 or −2 with appropriate reagents. The oxidation to +6 is easily accomplished by many oxidizing agents. For example, hydrogen peroxide oxidizes sulfurous acid:

$$H_2SO_3 + H_2O_2 \longrightarrow H_2SO_4 + H_2O$$

In the above reaction hydrogen peroxide is the oxidizing agent, since it oxidized the sulfur from a +4 to a +6 oxidation state.

In solution sulfurous acid, bisulfite ions, and sulfite ions are slowly oxidized by oxygen dissolved from the air. For this reason solutions containing bisulfite or sulfite ions contain some bisulfate and sulfate ions as well.

$$2\ H_2SO_3 + O_2 \longrightarrow 2\ H_2SO_4$$

$$2\ HSO_3^- + O_2 \longrightarrow 2\ HSO_4^-$$

$$2\ SO_3^{2-} + O_2 \longrightarrow 2\ SO_4^{2-}$$

Sulfur dioxide, sulfites, and bisulfites most commonly react as reducing agents. However, in some reactions these substances can function as oxidizing agents. An example is the reaction:

$$2\ H_2S + SO_2 \longrightarrow 2\ H_2O + 3\ S$$

The oxidation state of sulfur is −2 in $H_2S$, +4 in $SO_2$, and 0 in free sulfur. In the above reaction sulfur in $H_2S$ is oxidized from −2 to 0; sulfur in $SO_2$ is reduced from +4 to 0. Thus the $SO_2$ is the oxidizing agent (since it caused sulfur in $H_2S$ to be oxidized) and $H_2S$ is the reducing agent (since it caused sulfur in $SO_2$ to be reduced).

It is of interest that the foregoing reaction is used in petroleum refineries to (1) provide better fuels by removing objectionable sulfur compounds, (2) reduce air pollution by removing sulfur dioxide from refinery gases, and (3) yield a valuable by-product in the form of saleable sulfur.

## PROCEDURE

**Wear protective glasses.**

Assemble the apparatus shown in Figure 23.1, initially omitting the medicine dropper. Clamp tube A in an upright position on a ring stand. Place about 3 g of sodium bisulfite in tube A. Lubricate the hole in the stopper with glycerol. Draw as much conc. (12 M) hydrochloric acid into the medicine dropper as possible. Grip the medicine dropper by the glass part and insert it into the lubricated stopper, as shown. Refill the medicine dropper with acid as needed during the experiment.

### A. Solubility of Sulfur Dioxide in Water

Place the tip of delivery tube F into and near the bottom of a clean, dry test tube. Add 10 drops of the hydrochloric acid to the sodium bisulfite to generate sulfur dioxide. After 3 minutes remove the delivery tube from the test tube. Immediately cover the mouth of the test tube with your thumb and invert the tube into a beaker of tap water. Uncover the tube and allow it to remain in the beaker for several minutes, noting the water level in the tube from time to time. In the meantime continue with Part B.

### B. Reaction of Sulfur Dioxide with Water

Place about 5 mL of distilled water in a test tube; add a drop or two of concentrated hydrochloric acid to the sodium bisulfite in tube A and immediately insert delivery tube F into the distilled water. After the evolution of sulfur dioxide has subsided, withdraw the delivery tube. Test the solution in the tube with a strip of red and a strip of blue litmus paper.

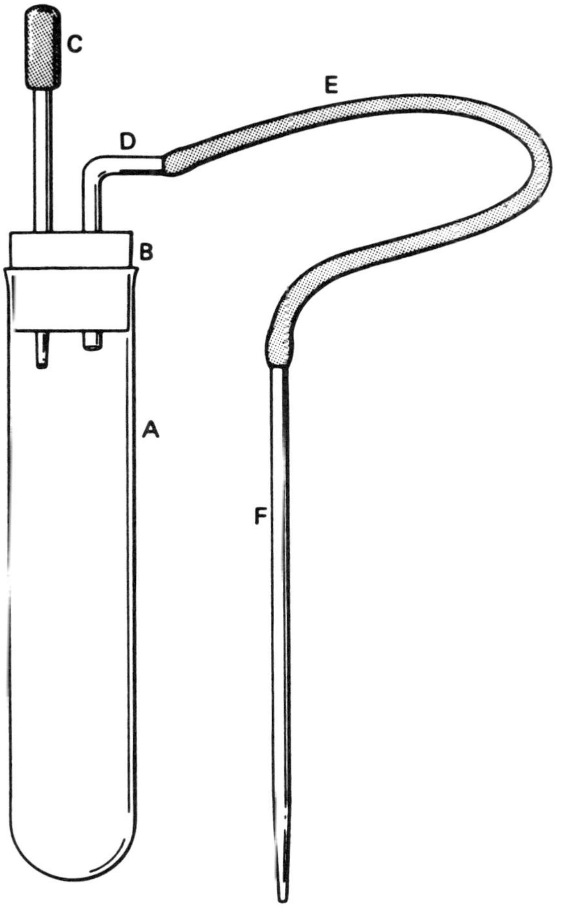


Figure 23.1. Generator for sulfur dioxide. A, 25 × 200 mm test tube (dry). B, No. 4 two-hole rubber stopper. C, Medicine dropper. D, Right-angle bend glass tubing. E, 25 cm of 3/16 in. (4.8 mm) rubber tubing. F, 6 mm glass tubing drawn to about 1 mm tip, about 15 cm overall length.

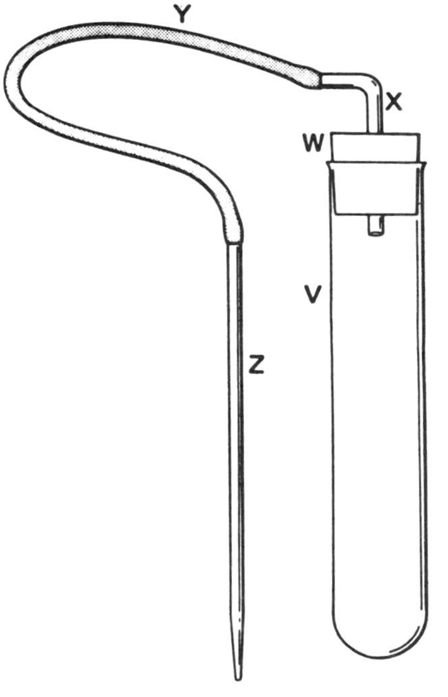


Figure 23.2. Generator for hydrogen sulfide. V, 18 × 150 mm test tube. W, No. 1 one-hole rubber stopper. X, Right-angle bend glass tubing. Y, 25 cm of 3/16 in. (4.8 mm) rubber tubing. Z, 6 mm glass tubing drawn to about 1 mm tip, about 15 cm overall length.

## C. Reactions of Sulfur Dioxide as a Reducing Agent

Arrange six test tubes in a rack and put about 5 mL of distilled water and 10 drops of dil. (6 M) hydrochloric acid into each tube. Then add reagents to the tubes, as follows, rinsing out the medicine dropper with distilled water between successive additions of reagents:

Tube 1: 10 drops 0.1 M iodine in potassium iodide solution.

Tube 2: 10 drops 0.1 M potassium chromate solution.

Tube 3: 10 drops 0.1 M potassium permanganate solution.

Tube 4: 10 drops 3 percent hydrogen peroxide solution.

Tube 5: 10 drops 0.1 M potassium chlorate solution.

Tube 6: 10 drops distilled water (control for comparison).

Begin the generation of sulfur dioxide by adding 3 drops of acid to tube A, and place delivery tube F into the iodine solution in tube 1. Continue to generate and deliver sulfur dioxide to tube 1 until the brown color of free iodine has disappeared. Repeat the sulfur dioxide addition process with the tube containing potassium chromate and with the tube containing potassium permanganate, until additional increments of the gas produce no further color changes in these solutions. **Note:** Add more HCL to tube A, as needed, to keep $SO_2$ bubbling into the solutions.

Many chemical reactions can be readily detected by the appearance of precipitates, evolution of gases, or color changes. But many other reactions produce no visible evidence of their progress, and we must rely on physical or chemical tests to tell whether a reaction has actually taken place. No color change is apparent when sulfur dioxide is added to tubes 4, 5, and 6. In order to be certain that enough sulfur dioxide is present to ensure the completion of any possible reactions, bubble sulfur dioxide into each tube for **1 minute.**

After adding sulfur dioxide to each tube in the series, add 10 drops of 0.1 M barium chloride solution to each tube and mix well. Allow the solution to stand for 2 or 3 minutes, then examine all tubes carefully to determine whether a precipitate has formed. If a precipitate has not formed in all of the tubes, heat all the tubes in a beaker of boiling water for 1 or 2 minutes, cool, and reexamine them. Record your observations in the table in Part C of the report form.

### D. Reaction of Sulfur Dioxide as an Oxidizing Agent

⚠ CAUTION: Hydrogen sulfide is very poisonous. **Work with this substance must be done in an adequate hood.**

Keep your sulfur dioxide generator intact. Assemble a hydrogen sulfide generator consisting of an 18 × 150 mm test tube with a one-hole stopper and delivery tube assembly, as shown in Figure 23.2. Put 0.5 to 1.0 g of iron(II) sulfide into this generator. Add about 1 mL of dil. (6 M) hydrochloric acid to the hydrogen sulfide generator tube and immediately replace the one-hole stopper. Add several drops of conc. hydrochloric acid to the sulfur dioxide generator. Now bubble both hydrogen sulfide and sulfur dioxide gases into about 5 mL of water in another test tube. Continue to bubble the gases through the water for about 2 minutes. Record your observations.

NAME ________________________

SECTION ________ DATE ________

INSTRUCTOR ____________________

## *REPORT FOR EXPERIMENT 23*

# Oxidation-Reduction—Sulfur Dioxide

### A. Solubility of Sulfur Dioxide in Water

Observation:

### B. Reaction of Sulfur Dioxide with Water

Observation:

### C. Reactions of Sulfur Dioxide as a Reducing Agent

| Substance tested | Evidence of reaction, if any, with $SO_2$ (before adding $BaCl_2$) | Relative amount of precipitate after $BaCl_2$ added<br><br>0 = None to slight<br>1 = Moderate to heavy |
|---|---|---|
| $I_2$ | | |
| $K_2CrO_4$ | | |
| $KMnO_4$ | | |
| $H_2O_2$ | | |
| $KClO_3$ | | |
| Water (control) | | |

**D. Reaction of Sulfur Dioxide as an Oxidizing Agent**

Observations:

## QUESTIONS AND PROBLEMS

1. What evidence did you see which indicates that sulfur dioxide is soluble in water?

2. Account for the color of litmus in a solution of sulfur dioxide in water.

3. Write a chemical equation to illustrate the reaction of sulfur dioxide with water.

4. Write balanced oxidation-reduction (redox) equations for each of the reactions occurring in the first five tubes of Part C. Refer to your text for an appropriate method of balancing redox equations. Assume reactants and products are as follows:

   (a) Sulfurous acid + Iodine + Water ⟶ Hydrogen iodide + Sulfuric acid

   (b) Sulfurous acid + Potassium chromate + Hydrochloric acid
   ⟶ Chromium(III) chloride + Sulfuric acid
   + Potassium chloride + Water

   (c) Sulfurous acid + Potassium permanganate ⟶ Manganese(II) sulfate + Sulfuric acid
   + Potassium sulfate + Water

   (d) Sulfurous acid + Hydrogen peroxide ⟶ Sulfuric acid + Water

   (e) Sulfurous acid + Potassium chlorate ⟶ Sulfuric acid + Potassium chloride

5. Write a balanced equation for the reaction of sulfur dioxide and hydrogen sulfide.

6. Referring to the reactions in Parts C and D, complete the following table by writing in the appropriate formulas and symbols.

| | Formula of Oxidizing Agent | Formula of Reducing Agent | Symbol of Element Oxidized | Symbol of Element Reduced |
|---|---|---|---|---|
| Part C | | | | |
| Tube 1 | | | | |
| Tube 2 | | | | |
| Tube 3 | | | | |
| Tube 4 | | | | |
| Tube 5 | | | | |
| Part D | | | | |

7. Write the formula of the precipitate that is formed after the barium chloride solution is added to the redox products in Part C.

____________

8. Sulfur dioxide arising from the combustion of sulfur containing fuels or from smelters is a major air pollutant. Explain, using appropriate equations, how this sulfur dioxide is responsible for the minute droplets of sulfuric acid solution in some smogs and for the sulfuric acid in acid rain.

# EXPERIMENT 24

# Halogens

## MATERIALS AND EQUIPMENT

Solids: light copper turnings, iodine crystals ($I_2$), magnesium strips, manganese dioxide ($MnO_2$), potassium bromide (KBr), potassium chloride (KCl), [C.P.], potassium iodide (KI), and steel wool. Liquids: bromine ($Br_2$), 1,1,1-trichloroethane ($CCl_3CH_3$), ethanol ($C_2H_5OH$) and concentrated sulfuric acid ($H_2SO_4$). Solutions: bromine water (saturated), iodine water (saturated), dilute (6 M) hydrochloric acid (HCl), and 0.1 M sodium nitrite ($NaNO_2$).

## DISCUSSION

The **halogens,** constituting Group VIIA of the periodic table, are a very reactive group of nonmetals. Fluorine, chlorine, bromine, and iodine–the first four members of the group–are of great economic value. Astatine, the fifth member, is a radioactive element that has been prepared only in minute quantities and will not be considered here. Chlorine and iodine are essential to our life processes, and fluorine is essential for sound teeth. Bromine, however, has no important physiological value to man. In this experiment we shall examine some properties and reactions of the halogens.

Because of their reactivity, halogens are never found in the free or uncombined state in nature. However, except for fluorine, they can be prepared in small quantities without difficulty, using simple equipment and readily available reagents. Free fluorine is a dangerous substance that requires specialized electrolytic apparatus for preparation. It is therefore not used for experimentation in general college chemistry laboratories. The free halogens are diatomic, their formulas being $F_2$, $Cl_2$, $Br_2$, and $I_2$.

All halogens combine directly with many other elements, both metals and nonmetals. Fluorine is the most reactive, followed in decreasing reactivity by chlorine, bromine, and iodine. All react directly with the more active metals to form salts collectively known as metal **halides.** The formulas of the halide ions are $F^-$, $Cl^-$, $Br^-$, and $I^-$.

The activity of free halogens as oxidizing agents decreases with increasing atomic number; fluorine is the most active oxidizing agent, and iodine is the least active. Conversely, iodide ion is the most active reducing agent (easiest to oxidize), and fluoride ion is the least active (most difficult to oxidize). Thus halogen atoms will oxidize halide ions of higher atomic number to the corresponding halogen atoms, and halide ions will reduce halogen atoms of lower atomic number to the corresponding halide ions.

Since fluoride ions and chloride ions are not easily oxidized, reasonably pure hydrogen fluoride and hydrogen chloride can be prepared by heating the corresponding metal halide with concentrated sulfuric acid. Hydrogen fluoride and hydrogen chloride are volatile and separate readily from the high boiling acid and nonvolatile salts. The equations for the reactions are:

$$NaF + H_2SO_4 \longrightarrow NaHSO_4 + HF\uparrow$$

$$NaCl + H_2SO_4 \longrightarrow NaHSO_4 + HCl\uparrow$$

CAUTION: Do not attempt to prepare hydrogen fluoride; it is **very poisonous.**

Because bromide and iodide ions are oxidized by concentrated sulfuric acid, neither HBr nor HI can be prepared in pure form by the method just described.

Free halogens can be prepared by a variety of methods. A common laboratory procedure is to oxidize the appropriate halide with manganese dioxide in the presence of concentrated sulfuric acid. This method is satisfactory for all halogens except fluorine.

$$4\ HX + MnO_2 \longrightarrow MnX_2 + X_2 + 2\ H_2O$$

X represents a halogen atom.

Chlorine, bromine, and iodine react with water, but the reactions are reversible. The general equation is

$$X_2 + HOH \rightleftarrows HX + HOX$$

Chlorine, for example, forms hydrochloric acid and hypochlorous acid:

$$Cl_2 + HOH \rightleftarrows HCl + HOCl$$

The hypochlorous acid formed is a weak acid, but it and its salts are strong oxidizing agents. Hydrochloric acid is a strong acid but is not an oxidizing agent.

If a metal, M, is brought into contact with a solution of a halogen in water, at least three different reactions are possible.

$$M + X_2 \longrightarrow MX_2$$

$$M + 2\ HX \longrightarrow MX_2 + H_2\uparrow$$

$$M + HOX + HX \longrightarrow MX_2 + H_2O$$

The reaction that will predominate in any specific instance depends on such factors as the nature of the metal and the halogen, temperature, and concentration of the reactants.

## PROCEDURE

NOTES:

1. **Wear protective glasses.**
2. CAUTION: The free halogens are irritating and toxic. Avoid skin contact with liquid and solid halogens and breathing of halogen vapors and gases.
3. Dispense bromine water in the fume hood.
4. Dispose of mixtures containing trichloroethane in the container provided.

### A. Physical Characteristics of Bromine and Iodine

1. Note the physical characteristics of the sample of free bromine in the fume hood. (Do not open the bottle.)

2. Note the physical characteristics of the free iodine in the bottle on the reagent shelf. (Do not allow the iodine to touch your skin.)

3. Arrange eight test tubes in a rack and add the following:

Tube 1: 3 mL water and 1 mL bromine water.

Tube 2: 20 drops trichloroethane, 3 mL water, and 1 mL bromine water.

Tube 3: 3 mL saturated iodine water.

Tube 4: 3 mL saturated iodine water and 20 drops of trichloroethane.

Tube 5: A crystal of iodine (no larger than a grain of rice) and 2 mL water.

Tube 6: A crystal of iodine (no larger than a grain of rice) and 2 mL trichloroethane.

Tube 7: A crystal of iodine (no larger than a grain of rice), 2 mL water, and about 0.1 g potassium iodide crystals.

Tube 8: A crystal of iodine (no larger than a grain of rice) and 2 mL ethanol.

Stopper and shake each tube vigorously; observe the colors and solubility characteristics of bromine and iodine in these mixtures. Complete the Report Sheet before discarding these eight mixtures.

## B. Preparation of Bromine and Iodine

1. Mix a small quantity (no larger than half the exposed part of the eraser on an ordinary lead pencil) of potassium bromide with an equal quantity of manganese dioxide. Transfer the mixture to a clean, **dry** test tube, and tap the tube lightly to prevent particles from adhering to the upper part of the tube. Add 3 drops of conc. sulfuric acid from a medicine dropper. Warm the mixture **very gently** (no more than 30 seconds) over a **low flame**; heat just enough to promote the reaction but not enough to drive fumes out of the tube. Record your observations on the report form.

2. Repeat the procedure above, using potassium iodide instead of potassium bromide.

## C. Reaction of Water Solutions of Bromine and Iodine with Metals

Set up a series of eight clean (but not necessarily completely dry) test tubes in a rack. Prepare two small wads each of steel wool, light copper turnings, and magnesium strips. Each wad should be **no larger** than the exposed rubber part of a pencil eraser. Using a stirring rod, push each metal wad to the bottom of a test tube. Arrange the tubes in order and add 3 mL of saturated bromine water or iodine water to each, as indicated below:

Tube 1: Steel wool and bromine water.

Tube 2: Steel wool and iodine water.

Tube 3: Copper turnings and bromine water.

Tube 4: Copper turnings and iodine water.

Tube 5: Magnesium strips and bromine water.

Tube 6: Magnesium strips and iodine water.

Tube 7: Bromine water (control for comparison).

Tube 8: Iodine water (control for comparison).

Loosely stopper the tubes containing bromine water. Examine each tube after approximately 5, 10 and 20 minutes, noting evidence of reactions such as color changes, evolution of gases, etc. In the meantime start Part D.

### D. Oxidation of Chloride and Bromide Ions

1. Place a strip of blue litmus paper on a glass plate and put a small crystal of potassium iodide immediately adjacent to the litmus paper. Into a clean, **dry** test tube, put a small amount of potassium chloride (no larger than half the exposed rubber part of a pencil eraser). Add 3 drops of conc. sulfuric acid from a medicine dropper, and allow the mixture to react for a few seconds. Now place the lip of the test tube almost against the crystal and tilt the tube at an angle of about 30 degrees. This will allow the gas evolved from the reaction to flow down and contact the crystal and the litmus paper. Note any evidence of reaction on the potassium iodide crystal and on the litmus paper.

2. Repeat Part D.1, adding an equal amount of manganese dioxide to the potassium chloride and mixing them in the test tube before you add the concentrated sulfuric acid.

3. Repeat Part D.1, but substitute potassium bromide for the potassium chloride in the test tube.

### E. Selective Oxidation of a Halide Ion

Place three clean test tubes in a rack. Add small quantities (no larger than a grain of rice) of potassium chloride crystals to the first tube, potassium bromide crystals to the second, and potassium iodide crystals to the third. Next add about 3 mL of distilled water to each tube and shake to dissolve the crystals. After the crystals are dissolved, add 5 drops of dil. (6 M) hydrochloric acid, 1 mL of 0.1 M sodium nitrite solution, and 1 mL of trichloroethane to each tube, and mix thoroughly. Note which of the halide ions was oxidized by the sodium nitrite.

NAME ____________________

SECTION ________ DATE ____________

INSTRUCTOR ____________________

# *REPORT FOR EXPERIMENT 24*

# Halogens

## A. Physical Characteristics of Bromine and Iodine

1. Describe the physical characteristics of bromine.

2. Describe the physical characteristics of iodine.

3. (a) Color of bromine in water: ____________

   (b) Color of bromine in trichloroethane: ____________

   (c) Color of iodine in water: ____________

   (d) Color of iodine in trichloroethane: ____________

   (e) Color of iodine in ethanol: ____________

4. Iodine is more soluble in trichloroethane than in water.

   (a) What evidence of this fact did you observe in tube 4 in Part A.3?

   (b) What evidence of this fact did you observe in Part A.3 when comparing tubes 5 and 6? (Color intensity is not sufficient evidence in this case.)

5. In which solvent, water or water plus potassium iodide, does iodine have the greater solubility?

   ____________

6. In which solvent, water or ethanol, does iodine have the greater solubility?

   ____________

## B. Preparation of Bromine and Iodine

1. Evidence that bromine was produced.

2. Evidence that iodine was produced.

## C. Reaction of Water Solutions of Bromine and Iodine with Metals

| Metal | Description of Changes Observed (if any) | |
|---|---|---|
| | Bromine Water | Iodine Water |
| Iron (steel wool) | | |
| Copper turnings | | |
| Magnesium strips | | |
| Control | | |

**D. Oxidation of Chloride and Bromide Ions**

| Reactants | Effect of Gaseous Product(s) on: | |
|---|---|---|
| | Blue Litmus | KI Crystal |
| 1. $KCl + H_2SO_4$ | | |
| 2. $KCl + MnO_2 + H_2SO_4$ | | |
| 3. $KBr + H_2SO_4$ | | |

**E. Selective Oxidation of a Halide Ion**

1. Which of the halide ions, if any, was oxidized by sodium nitrite? Cite evidence.

## QUESTIONS AND PROBLEMS

1. Part B.1. Write a balanced equation for the reaction that occurred when potassium bromide, manganese dioxide, and sulfuric acid were heated. Assume the products are bromine, potassium sulfate, manganese(II) sulfate, and water.

2. Part B.2. Write a balanced equation for the preparation of iodine from potassium iodide, manganese dioxide, and sulfuric acid.

3. Part C. What evidence, other than color changes, did you observe that the reaction of bromine with magnesium was different from the reaction of bromine with copper?

4. Part C(a) Write the equation for the reaction between bromine water and copper. The reactants in the equation will be copper, hydrobromic acid, and hypobromous acid.

(b) Explain why, on the basis of the activity (electromotive) series of the metals, hydrobromic acid alone will not react with copper.

5. Part D(a) Was hydrogen chloride produced when sulfuric acid and potassium chloride reacted? Cite evidence.

(b) Was bromine produced when sulfuric acid and potassium bromide reacted? Cite evidence.

(c) Was chlorine produced when sulfuric acid, manganese dioxide, and potassium chloride reacted? Cite evidence.

6. Part D. Write equations for the reactions of (a) sulfuric acid and potassium chloride and (b) chlorine and potassium iodide.

(a)

(b)

7. Part E. Balance the following redox equation:

$$KI + \quad HNO_2 + \quad HCl \longrightarrow I_2 + \quad NO + \quad KCl + \quad H_2O$$

# EXPERIMENT 25

# Chemical Equilibrium—Reversible Reactions

## MATERIALS AND EQUIPMENT

Solid: ammonium chloride ($NH_4Cl$). Solutions: saturated ammonium chloride, concentrated (15 M) ammonium hydroxide ($NH_4OH$), 0.1 M cobalt(II) chloride ($CoCl_2$), 0.1 M iron(III) chloride ($FeCl_3$), concentrated (12 M) hydrochloric acid (HCl), dilute (6 M) nitric acid ($HNO_3$), phenolphthalein, 0.1 M potassium chromate ($K_2CrO_4$), 0.1 M potassium thiocyanate (KSCN), 0.1 M silver nitrate ($AgNO_3$), saturated sodium chloride (NaCl), 10 percent sodium hydroxide (NaOH), and dilute (3 M) sulfuric acid ($H_2SO_4$).

## DISCUSSION

In many chemical reactions the reactants are not totally converted to the products because of a reverse reaction; that is, because the products react to form the original reactants. Such reactions are said to be **reversible** and are indicated by a double arrow ($\rightleftarrows$) in the equation. The reaction proceeding to the right is called the **forward reaction**; that to the left, the **reverse reaction.** Both reactions occur simultaneously.

Every chemical reaction proceeds at a certain rate or speed. The rate of a reaction is variable and depends on the concentrations of the reactants and the conditions under which the reaction is conducted. When the rate of the forward reaction is equal to the rate of the reverse reaction, a condition of **chemical equilibrium** exists. At equilibrium the products react at the same rate they are produced. Thus the concentrations of substances in equilibrium do not change, but both reactions, forward and reverse, are still occurring.

The principle of Le Chatelier relates to systems in equilibrium and states that when the conditions of a system in equilibrium are changed the system reacts to counteract the change. In this experiment we will observe the effect of changing the concentration of one or more substances in a chemical equilibrium. Consider the hypothetical equilibrium system

$$A + B \rightleftarrows C + D$$

When the concentration of any one of the species in this equilibrium is changed, the equilibrium is disturbed. Changes in the concentrations of all the other substances will occur to establish a new position of equilibrium. For example, when the concentration of B is increased, the rate of the forward reaction increases, the concentration of A decreases, and the concentrations of C and D increase. After a period of time the two rates will become equal and the system will again be in equilibrium. The following statements indicate how the equilibrium will shift when the concentrations of A, B, C, and D are changed.

An increase in the concentration of A or B will cause the equilibrium to shift to the right.

An increase in the concentration of C or D will cause the equilibrium to shift to the left.

A decrease in the concentrations of A or B will cause the equilibrium to shift to the left.

A decrease in the concentrations of C or D will cause the equilibrium to shift to the right.

Evidence of a shift in equilibrium by a change in concentration can easily be observed if one of the substances involved in the equilibrium is colored. The appearance of a precipitate or the change in color of an indicator can sometimes be used to detect a shift in equilibrium.

Net ionic equations for the equilibrium systems to be studied are given below. These equations will be useful for answering the questions in the Report Sheet.

**Saturated Sodium Chloride Solution**

$$\text{NaCl (solid)} \underset{}{\overset{H_2O}{\rightleftarrows}} Na^+ + Cl^-$$

**Saturated Ammonium Chloride Solution**

$$NH_4Cl \text{ (solid)} \overset{H_2O}{\rightleftarrows} NH_4^+ + Cl^-$$

**Iron(III) Chloride plus Potassium Thiocyanate**

The equilibrium involves the following ions in solution:

$$Fe^{3+} + SCN^- \rightleftarrows Fe(SCN)^{2+}$$

Pale yellow Colorless Red

**Potassium Chromate with Nitric and Sulfuric Acids**

The equilibrium involves the following ions in solution:

$$2\,CrO_4^{2-} + 2\,H^+ \rightleftarrows Cr_2O_7^{2-} + H_2O$$

Yellow Orange

**Cobalt(II) Chloride Solution**

The equilibrium involves the following ions in solutions:

$$Co(H_2O)_6^{2+} + 4\,Cl^- \rightleftarrows CoCl_4^{2-} + 6\,H_2O$$

Pink Blue

**Ammonia Solution**

$$NH_3 + H_2O \rightleftarrows NH_4^+ + OH^-$$

## PROCEDURE

**Wear protective glasses.**

NOTE: Record observed evidence of equilibrium shifts as each experiment is done.

### A. Saturated Sodium Chloride Solution

Add a few drops of conc. hydrochloric acid to 2 to 3 mL of saturated sodium chloride solution in a test tube, and note the results.

### B. Saturated Ammonium Chloride Solution

Repeat Part A, using saturated ammonium chloride solution instead of sodium chloride solution.

### C. Iron(III) Chloride plus Potassium Thiocyanate

Prepare a stock solution to be tested by adding 2 mL each of 0.1 M iron(III) chloride and 0.1 M potassium thiocyanate solutions to 100 mL of distilled water and mix. Pour about 5 mL of this stock solution into each of four test tubes.

1. Use the first tube as a control for comparison.

2. Add about 1 mL of 0.1 M iron(III) chloride solution to the second tube and observe the color change.

3. Add about 1 mL of 0.1 M potassium thiocyanate solution to the third tube and observe the color change.

4. Add 0.1 M silver nitrate solution dropwise (less than 1 mL) to the fourth tube until almost all the color is discharged. The white precipitate formed consists of both AgCl and AgSCN. Pour about half the contents (including the precipitate) into another tube. Add 0.1 M potassium thiocyanate solution dropwise (1 to 2 mL) to one tube and 0.1 M iron(III) chloride solution (1 to 2 mL) to the other. Observe the results.

### D. Potassium Chromate with Nitric and Sulfuric Acids

Pour about 2 to 3 mL of 0.1 M potassium chromate solution into each of two test tubes. Add 2 drops of dil. (6 M) nitric acid to one tube. 2 drops of dil. (3 M) sulfuric acid to the other, and note the color change. Now add 10 percent sodium hydroxide solution dropwise to each tube until the original color of potassium chromate is restored.

### E. Cobalt(II) Chloride Solution

Place about 2 mL (no more) of 0.1 M cobalt(II) chloride solution into each of three test tubes.

1. To one tube add about 3 mL of conc. hydrochloric acid dropwise and note the result. Now add water dropwise to the solution until the original color (reverse reaction) is evident.

2. To the second tube add about 1.5 g of solid ammonium chloride and shake to make a saturated salt solution. Compare the color with the solution in the third tube (control). Place the second and third tubes (unstoppered) in a beaker of boiling water, shake occasionally, and note the results. Cool both tubes under tap water until the original color (reverse reaction) is evident.

### F. Ammonia Solution

Prepare an ammonia stock solution by adding 4 drops of conc. ammonium hydroxide and 3 drops of phenolphthalein to 100 mL of tap water and mix. Pour about 5 mL of this stock solution into each of two test tubes.

1. Dissolve a very small amount of solid ammonium chloride in the stock solution in the first tube and observe the result.

2. Add a few drops of dil. (6 M) hydrochloric acid to the stock solution in the second tube. Mix and observe the result.

NAME______________________________

SECTION__________ DATE ____________

INSTRUCTOR________________________

## *REPORT FOR EXPERIMENT 25*

# Chemical Equilibrium—Reversible Reactions

Refer to equilibrium equations on page 194 when answering these questions.

**A. Saturated Sodium Chloride**

1. What is the evidence for a shift in equilibrium?

2. What ion caused the equilibrium to shift? __________________

3. In which direction did the equilibrium shift? __________________

4. If solid sodium hydroxide were added to neutralize the hydrochloric acid, would this reverse the reaction and cause the precipitated sodium chloride to redissolve? Explain.

**B. Saturated Ammonium Chloride**

1. What is the evidence for a shift in equilibrium?

2. In which direction did the equilibrium shift? __________________

3. What ion caused the equilibrium to shift? __________________

**C. Iron(III) Chloride plus Potassium Thiocyanate**

1. What is the evidence for a shift in equilibrium when iron(III) chloride is added to the stock solution?

2. What is the evidence for a shift in equilibrium when potassium thiocyanate is added to the stock solution?

3. (a) What is the evidence for a shift in equilibrium when silver nitrate is added to the stock solution? (The formation of a precipitate is not the evidence since the precipitate is not one of the substances in the equilibrium.)

3. (b) The change in concentration of which ion in the equilibrium caused this equilibrium shift?

____________________

(c) Write a net ionic equation to illustrate how this concentration change occurred.

(d) When the mixture in C.4 was divided and further tested, what evidence showed that the mixture still contained $Fe^{3+}$ ions in solution?

**D. Potassium Chromate with Nitric and Sulfuric Acids**

1. What was the evidence for a shift in equilibrium when the nitric acid and the sulfuric acid were added to the potassium chromate solution?

2. Explain how sodium hydroxide caused the equilibrium to shift back again.

**E. Cobalt(II) Chloride Solution**

1. What was the evidence for a shift in equilibrium when conc. hydrochloric acid was added to the cobalt(II) chloride solution?

2. (a) Write the equilibrium equation for this system.

(b) State whether the concentration of each of the following substances was increased, decreased, or unaffected when the conc. hydrochloric acid was added to cobalt chloride solution:

$Co(H_2O)_6^{2+}$ ____________________, $Cl^-$ ____________________, $CoCl_4^{2-}$ ____________________

3. (a) What did you observe when ammonium chloride was added to cobalt(II) chloride solution?

(b) What did you observe when this mixture was heated?

3. (c) Explain why heating the mixture caused the equilibrium to shift.

(d) What did you observe when the mixture was cooled?

(e) Explain why cooling the mixture caused the equilibrium to shift.

## F. Ammonia Solution

1. What is the evidence for a shift in equilibrium when ammonium chloride was added to the stock solution?

2. Explain, in terms of the equilibrium, the results observed when hydrochloric acid was added to the stock solution.

3. State whether the concentration of each of the following was increased, decreased, or was unaffected when dilute hydrochloric acid was added to the ammonia stock solution:

   $NH_3$ ____________, $NH_4^+$ ____________, $OH^-$ ____________, Pink color ____________

4. (a) In which direction would the equilibrium shift if sodium hydroxide were added to the ammonia stock solution?

   (b) Would the sodium hydroxide tend to decrease the color intensity? Explain.

5. Would boiling the ammonia solution have any effect on the equilibrium? Explain.

# EXPERIMENT 26

# Heat of Reaction

## MATERIALS AND EQUIPMENT

Solid: ammonium chloride ($NH_4Cl$). Solutions: concentrated (12 M) hydrochloric acid (HCl), concentrated (16 M) nitric acid ($HNO_3$), 10 percent sodium hydroxide (NaOH), and concentrated (18 M) sulfuric acid ($H_2SO_4$). A styrofoam cup; a thermometer.

## DISCUSSION

All chemical reactions involve a heat effect. If the reaction liberates heat it is called an **exothermic reaction.** If the reaction consumes heat it is called an **endothermic reaction.** These heat effects are the result of the formation or breaking of chemical bonds. Energy is required to break bonds, and energy is liberated in the formation of chemical bonds. The **heat of reaction** is the net heat effect for the reaction and is often the combination of several heat effects.

This experiment will investigate the heat of reaction of three types of reactions: the hydration of a liquid, the dissolving of a solid, and neutralization (the reaction of an acid with a base). The quantity of heat will not be measured in any of the experiments. Instead temperature changes will be observed as indications of heat effects.

### Hydration of Concentrated Sulfuric Acid

A strong heat effect is observed when conc. sulfuric acid is added to water. This reaction is so exothermic that it can lead to dangerous spattering. The general rule "add the acid to the water" will help avoid the danger.

The reaction involved is the hydration of sulfuric acid molecules to form $H_2SO_4 \cdot H_2O$ and $H_2SO_4 \cdot 2H_2O$. The formation of the bonds between the acid and the water is the primary cause of the heat liberated. The usual commercial concentrated sulfuric acid contains less than 5 percent water, so some of the acid molecules are already hydrated.

### Dissolving of Ammonium Chloride

When a salt dissolves there are two major competing heat effects. The energy required to break the ionic bonds of the crystal lattice is called the **lattice energy.** As the ions are freed from the crystal lattice they become hydrated, liberating energy. This energy is known as the **hydration energy.** If the lattice energy is greater than the hydration energy the reaction will be endothermic. If the hydration energy is greater than the lattice energy the reaction will be exothermic.

### Neutralization Reactions

When hydrochloric acid is added to sodium hydroxide solution, heat is evolved. Since the reaction of an acid with a base is called neutralization, this heat is called the **heat of neutralization.** The molecular equation for this reaction is:

$$HCl + NaOH \longrightarrow H_2O + NaCl + Heat$$

Since all the substances are soluble, and all of them except water are strong electrolytes, the total ionic equation is:

$$H^+ + Cl^- + Na^+ + OH^- \longrightarrow H_2O + Na^+ + Cl^- + Heat$$

Therefore the net ionic equation is:

$$H^+ + OH^- \longrightarrow H_2O + Heat$$

According to the net ionic equation, a neutralization reaction between a strong acid and a strong base is the reaction of the hydrogen ions ($H^+$) and the hydroxide ions ($OH^-$) to form water. Since this reaction is common to all strong acids and bases, and since a definite amount of heat is liberated for each mole of reacting hydrogen ion and hydroxide ion, it should be possible to use different strong acids with a fixed amount of base and obtain the same quantity of heat for each acid neutralized. This experiment will investigate this hypothesis.

Each neutralization will involve the same total volume of liquid. Thus if the same quantity of heat actually is produced, the temperature increase should be the same for each neutralization. This assumes a constant heat capacity for the various solutions used, which is not precisely true, but is satisfactory for the purpose of this experiment. Duplicate runs are made for each acid used. Because of experimental errors, variations occur in the temperature increases obtained in duplicate runs with the same acid. Thus the variations in temperature changes obtained with different acids should be of the same order of magnitude, if our hypothesis of identical reactions is correct. It is necessary to use dilute acids and bases to minimize heats of hydration.

In all the experimental runs the moles of hydroxide ion will be the limiting reactant. Thus there will always be more moles of hydrogen ion available than of hydroxide ion. The excess moles of hydrogen ion will remain unreacted.

## PROCEDURE

NOTES:

1. CAUTION: **Wear protective glasses.**
2. Record all temperatures to the nearest 0.1°C.

### A. Hydration of Concentrated Sulfuric Acid

Pour 44 mL of tap water into a styrofoam cup. Read and record the temperature of the water. Measure 4.0 mL of conc. sulfuric acid and pour it into the water in the styrofoam cup. Stir with the thermometer and record the maximum temperature reached. Rinse off any acid that runs down the side of the bottle and rinse out the graduated cylinder. Pour this dilute sulfuric acid solution from the styrofoam cup into a 125 mL Erlenmeyer flask and save for Part C.

## B. Dissolving of Ammonium Chloride

Rinse the styrofoam cup and pour 20 mL of tap water into it. Read and record the temperature of the water. Weigh approximately 3 g of ammonium chloride crystals and pour them into the styrofoam cup. Stir with the thermometer and record the minimum temperature reached.

## C. Neutralization Reactions

You will need four stock solutions. One, the NaOH stock solution, requires a 250 mL Erlenmeyer flask; the other three can be in either 250 or 125 mL flasks. Prepare these stock solutions by adding the acid or base to water as required and mixing. Label each flask for identification.

| | | |
|---|---|---|
| Flask 1 (250 mL) | NaOH stock solution: | 65 mL $H_2O$ + 65 mL 10 percent NaOH. |
| Flask 2 | HCl stock solution: | 75 mL $H_2O$ + 25 mL concentrated HCl. |
| Flask 3 | $HNO_3$ stock solution: | 21 mL $H_2O$ + 5 mL conc. $HNO_3$ |
| Flask 4 | $H_2SO_4$ stock solution: | $H_2SO_4$ solution prepared in Part A. |

After the four stock solutions are prepared, carry out the five neutralizations using the specified amounts of stock solutions and water. Make a duplicate run for each neutralization. Use the following six-step procedure for each neutralization:

Step 1: Pour the measured volumes of acid and water into the styrofoam cup.

Step 2: Measure out the sodium hydroxide solution.

Step 3: Read and record the temperature of the water-acid mixture.

Step 4: Add the base, stir with the thermometer, and record the maximum temperature reached.

Step 5: Calculate the temperature change.

Step 6: Empty the styrofoam cup; rinse the cup and the graduated cylinders with water.

The amounts of the prepared stock solutions to be used for each neutralization are as follows:

Neutralization 1. 10 mL HCl, 20 mL $H_2O$, 10 mL NaOH

Neutralization 2. 10 mL $HNO_3$, 20 mL $H_2O$, 10 mL NaOH

Neutralization 3. 10 mL $H_2SO_4$, 20 mL $H_2O$, 10 mL NaOH

Neutralization 4. 20 mL HCl, 10 mL $H_2O$, 10 mL NaOH

Neutralization 5. 10 mL HCl, 10 mL $H_2O$, 20 mL NaOH

NAME ______________________

SECTION ________ DATE __________

INSTRUCTOR ______________________

*REPORT FOR EXPERIMENT 26*

# Heat of Reaction

## A. Hydration of Concentrated Sulfuric Acid

1. Temperature of water. ______________

   Temperature of water after adding acid. ______________

   Temperature change. ______________

2. Is this an endothermic or exothermic reaction? ______________

3. In this reaction what is the major change, in terms of bonds made or broken, that causes the heat effect?

4. If you had used dilute sulfuric acid rather than concentrated sulfuric acid, would you expect the temperature change to be greater or less? ______________
   Why?

## B. Dissolving of Ammonium Chloride

1. Temperature of water. ______________

   Temperature of water after adding salt. ______________

   Temperature change. ______________

2. Is this an endothermic or exothermic reaction? ______________

3. In this reaction what is the major change, in terms of bonds made or broken, that causes the heat effect?

4. What other heat effect is present in this reaction, acting in the opposite direction?

5. If you had used 40 mL of water and 6 g of ammonium chloride, rather than the 20 mL and 3 g in the experiment, would you expect to get a larger, smaller, or identical temperature change? ______________
Why?

## C. Neutralization Reactions

**Data Table**

| | | Temp. Before Adding NaOH | Temp. After Adding NaOH | Temp. Change |
|---|---|---|---|---|
| 1 10 mL HCl, 20 mL $H_2O$, 10 mL NaOH | Run 1 | ______ | ______ | ______ |
| | Run 2 | ______ | ______ | ______ |
| | Average | XXXX | XXXX | ______ |
| 2. 10 mL $HNO_3$, 20 mL $H_2O$, 10 mL NaOH | Run 1 | ______ | ______ | ______ |
| | Run 2 | ______ | ______ | ______ |
| | Average | XXXX | XXXX | ______ |
| 3. 10 mL $H_2SO_4$, 20 mL $H_2O$, 10 mL NaOH | Run 1 | ______ | ______ | ______ |
| | Run 2 | ______ | ______ | ______ |
| | Average | XXXX | XXXX | ______ |
| 4. 20 mL HCl, 10 mL $H_2O$, 10 mL NaOH | Run 1 | ______ | ______ | ______ |
| | Run 2 | ______ | ______ | ______ |
| | Average | XXXX | XXXX | ______ |
| 5. 10 mL HCl, 10 mL $H_2O$, 20 mL NaOH | Run 1 | ______ | ______ | ______ |
| | Run 2 | ______ | ______ | ______ |
| | Average | XXXX | XXXX | ______ |

## QUESTIONS AND PROBLEMS

The following pertain to Part C:

1. (a) How do the average temperature changes observed in Parts 1, 2, and 3 compare?

   (b) What do you conclude from this?

2. (a) How does the average temperature change in Part 4 compare with that in Parts 1, 2, and 3?

   (b) What do you conclude from this?

3. (a) How does the average temperature change in Part 5 compare with that in Part 1?

   (b) What do you conclude from this?

# EXPERIMENT 27

# Boiling Points and Melting Points

## MATERIALS AND EQUIPMENT

Solids: Boiling stones, benzoic acid ($C_6H_5COOH$), benzophenone ($C_6H_5COC_6H_5$), cholesterol ($C_{27}H_{45}OH$), diphenylacetic acid ($C_{14}H_{12}O_2$), trans-cinnamic acid ($C_9H_8O_2$), urea ($CH_4N_2O$), 50% urea-50% trans-cinnamic acid, 50% diphenylacetic acid-50% cholesterol, and an unknown. Liquids: Isopropyl alcohol ($CH_3CHOHCH_3$), ethyl alcohol ($C_2H_5OH$), and methyl alcohol ($CH_3OH$). Oil bath (mineral or vegetable oil), wire stirrer, 0-250°C thermometer, capillary tubes (sealed at one end), rubber bands cut from 3/16 inch rubber tubing.

## DISCUSSION

Organic chemistry is the branch of chemistry that deals with carbon-containing molecules. Carbon bonds not only to other atoms but to itself forming multiple chains and rings. This feature of carbon allows almost an unlimited number of organic compounds to exist. Organic chemists are faced with the difficult task of identifying and categorizing this vast array of molecules. Each organic compound has many chemical and physical properties associated with its structure. If enough of its properties are identified, we are able to say that we are dealing with one specific compound and not another.

This experiment will introduce two physical properties, the boiling point and the melting point, that organic chemists use to identify organic molecules. The boiling point of a liquid is the temperature at which the vapor pressure of the liquid equals the surrounding atmospheric pressure. Since the boiling point is dependent upon atmospheric pressure, a liquid will boil at different temperatures depending on its height above sea level. Consequently, water boils at a lower temperature in the mountains than it does at the sea shore.

The second property we will measure is the melting point of a substance. The melting point is the temperature at which the solid becomes a liquid. Unlike the boiling point, normal atmospheric pressures do not significantly affect the melting point.

Most pure substances have sharp melting points. If a substance is made impure by mixing it with another solid chemical, the temperature at which the mixture melts is no longer sharp but is broadened over a range. Then the melting temperature is known as the "melting point range." Mixtures also have melting point ranges that are usually lower than the melting points of the pure components. This broadening and lowering of the melting point can be used to identify unknown substances. Two chemicals can have the same melting point, e.g., urea and trans-cinnamic acid. Thus, if we have an unknown sample that is either urea or trans-cinnamic acid, and mix it with urea and obtain a sharp melting point, we know the unknown must be urea. But, if the unknown when mixed with urea gives a broad and lowered melting point range, then the unknown must be trans-cinnamic acid. This procedure is known as the "method of mixed melting points."

## PROCEDURE

**Wear protective glasses.**

CAUTION: All three alcohols are flammable. The quantities used are small, but be sure to keep the flame as far as possible from the mouth of the test tube (your neighbor's test tube also!).

### A. Boiling Points

1. Assemble the apparatus illustrated in Figure 27.1.
2. Determine the boiling point of these three liquids.
   (a) methyl alcohol (methanol)
   (b) ethyl alcohol (ethanol)
   (c) isopropyl alcohol (2-propanol)
3. Pour 2 mL of methanol into a clean dry test tube, add two small boiling stones, and secure the tube to a ring stand so that
   (a) the surface of the methanol is slightly below the oil level and
   (b) the tip of the thermometer is about 1 cm above the surface of the methanol (see Figure 27.1).
4. Heat the mineral oil. Check the temperature of the oil bath with a second thermometer. Do not allow the bath's temperature to go above 80°C for this methanol sample. As the sample boils, the vapor will cause the temperature reading of the thermometer to increase until it remains constant at one value. This constant value is the boiling point of the sample. Record this temperature.
5. Repeat Steps 3 and 4 for the other samples. Do not allow the temperature of the bath to exceed 90°C for the ethanol sample. Allow the bath to cool to about 75°C before you place the third sample in the oil. For the 2-propanol sample, do not let the temperature of the bath exceed 100°C.

### B. Melting Points

You will measure the melting points of several pure compounds and the melting point ranges for two mixtures. You also will determine the identity of an unknown solid by the method of mixed melting points.

1. Assemble the apparatus shown in Figure 27.2. (Other methods may be used as indicated by your instructor.)

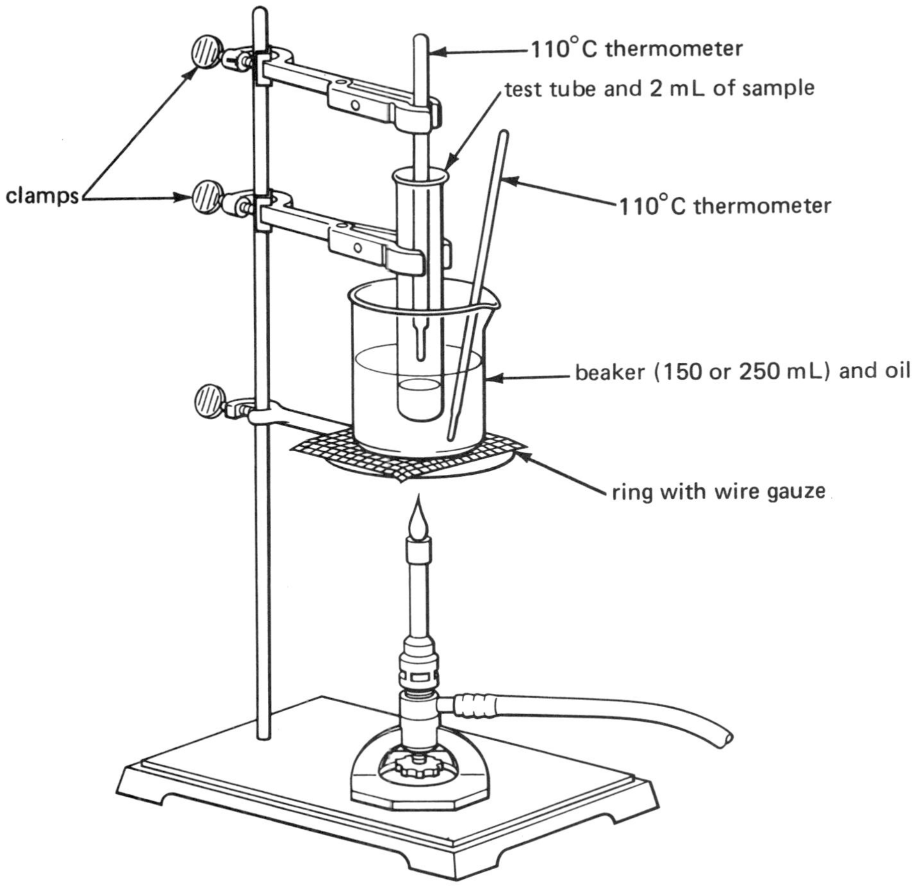


**Figure 27.1. Boiling Point Apparatus.**

2. Measure the melting points (or ranges) for the following compounds and mixtures:

   (a) benzophenone
   (b) benzoic acid
   (c) urea
   (d) trans-cinnamic acid
   (e) diphenylacetic acid
   (f) cholesterol
   (g) 50%urea-50% trans-cinnamic acid
   (h) 50% diphenylacetic acid-50% cholesterol

3. You will need a separate capillary tube (sealed at one end) for each sample. To fill the capillary tube, make a small pile of the solid on a watch glass and force the open end of the capillary through the crystals to get some of the sample into the tube. Then tap the sealed end of the tube lightly on the table top to get the sample down into the sealed end. Repeat this procedure until you have about 0.5 cm of the sample in the capillary.

4. Use a small rubber band to attach the capillary tube to the thermometer. The bottom of the capillary tube should be level with the bottom of the thermometer.

5. Place the thermometer and capillary tube in the oil bath as shown in Figure 27.2.

6. It is important to stir the oil to produce a uniform temperature. Use a wire that has a loop at one end and a handle at the other (Figure 27.2). Stir the oil by raising and lowering the wire loop as you slowly heat the oil.

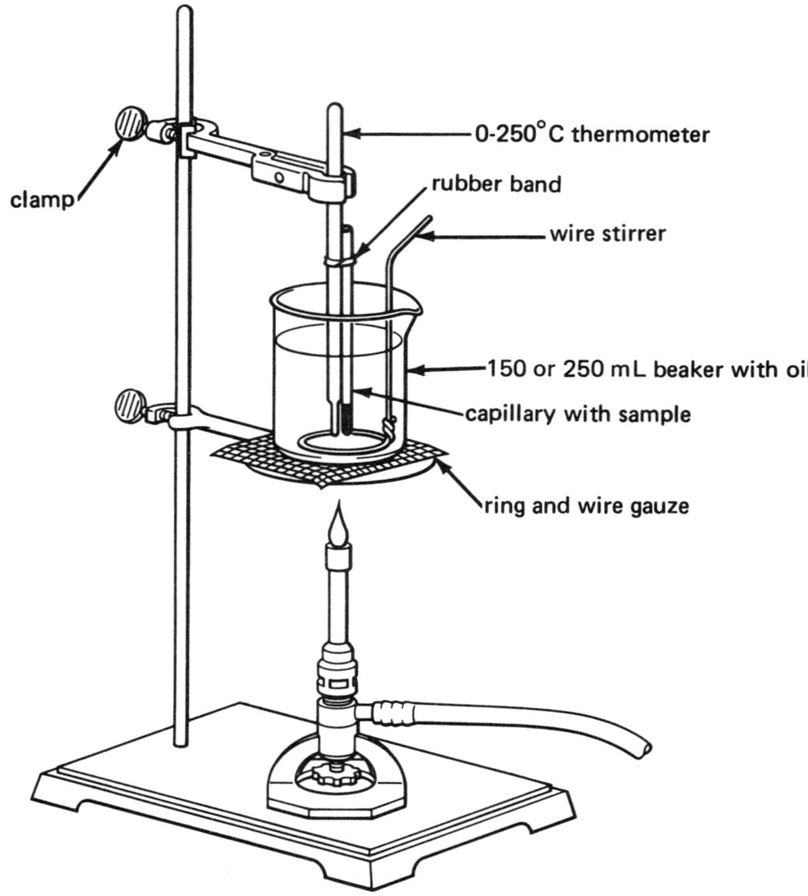


**Figure 27.2. Melting Point Apparatus. (Other types are available)**

7. Watch **both** the temperature and the crystals as you slowly heat the oil bath. Record the temperature when the crystals first begin to melt and also the temperature at which the last crystals melt. The melting process may occur over a range of several degrees. (Be sure that you stir the oil constantly.)

8. Repeat Steps 3-7 for the remaining samples. Be sure the oil bath has cooled below the expected melting point of your next sample.

9. Obtain an unknown sample from your instructor.

10. Determine the melting point range for your unknown. You should now have a good idea as to its identity by comparing its melting point to those of the knowns that you have already determined.

11. Determine by the "method of mixed melting points" if your unknown is what you believe it to be. Record your results. (Be sure your sample and the known are **very thoroughly mixed** before determining the mixed melting point range.)

NAME ____________

SECTION ________ DATE ________

INSTRUCTOR ____________

## *REPORT FOR EXPERIMENT 27*

# Boiling Points and Melting Points

For this report you will need to obtain literature values (from Handbooks) for the accepted boiling and melting points.

### A. Boiling Points

| Sample | Handbook values of boiling points, °C | Experimental values of boiling points, °C |
|---|---|---|
| Methyl Alcohol | | |
| Ethyl Alcohol | | |
| Isopropyl Alcohol | | |

### B. Melting Point Ranges

| Sample | Handbook values of melting points,°C | Experimental values melting point range, °C |
|---|---|---|
| Benzophenone | | |
| Benzoic acid | | |
| Urea | | |
| Trans-cinnamic acid | | |
| Diphenylacetic acid | | |
| Cholesterol | | |
| 50% urea and 50% trans-cinnamic acid | | |
| 50% cholesterol and 50% diphenlyacetic acid | | |
| Unknown | none | |

Results for the "method of mixed melting points":

(a) With what compound did you mix your unknown? ____________

(b) Observed melting point range ____________

(c) What is your unknown number? ____________

(d) What is the name of your unknown? ____________

## QUESTIONS AND PROBLEMS

1. Why must you allow the oil bath to cool between melting point determinations?

2. (a) If you performed a melting point experiment on the same compound in San Francisco and on top of Mt. Everest, would your results differ? Explain.

   (b) If you performed a boiling point experiment on the same compound in San Francisco and on top of Mt. Everest, would your results differ? Explain.

3. Cocaine melts at 98°C and glucose melts at 146°C. As a chemist for the government it is your task to quickly identify the contents of three vials. One contains pure cocaine, another pure glucose, and the third a mixture of cocaine and glucose. How would you accomplish your assignment?

4. Suppose you determine the melting point of cortisone (a hormone) to be 230°C. Your two neighbors obtain values of 226°C and 233°C. Why might their values differ from your value?

# EXPERIMENT 28

# Hydrocarbons

## MATERIALS AND EQUIPMENT

Solid: calcium carbide (about 3/8 in. lumps) ($CaC_2$). Liquids: pentene (amylene) ($C_5H_{10}$), benzene ($C_6H_6$), heptane ($C_7H_{16}$), and kerosene. Solutions: 5% bromine ($Br_2$) in 1,1,1-trichloroethane ($CCl_3CH_3$) and 0.1 M potassium permanganate ($KMnO_4$). Wooden splints.

### Hydrocarbons

Hydrocarbons are organic compounds made up entirely of carbon and hydrogen atoms. Their principal natural sources are coal, petroleum, and natural gas. Hydrocarbons are grouped into several series by similarity of molecular structure. Some of these are the alkanes, alkenes, alkynes, and aromatic hydrocarbons.

### Alkanes

Also known as the paraffins or **saturated hydrocarbons,** the **alkanes** are straight- or branched-chain hydrocarbons having only single bonds between carbon atoms. They are called saturated hydrocarbons because all their carbon-carbon bonds are single bonds.

The first 10 members of the alkane series and their molecular formulas are listed below:

| | | | |
|---|---|---|---|
| Methane | $CH_4$ | Hexane | $C_6H_{14}$ |
| Ethane | $C_2H_6$ | Heptane | $C_7H_{16}$ |
| Propane | $C_3H_8$ | Octane | $C_8H_{18}$ |
| Butane | $C_4H_{10}$ | Nonane | $C_9H_{20}$ |
| Pentane | $C_5H_{12}$ | Decane | $C_{10}H_{22}$ |

Like most organic substances, the alkanes are combustible. The products of their complete combustion are carbon dioxide and water. The reactions of the alkanes are of the substitution type; that is, some atom or group of atoms is substituted for one or more of the hydrogen atoms in the hydrocarbon molecule. For example, in the bromination of methane, a bromine atom is substituted for a hydrogen atom. This reaction does not occur appreciably in the dark at room temperature but is catalyzed by ultraviolet light. The equation is:

$$\underset{\text{Methane}}{CH_4} + Br_2 \xrightarrow[\text{light}]{\text{Ultraviolet}} \underset{\text{Methyl bromide}}{CH_3Br} + HBr$$

### Alkenes

Also known as the **olefins**, the **alkenes** are a series of straight- or branched-chain hydrocarbons containing a carbon-carbon double bond in their structures. They are considered to be **unsaturated**

**hydrocarbons.** The first two members of the series are ethylene ($C_2H_4$) and propylene ($C_3H_6$). Their structural and condensed structural formulas are:

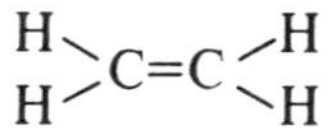

$CH_2{=}CH_2$

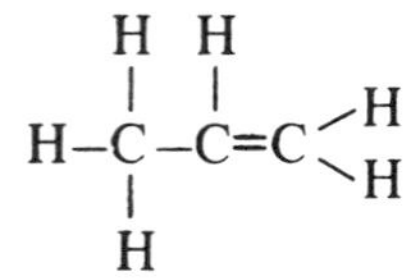

$CH_3CH{=}CH_2$

Ethylene (ethene) — Propylene (propene)

The functional group of this series is the carbon-carbon double bond (C=C); it is a point of high reactivity. Alkenes undergo addition-type reactions; that is, other groups are added to the double bond, causing the molecule to become saturated. For example, when hydrogen is added, one H atom from $H_2$ is added to each carbon atom of the double bond to saturate the molecule, forming an alkane:

$$CH_3CH{=}CH_2 + H_2 \xrightarrow[\text{Heat and pressure}]{\text{Ni Catalyst}} CH_3CH_2CH_3$$

Propylene — Propane

When a halogen such as bromine is added, one Br atom from $Br_2$ is added to each carbon atom of the double bond to saturate the molecule:

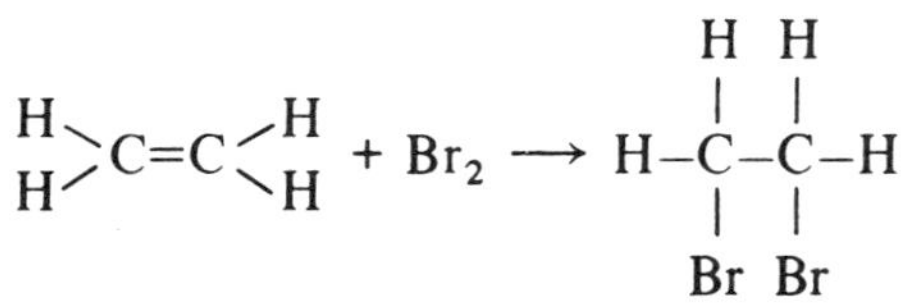

1,2-Dibromoethane
(Ethylene dibromide)

Evidence that bromine has reacted is the disappearance of the red-brown color of free bromine. Other reactions of olefins also show the increased reactivity of the alkenes over the alkanes.

Unsaturated hydrocarbons can be oxidized by potassium permanganate. This reaction is known as the Baeyer test for unsaturation. Evidence that reaction has occurred is the rapid disappearance (within a few seconds) of the purple color of the permanganate ion. The resulting reaction products will not be colorless. Potassium permanganate is a very strong oxidizing agent and gives similar results when reacted with other oxidizable substances, such as alcohols.

### Alkynes

Also called the **acetylenes,** the alkynes are another class of **unsaturated hydrocarbons,** but they contain a carbon-carbon triple bond in their structures. The first two members of this series are acetylene (ethyne) and propyne:

| $H{-}C{\equiv}C{-}H$ | $CH_3C{\equiv}CH$ |
|---|---|
| Acetylene | Propyne |

Acetylene is the most important member of this series and can be prepared from calcium carbide and water. The equation for this reaction is

$$CaC_2 + 2\,H_2O \longrightarrow CH{\equiv}CH\uparrow + Ca(OH)_2$$

Mixtures of acetylene and air are explosive. The alkynes undergo addition-type reactions similar to those of the alkenes.

## Aromatic Hydrocarbons

The parent substance of this class of hydrocarbons is benzene ($C_6H_6$). From its formula benzene appears to be a highly unsaturated molecule; the corresponding six-carbon alkane contains 14 hydrogen atoms per molecule ($C_6H_{14}$). However, the chemical reactions of benzene show that its behavior is like that of the saturated hydrocarbons in many respects. Its reactions are primarily of the substitution type.

The carbon atoms in a benzene molecule are arranged in a six-membered ring structure, with one hydrogen atom attached to each carbon atom. The following diagrams represent the benzene molecule; in the second and third structures it is understood that a carbon and a hydrogen atom are present at each corner of the hexagon or benzene ring:

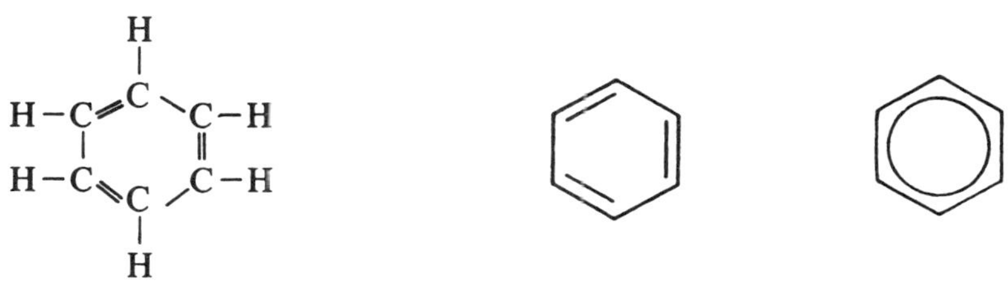

## PROCEDURE

**Wear protective glasses.**

In the reactions below, heptane will be used to represent the saturated hydrocarbons; pentene (amylene), the unsaturated hydrocarbons; and benzene, the aromatic hydrocarbons.

CAUTION: Because hydrocarbons are extremely flammable, do not handle them near open flames. Discard any waste and reaction products as directed by your instructor.

### A. Combustion

Obtain about 1 mL (no more) of heptane in an evaporating dish and start it burning by carefully bringing a lighted match or splint to it. Repeat with an equally small volume of pentene.

### B. Reaction with Bromine

CAUTION: **Dispense bromine solution under the hood**, and be especially careful not to spill bromine on your hands.

Take three clean, **dry** test tubes. Place about 1 mL of heptane in the first tube, 1 mL of pentene in the second, and 1 mL of benzene in the third. Add 3 drops of 5 percent bromine in trichloroethane solution to each sample; stopper the tubes and note the results. Any tube that still shows bromine color after 1 minute should be exposed to sunlight or to a strong electric light for an additional 2 minutes.

### C. Reaction with Potassium Permanganate

The Baeyer test for unsaturation in hydrocarbons involves the reaction of hydrocarbons with potassium permanganate solution. Evidence that reaction has occurred is the rapid disappearance (within a few seconds) of the purple color of the permanganate ion. Potassium permanganate is a very strong oxidizing agent and gives similar results when reacted with other oxidizable substances, such as alcohols.

Add 2 drops of potassium permanganate solution to about 1 mL each of heptane, pentene, and benzene in test tubes. Mix and note the results.

### D. Kerosene

Determine which class of hydrocarbons (alkanes, alkenes, or aromatic) kerosene belongs to by reacting it with bromine and with potassium permanganate, as in Tests B and C.

### E. Acetylene

In this part of the experiment you will prepare acetylene and test its combustibility.

Fill a 400 mL beaker nearly full of tap water. Fill three test tubes (18 X 150 mm) with water as follows: **Tube (1)** completely full; **Tube (2)** 15 mL; and **Tube (3)** 6 mL.

Obtain a small lump of calcium carbide from the reagent bottle and drop it into the beaker of water. Place your thumb over the full test tube of water (**Tube 1**) and invert it in the beaker. Hold the tube over the bubbling acetylene and, when it is full of gas, stopper it while the tube is still under the water. Displace the water in the other two tubes in the same manner; **stopper them immediately after the water is displaced.**

Test the contents of each tube as follows:

**Tube 1.** Bring the mouth of the tube to the burner flame as you remove the stopper. After the acetylene ignites, tilt the mouth of the tube up and down.

**Tube 2.** Bring the mouth of the tube to the burner flame as you remove the stopper.

⚠ **Tube 3. Wrap the tube in a towel** and bring the mouth of the tube to the burner flame as you remove the stopper. This sample is the most highly explosive of the three samples tested.

### F. Solubility Tests

Test the solubility of heptane, pentene, and benzene in water by adding 1 mL (or less) of each hydrocarbon to about 5 mL portions of water. Shake each mixture for a few seconds and note whether they are soluble. For any that are not soluble note the relative density of the hydrocarbon with respect to water.

Test the miscibility of these three hydrocarbons with each other by mixing about 1 mL of each in a **dry** test tube.

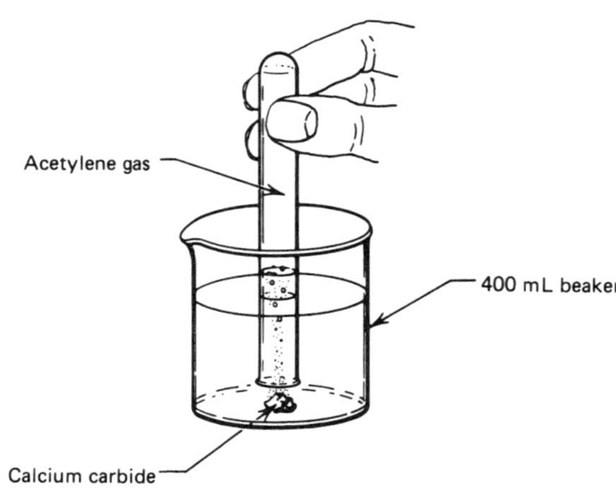


**Figure 28.1. Collecting acetylene from calcium carbide-water reaction**

NAME ______________

SECTION ________ DATE ________

# *REPORT FOR EXPERIMENT 28*

INSTRUCTOR ______________

# Hydrocarbons

## A. Combustion

1. Describe the combustion characteristics of heptane and pentene.

2. (a) Write a balanced equation to represent the complete combustion of heptane.

   (b) How many moles of oxygen are needed for the combustion of 1 mole of heptane in this reaction?

   ________

## B. and C. Reaction with Bromine and Potassium Permanganate

Data Table: Place an X in the column where a reaction was observed.

| | Heptane (Saturated Hydrocarbon) | Pentene (Unsaturated Hydrocarbon) | Benzene (Aromatic Hydrocarbon) |
|---|---|---|---|
| Immediate reaction with $Br_2$ (without exposure to light) | | | |
| Slow reaction with $Br_2$ (or only after exposure to light) | | | |
| Reaction with $KMnO_4$ | | | |

1. Which of the three hydrocarbons reacted with bromine (without exposure to light)?

2. Write an equation to illustrate how heptane reacts with bromine when the reaction mixture is exposed to sunlight.

3. Write an equation to illustrate how pentene reacts with bromine. Assume the pentene is $CH_3CH_2CH{=}CHCH_3$ and use structural formulas.

4. Which of the three hydrocarbons gave a positive Baeyer test?

## D. Kerosene

1. (a) Did you observe any evidence of reaction with bromine before exposure to light? If so, describe.

   (b) Did you observe any evidence of reaction with bromine after exposure to light? If so, describe.

2. Did you observe any evidence of reaction with potassium permanganate? If so, describe.

3. Based on these tests (bromine and potassium permanganate), to which class of hydrocarbon does kerosene belong?

____________

## E. Acetylene

1. Describe the combustion characteristics of acetylene:

   (a) Tube 1.

   (b) Tube 2.

   (c) Tube 3.

2. Write an equation for the complete combustion of acetylene.

**F. Solubility Tests**

1. Which of the three hydrocarbons are soluble in water?

2. From your observations what do you conclude about the density of hydrocarbons with respect to water?

## QUESTIONS AND PROBLEMS

1. Do you expect that acetylene would react with bromine without exposure to light? Explain your answer.

2. Write structural formulas for the three different isomers of pentane, all having the molecular formula $C_5H_{12}$.

3. Write structural formulas for (a) ethylene, (b) propylene, and (c) the three different isomeric butylenes ($C_4H_8$).

# EXPERIMENT 29

# Alcohols, Esters, Aldehydes, and Ketones

## MATERIALS AND EQUIPMENT

Solids: Copper wire (No. 18, with spiral); salicylic acid [$C_6H_4(COOH)(OH)$]. Liquids: acetic acid, glacial ($CH_3COOH$); acetone [$(CH_3)_2C{=}O$]; ethyl alcohol ($C_2H_5OH$); isoamyl alcohol ($C_5H_{11}OH$); isopropyl alcohol (iso-$C_3H_7OH$); methyl alcohol ($CH_3OH$). Solutions: dilute ammonium hydroxide ($NH_4OH$), 10 percent glucose ($C_6H_{12}O_6$), 10 percent formaldehyde ($H_2C{=}O$), 0.1 M potassium permanganate ($KMnO_4$), 0.1 M silver nitrate ($AgNO_3$), 10 percent sodium hydroxide (NaOH), and dilute and concentrated sulfuric acid ($H_2SO_4$).

## DISCUSSION

In this experiment we will examine some of the properties and characteristic reactions of four classes of organic compounds: alcohols, esters, aldehydes, and ketones.

### Alcohols

The formulas of alcohols may be derived from alkane hydrocarbon formulas by replacing a hydrogen atom with a hydroxyl group (OH). In the resulting alcohols the OH group is bonded to the carbon atom by a covalent bond and is not an ionizable hydroxide group. Examples follow:

| Alkane | Alcohol | Name of Alcohol |
|---|---|---|
| $CH_4$ | $CH_3OH$ | Methyl alcohol (Methanol) |
| $CH_3CH_3$ | $CH_3CH_2OH$ | Ethyl alcohol (Ethanol) |
| $CH_3CH_2CH_3$ | $CH_3CH_2CH_2OH$ | n-Propyl alcohol (1-Propanol) |
| $CH_3CH_2CH_3$ | $CH_3CH(OH)CH_3$ | Isopropyl alcohol (2-Propanol) |

Thus there is an entire series of alcohols. The functional group of the alcohols is the hydroxyl group, OH.

### Esters

This class of organic compounds may be formed by reacting alcohols with organic acids. Esters generally have a pleasant odor; many of them occur naturally, being found mainly in fruits and fatty material.

Methyl acetate will be used as an example illustrating the formation of an ester. When acetic acid and methyl alcohol are reacted together, using sulfuric acid as a catalyst, a molecule of water is split out between a molecule of the acetic acid and a molecule of the alcohol, forming the ester. The equation is

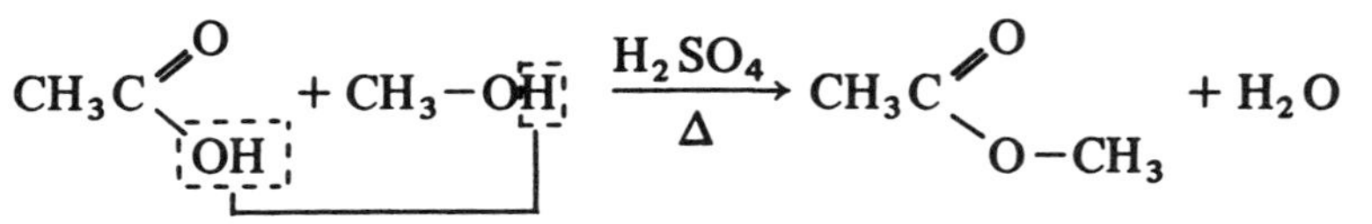

Acetic acid    Methyl alcohol    Methyl acetate

The functional group characterizing organic acids is

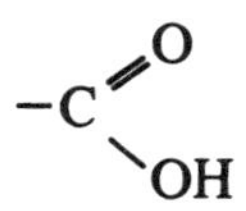

or –COOH

It is called the **carboxyl group.**

Esters are named in the following manner. The first part of the name is taken from the name of the alcohol, the second part is derived by adding the suffix **ate** to the identifying stem of the acid. Thus **acetic** becomes **acetate**, and the name of the ester derived from methyl alcohol and acetic acid is methyl acetate.

### Aldehydes and Ketones

The functional groups of the aldehydes and ketones are

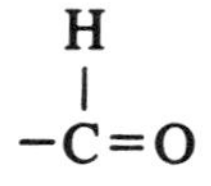

Aldehyde

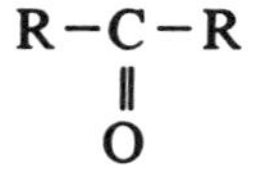

Ketone

Aldehydes and ketones may be obtained by oxidizing alcohols. One major difference between aldehydes and ketones is that aldehydes are very easily oxidized to acids, but ketones are not easily further oxidized. Thus aldehydes are good reducing agents. Chemical reactions for distinguishing aldehydes and ketones are based on this difference.

| Alcohol | Aldehyde | | Ketone |
|---|---|---|---|
| $CH_3OH$ | $H-\underset{\vert}{\overset{H}{C}}=O$ | Formaldehyde (Methanal) | — |
| $CH_3CH_2OH$ | $CH_3\overset{H}{\underset{}{C}}=O$ | Acetaldehyde (Ethanal) | — |
| $CH_3CH_2CH_2OH$ | $CH_3CH_2\underset{H}{C}=O$ | Propionaldehyde (Propanal) | — |
| $CH_3\underset{OH}{C}HCH_3$ | — | — | $CH_3\underset{O}{\overset{}{C}}CH_3$ Acetone (Propanone) |

## PROCEDURE

**Wear protective glasses.**

### A. Combustion of Alcohols

Obtain about 1 mL (no more) of methyl alcohol in an evaporating dish and ignite the alcohol with a match or burning splint. Repeat with equally small volumes of ethyl alcohol and isopropyl alcohol.

### B. Oxidation of Alcohols

1. **Oxidation with Potassium Permanganate.** Mix 3 mL of methyl alcohol with 12 mL of water and divide the solution into three equal portions, placing them in three test tubes. To a fourth tube add 5 mL water. Add 1 drop of 10 percent sodium hydroxide to the first tube and 1 drop of dilute sulfuric acid to the second. Now add 1 drop of potassium permanganate solution to each of the four tubes. Mix and note how long it takes for reaction to occur in each of the first three tubes, using the fourth tube as a reference tube. Disappearance of the purple permanganate color is evidence of reaction. (Patience, some reactions take a long time.)

Repeat this oxidation procedure, using isopropyl alcohol instead of methyl alcohol.

2. **Oxidation with Copper(II) Oxide.** Put about 2 mL of methyl alcohol in a test tube. Obtain from the reagent shelf about a 20 cm piece of copper wire with a four- or five-turn spiral at one end. Warm the alcohol slightly to promote alcohol vapors in the tube. Heat the copper spiral in the hottest part of the burner flame to get a good copper(II) oxide coating. Do not overheat the copper or it will melt. While the copper spiral is very hot, lower it part way into the tube (not to the liquid) and note the results. Heat the wire again and lower it into the tube, finally dropping it into the liquid alcohol. Remove the wire and **gently waft** the vapors from the tube to your nose to detect the odor of formaldehyde resulting from the oxidation of the methyl alcohol. **Return the copper wire to the reagent shelf.**

### C. Formation of Esters

Take three test tubes and mix the following reagents in them:

Tube 1: 3 mL ethyl alcohol, 0.5 mL glacial acetic acid, and 10 drops of concentrated sulfuric acid.

Tube 2: 3 mL isoamyl alcohol, 0.5 glacial acetic acid, and 10 drops concentrated sulfuric acid.

Tube 3: Salicylic acid crystals (about 1 cm deep in the tube), 2 mL methyl alcohol, and 10 drops concentrated sulfuric acid.

Heat the tubes by placing them in boiling water for 3 minutes.

Products formed:

Tube 1: Ethyl acetate.

Tube 2: Isoamyl acetate.

Tube 3: Methyl salicylate.

After heating, pour a small amount of each product onto a piece of filter paper and **carefully** smell it and describe the odor.

### D. Tollens' Test for Aldehydes

This test is based on the ability of the aldehyde group to reduce silver ion in solution, forming either a black deposit of free silver or a silver mirror. The aldehyde group is oxidized to an acid in the reaction. Tollens' reagent is made by reacting silver nitrate solution with dilute ammonium hydroxide. Rinse all glass equipment with distilled water before use.

**Carefully** clean three test tubes. To 8 mL of 0.1 M silver nitrate solution in one of these tubes, add dilute ammonium hydroxide 1 drop at a time until the brown precipitate of silver oxide that is formed just dissolves (mix after each drop is added). Now add 7 mL of distilled water, mix, and divide the solution (Tollens' reagent) equally among the three test tubes.

To Tube 1, add 2 drops of 10 percent formaldehyde solution and mix. To Tube 2, add 2 drops of acetone and mix. To Tube 3, add 5 drops of 10 percent glucose solution and mix. Allow the tubes to stand undisturbed and note the results. The solution containing the glucose may take 10 to 15 minutes to react.

NAME ____________________

SECTION __________ DATE __________

## *REPORT FOR EXPERIMENT 29*

INSTRUCTOR ____________________

# Alcohols, Esters, Aldehydes, and Ketones

### A. Combusion of Alcohols

1. Compare the combustion characteristics of methyl, ethyl, and isopropyl alcohols, in terms of color and luminosity of their flames.

2. What type of flame would you predict for the combustion of amyl alcohol ($C_5H_{11}OH$)?

3. Write and balance the equation for the complete combustion of ethyl alcohol.

### B. Oxidation of Alcohols

1. **Oxidation with Potassium Permanganate**

(a) Time required for oxidation of methyl alcohol by potassium permanganate:

Tube 1: Alkaline solution. __________

Tube 2: Acid solution. __________

Tube 3: Neutral solution. __________

(b) Time required for oxidation of isopropyl alcohol by potassium permanganate:

Tube 1: Alkaline solution. __________

Tube 2: Acid solution. __________

Tube 3: Neutral solution. __________

(c) Balance the equation for the oxidation of methyl alcohol:

$$CH_3OH + \quad KMnO_4 \longrightarrow \quad H_2C{=}O + \quad KOH + \quad H_2O + \quad MnO_2$$

2. **Oxidation with Copper(II) Oxide**

(a) Write the equation for the oxidation reaction that occurred on the copper spiral when it was heated.

(b) What evidence of oxidation or reduction did you observe when the heated spiral was lowered into methyl alcohol vapors?

(c) Write and balance the oxidation-reduction equation between methyl alcohol and copper(II) oxide.

**C. Formation of Esters**

1. Describe the odor of:

(a) Ethyl acetate.

(b) Isoamyl acetate.

(c) Methyl salicylate.

2. Write an equation to illustrate the formation of ethyl acetate from ethyl alcohol and acetic acid.

3. (a) The formula for isoamyl alcohol is $CH_3CH(CH_3)CH_2CH_2OH$. Write the formula for isoamyl acetate.

(b) The formula for salicylic acid is

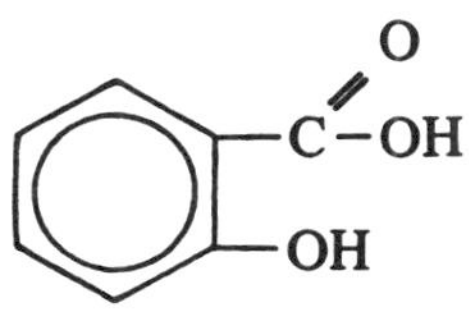

Write the formula for methyl salicylate.

**D. Tollens' Test for Aldehydes**

1. How is a positive Tollens' test recognized?

2. Which of the substances tested gave a positive Tollens' test?

3. Circle the formula(s) of the compounds listed that will give a positive Tollens' test:

$CH_3OH$, $C_2H_5OH$, $CH_3\overset{\overset{\displaystyle H}{|}}{C}{=}O$, $CH_3\underset{\underset{\displaystyle O}{\|}}{C}CH_2CH_3$, $Na_2CO_3$

4. Write the formula for the oxidation product formed from formaldehyde in the Tollens' test.

## QUESTIONS AND PROBLEMS

1. There are four butyl alcohols of formula $C_4H_9OH$. Write their structural formulas.

2. Write the name of the ester that can be derived from the following pairs of acids and alcohols:

| Alcohol | Acid | Ester |
|---|---|---|
| Methyl alcohol | Acetic acid | |
| Ethyl alcohol | Formic acid | |
| Isopropyl alcohol | Butyric acid | |

# EXPERIMENT 30

# Esterification—Distillation: Synthesis of n-Butyl Acetate

## MATERIALS AND EQUIPMENT

Solids: Magnesium sulfate, anhydrous ($MgSO_4$), boiling stones, cotton. Liquids: n-butyl alcohol ($CH_3CH_2CH_2CH_2OH$), glacial acetic acid ($HC_2H_3O_2$), concentrated sulfuric acid ($H_2SO_4$). Solutions: saturated sodium bicarbonate ($NaHCO_3$). Reflux and distillation equipment (to be used and returned at the end of the laboratory period): 100 mL or 250 mL round bottom distilling flask, distilling column, distillation take-off head, condenser, 200° or 250°C thermometer, 250 mL separatory funnel.

## DISCUSSION

Carboxylic acids react with alcohols to form esters through a condensation reaction known as esterification. The general equation for the reaction is

$$\underset{\text{Acid}}{R-\overset{\overset{\large O}{\|}}{C}-OH} + \underset{\text{Alcohol}}{R'OH} \rightleftarrows \underset{\text{Ester}}{R-\overset{\overset{\large O}{\|}}{C}-O-R'} + H_2O$$

The reaction proceeds slowly, usually requiring many hours to reach equilibrium. However, when the reaction is catalyzed with a small amount of a mineral acid, such as sulfuric acid, equilibrium is established in a few hours.

According to Le Chatelier's principle, the yield of ester may be increased by increasing the concentration of either reactant. In this experiment, an excess of acid is used.

Since reactions go faster at higher temperatures, the esterification is conducted at the boiling point of the reaction mixture. This is accomplished by refluxing, that is, by boiling the mixture, condensing the vapor in a water-cooled condenser, and returning the liquid to the reaction flask. Figure 30.1 illustrates the apparatus setup for refluxing.

In this experiment you will prepare n-butyl acetate by refluxing n-butyl alcohol with an excess of acetic acid and a small amount of sulfuric acid as catalyst.

$$CH_3COOH + CH_3CH_2CH_2CH_2OH \overset{H^+}{\rightleftarrows} CH_3\overset{\overset{\large O}{\|}}{C}-OCH_2CH_2CH_2CH_3 + H_2O$$

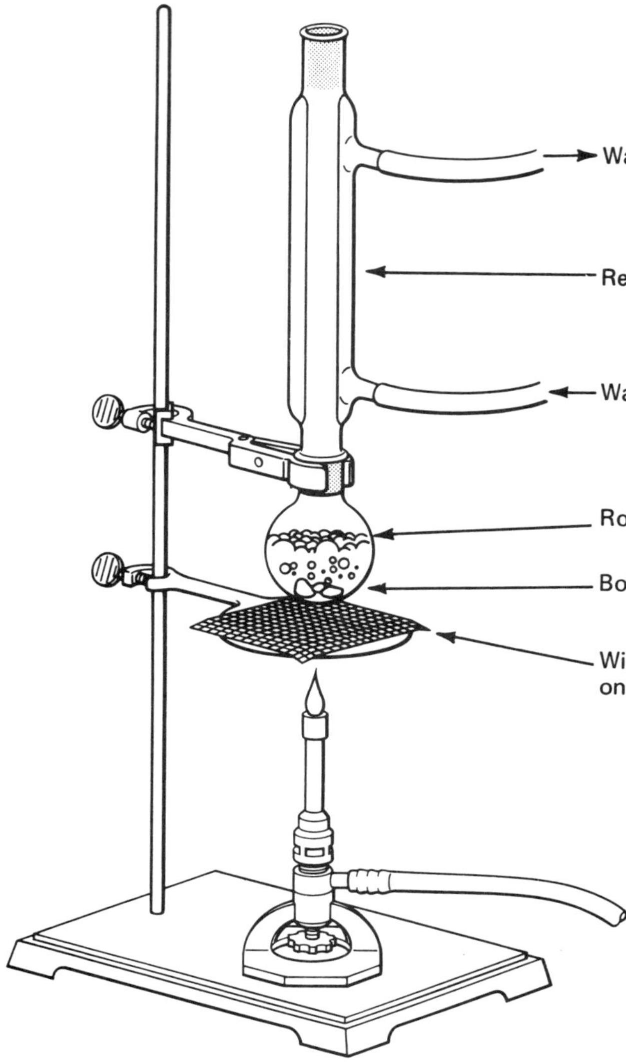


**Figure 30.1. Apparatus for refluxing a reaction mixture.**

Since the mixture is refluxed for only 1 hour, equilibrium will not be reached. The excess acetic acid and unreacted alcohol are separated from the product by extracting them with water using a separatory funnel. In this extraction, we take advantage of the relative insolubility (see Table 30.1) of n-butyl acetate in water. To remove the last traces of acid, the product is washed with sodium bicarbonate solution. This is followed by a water wash to remove sodium bicarbonate, since esters are readily hydrolyzed in alkaline solutions. The ester is finally dried with anhydrous magnesium sulfate and distilled. The drying agent removes water that is dissolved in the ester by forming the hydrate $MgSO_4 \cdot 7H_2O$.

**Table 30.1 Physical properties of the organic reactants and products in this experiment.**

| | Molar Mass g/mole | Density g/mL | Boiling Point, °C | Solubility g/100g $H_2O$ |
|---|---|---|---|---|
| Acetic acid | 60.0 | 1.05 | 118 | Infinite |
| n-Butyl alcohol | 74.0 | 0.81 | 117 | $9^{15°}$ |
| n-Butyl acetate | 116.0 | 0.88 | 126 | $0.7^{20°}$ |

### Distillation

Distillation is a widely used technique for separation and for purification of compounds. The separation depends on the differences in vapor pressure of the components in a solution. As the temperature of a liquid is raised, the vapor pressure increases until it becomes equal to the pressure of the atmosphere above the liquid. At that temperature the liquid boils. The temperature where the vapor pressure of the liquid equals the atmospheric pressure is called the boiling point of the liquid. The boiling point of a pure liquid is constant and, at 1 atmosphere pressure, is known as the normal boiling point.

Vapor pressure and boiling point are inversely related. A liquid that has a high vapor pressure at room temperature will have a lower boiling point than a liquid with a low vapor pressure at room temperature. Both, of course, will have a vapor pressure of 760 torr (mm Hg) at their respective normal boiling points. Compare the examples given below.

| | Boiling point, °C | Vapor pressure, torr (20°C) |
|---|---|---|
| Ethyl ether | 34.6 | 442 |
| Water | 100 | 17.5 |
| Acetic acid | 118 | 11.7 |

When a solution containing two miscible liquid compounds is boiling, the vapor above the solution will contain molecules of both compounds. However, this vapor is richer in the lower boiling compound, that is, the compound with the higher vapor pressure. When this vapor is condensed to a liquid and revaporized a number of consecutive times in a fractional distillation column, it continually becomes richer in the lower boiling compound and is eventually separated (distilled) as a purified or pure compound. With good control of heat input and the rate of distillation, the temperature at the top of the distilling column will remain fairly constant at the boiling point of the first compound being distilled. After the first compound has been removed from the solution, the temperature will rise to the boiling point of the second compound and it can then be collected by continued distillation.

### Use of the Separatory Funnel

The separatory funnel (Figure 30.2) is designed for separating immiscible liquids. Thus, it is narrow at the base, near the stopcock, so that the interface between the immiscible liquid layers is confined in a small space for accurate phase separation. The separatory funnel is used for liquid-liquid extractions, that is, extracting a substance from one liquid phase into another liquid phase. This is accomplished by placing both liquids in the funnel, mixing thoroughly by shaking, and allowing the two liquid layers to separate. The lower layer is then drawn off through the stopcock. Multiple extractions can be done without removing the upper layer from the funnel by adding more extracting solvent and repeating the process.

The location of the two liquid layers depends on the relative densities of the solutions. The solution that has the higher density will be the lower layer. When the densities of the two immiscible liquids are close to each other, separation after shaking is often slow, and even difficult because of the formation of emulsions.

The extraction process is done with the separatory funnel in the inverted position. The extracting solvent and the solution to be extracted are placed into the funnel and the stopper is

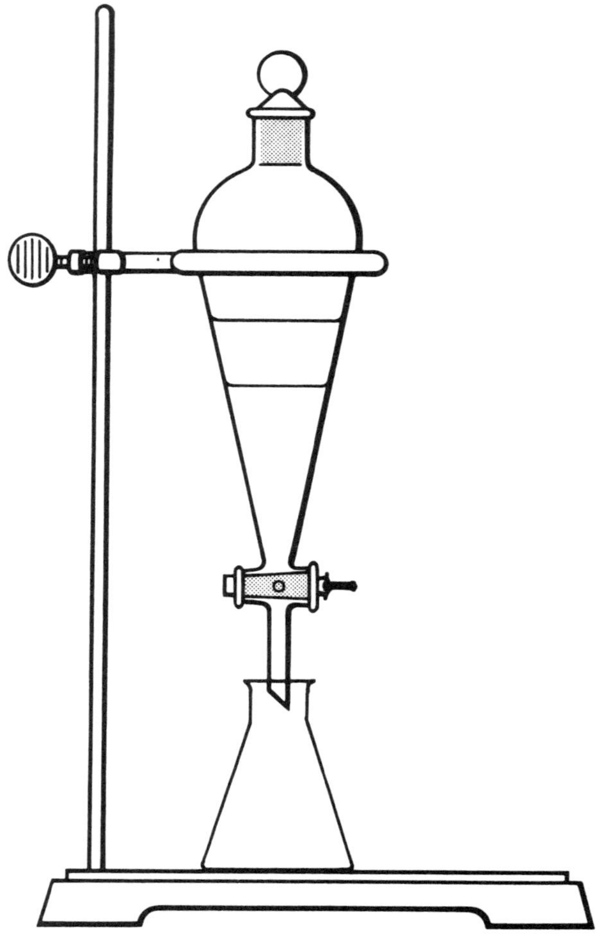

**Figure 30.2 Separatory funnel containing two liquid phases.**

inserted. (Be certain that the stopcock is closed when adding these liquids.) With the forefinger firmly on top of the stopper, the separatory funnel is inverted and the stopcock opened to relieve any pressure that may have built up inside the funnel. The stopcock is then closed and the contents shaken vigorously for about 40 seconds, stopping occasionally to relieve any pressure built up by carefully opening the stopcock. (**Be careful not to point the separatory funnel at yourself or at anyone else when you open the stopcock.**) Close the stopcock and place the separatory funnel upright in the iron ring for the separation of the two liquid phases.

## PROCEDURE

**Wear protective glasses.**

### A. Preparation of n-butyl acetate

Assemble the apparatus in Figure 30.1 for refluxing according to the following directions. If your equipment has ground glass joints, lightly grease the male joint before putting the apparatus together.

1. Clamp a 100 mL (or 200 mL) round bottom flask in place above the wire gauze.
2. Add 27.5 mL (22.3 g) of n-butyl alcohol and, with the same graduated cylinder measure and add 40.0 mL (42.0 g) of glacial acetic acid. Swirl the flask to mix the reagents.

3. Slowly add 1 mL of concentrated $H_2SO_4$ and swirl to mix the reagents thoroughly.
4. Add 3 boiling stones.
5. Place the reflux condenser on the flask and start a slow stream of water flowing through the condenser.
6. Heat the mixture with a small flame. Adjust the flame to control the rate of boiling so that the vapor is condensing no further than one-fourth the way up in the condenser. Continue boiling and refluxing for 1 hour.
7. During the 1-hour reflux time, get the equipment ready to separate the product.
8. After the mixture has refluxed for 1 hour, turn off the burner and allow the mixture to cool for about 5 minutes. Then cool the mixture further in an ice-water bath.

**B. Separation of n-butyl acetate**

1. Pour the cold reaction mixture through a funnel into a 250 mL separatory funnel.
2. Add 100 mL of water to the separatory funnel and close it with the stopper.
3. Place your forefinger firmly on top of the stopper, invert the separatory funnel and open the stopcock to relieve any pressure that may have built up inside the funnel. Close the stopcock.
4. With the funnel still inverted and held in both hands, shake the contents vigorously for about 40 seconds, stopping occasionally to relieve any pressure build-up by carefully opening the stopcock.
5. Close the stopcock and place the separatory funnel upright in the iron ring for separation of the aqueous layer from the ester layer.
6. After the two liquid layers have separated, remove the stopper and withdraw the lower layer through the stopcock, stopping just short of removing all of the aqueous layer. The ester should still be in the separatory funnel.
7. Add another 100 mL of water to the separatory funnel and repeat Steps B. 3 through 6.
8. Add 25 mL of saturated $NaHCO_3$ solution and repeat Steps B. 3 through 6. If the solution is still acidic at this time, carbon dioxide will be formed and pressure will build up inside the funnel. Test the aqueous solution being removed from the funnel with red and blue litmus paper. If this solution is acidic, wash with another 15 mL of the $NaHCO_3$ solution.
9. Add 25 mL of water to the funnel and repeat Steps B. 3 through 6. This time carefully remove all the aqueous layer through the stopcock.
10. Pour the crude n-butyl acetate out the top of the separatory funnel into a clean, dry, 125 mL Erlenmeyer flask. Add 2.5 g of anhydrous $MgSO_4$ powder, close the flask with a cork stopper, and swirl the contents several times to bring the liquid in contact with the drying agent ($MgSO_4$).
11. Label and store the flask in your laboratory locker for distillation at the next laboratory period.

## C. Distillation of n-butyl acetate

NOTE: All equipment used for the distillation must be **dry**.

1. Assemble the distillation apparatus shown in Figure 30.3 using a 100 mL distilling flask. If ground glass equipment is used, lightly grease the male joints. Be sure that all joints fit tightly and that the distilling flask and condenser are clamped to the ring stand. A 150°C (or greater) thermometer is needed. The height of the thermometer is adjusted so that the mercury bulb is just below the side arm of the distillation head. After attaching the rubber tubing to the condenser, turn on the water so that a slow stream is flowing through the condenser.

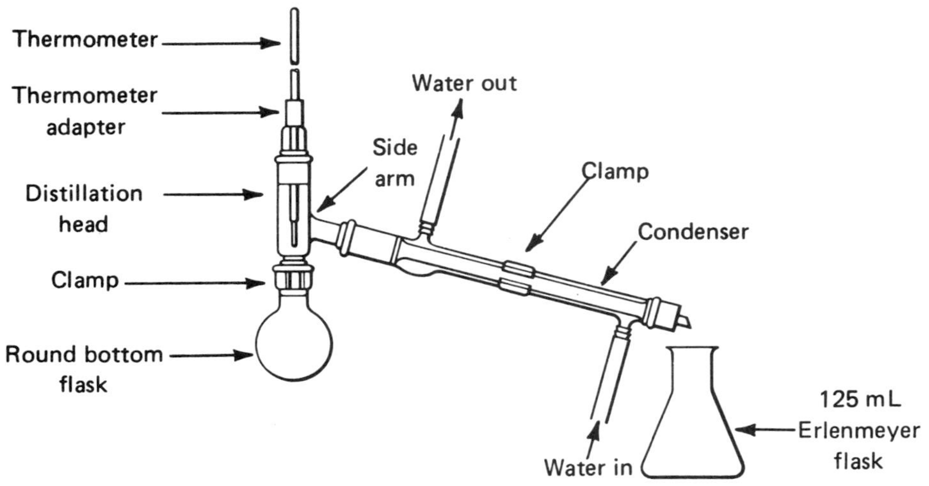


**Figure 30.3. Simple distillation apparatus.**

2. Pour the dried product into the round bottom flask through a funnel that is fitted with a very small plug of cotton just enough to keep the drying agent from entering the flask.
3. Add 3 boiling stones.
4. Heat the flask with a small flame. Adjust the heat input so that the rate of distillation is about 1 drop of distillate per second. (An ideal distillation is at equilibrium when vapor is condensing and liquid is dripping off of the thermometer back into the distilling flask.)
5. Collect the first distillate up to 122°C in a small Erlenmeyer flask. The first distillate that comes over may be cloudy because of traces of water in the system.
6. At 122°C replace the Erlenmeyer flask with a weighed, dry, 50 mL graduated cylinder. Collect all the distillate (n-butyl acetate) that comes over up to 127°C. Record the boiling range of the ester collected.

   ⚠ CAUTION: DO NOT heat the distilling flask to dryness.

7. Turn off the burner when the temperature exceeds 127°C or when the volume in the distilling flask decreases to 2 or 3 mL.
8. Weigh the graduated cylinder. Record the volume and the mass of the n-butyl acetate in the data table.
9. Hand in or dispose of the n-butyl acetate as directed by your instructor.

NAME ____________________

SECTION __________ DATE __________

INSTRUCTOR ____________________

## *REPORT FOR EXPERIMENT 30*

# Esterification—Distillation: Synthesis of n-Butyl Acetate

**Data Table**

| | Volume mL | Mass g | Molar Mass g/mole | Moles | Boiling Range |
|---|---|---|---|---|---|
| n-Butyl alcohol | | | | | — |
| Acetic acid | | | | | — |
| n-Butyl acetate | | | | | |

### CALCULATIONS

**Show calculation setups and answers.**

1. Calculate the limiting reactant in this reaction.

____________________

2. Calculate the theroretical yield of n-butyl acetate.

____________________

3. Calculate the percentage yield of n-butyl acetate.

____________________

## QUESTIONS AND PROBLEMS

1. Why was an excess of acetic acid used in this synthesis?

2. At approximately what temperature was the solution boiling during the reflux period?

____________

3. Explain why it is possible to separate n-butyl acetate from excess and unreacted acetic acid and n-butyl alcohol.

4. Why is the n-butyl acetate the top layer of the two liquid phases in the separatory funnel?

5. The volume of water and $NaHCO_3$ solution used to wash the n-butyl acetate was 250 mL. Calculate the maximum loss of n-butyl acetate due to its solubility in this volume of water.

____________

6. Write an equation to show how anhydrous $MgSO_4$ behaves as a drying agent. Name the product.

____________

7. Why can a simple distillation setup be used to distill the n-butyl acetate as prepared in this experiment?

8. Methyl salicylate (oil of wintergreen) is an ester and can be prepared from salicylic acid and methyl alcohol by the same method used in this experiment. Write an equation for this reaction.

# EXPERIMENT 31

# Synthesis of Aspirin

## MATERIALS AND EQUIPMENT

Solids: Salicylic acid ($HO-C_6H_4-COOH$). Liquids: Acetic anhydride [$(CH_3CO)_2O$], ethyl alcohol ($C_2H_5OH$), mineral oil, 85% phosphoric acid ($H_3PO_4$). Buchner funnel and suction flask, melting point tubes, filter paper to fit the Buchner funnel.

## DISCUSSION

Aspirin (acetyl salicylic acid, or A.S.A.) is the drug that is most widely used for self-medication. The familiar aspirin tablet contains 5 grains (about 325 milligrams) of acetyl salicylic acid and a small amount of an inert binding material such as starch. More than four million pounds of aspirin is manufactured each year in the United States. As a drug, aspirin has analgesic, antipyretic, and anti-inflammatory properties; that is, it can relieve pain, lower fever, and reduce inflammation.

Aspirin belongs to a group of drugs called salicylates because of their structural relationship to salicylic acid (SA),

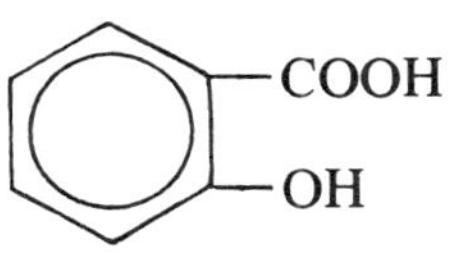


Salicylic acid is both an aromatic carboxylic acid and a phenol. Aspirin is represented by this structural formula:

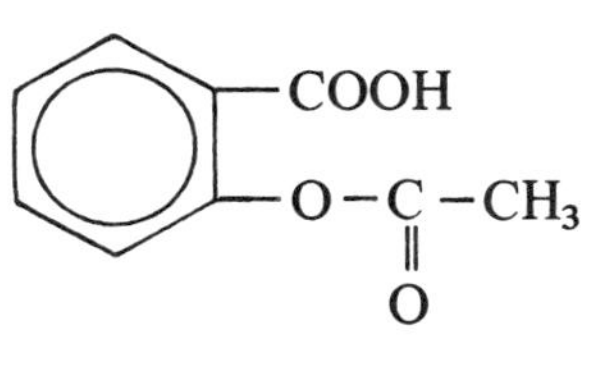


Acetyl salicylic acid (Aspirin)

This formula shows that aspirin is an ester formed between acetic acid and the phenol −OH group of salicylic acid. Aspirin is a weak acid because of its carboxyl group. Acetyl salicylic acid is practically insoluble in cold water, but its sodium salt is soluble in water.

Although it is clearly a derivative of salicylic and acetic acids, aspirin usually is not made from these substances. It can be prepared by reacting either acetyl chloride or acetic anhydride with salicylic acid. Generally, acetic anhydride, the acid anhydride of acetic acid, is used in the synthesis. Phosphoric acid catalyzes the reaction.

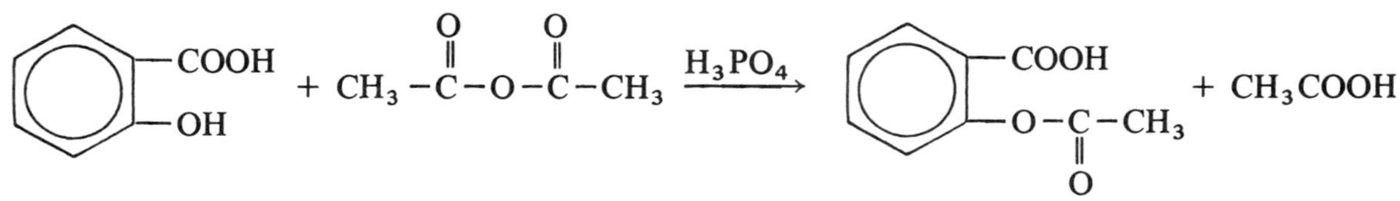

| Salicylic acid | Acetic anhydride | Acetyl salicylic acid | Acetic acid |
|---|---|---|---|
| Molar mass = 138.1 | Molar mass = 102.1 | Molar mass = 180.2 | Molar mass = 60.06 |

Aspirin can be recovered from the reaction mixture by adding cold water and filtering out the precipitated aspirin. The crude aspirin crystals are contaminated with small amounts of impurities, chiefly acetic and phosphoric acids. These impurities can be removed by recrystallization of the aspirin. To accomplish this, the crude aspirin is dissolved in hot alcohol, water is added, and the mixture is cooled. The crystals that form are filtered from the solution and dried to yield a product of high purity.

### Yields in Organic Synthesis

In an organic synthesis the actual amount of finished product is almost always less—often a great deal less—than the amount theoretically obtainable from the reactants used. Incomplete reactions or side reactions—that is, reactions that do not produce the desired product—can reduce the amount of product obtained. In addition, losses invariably occur in the recovery and purification steps (crystallization and recrystallization in this case). Thus, the percent of the theoretical amount of product actually obtained, or percentage yield, is a number that can be used to judge the success of a synthesis. It is calculated by this formula

$$\text{Percentage yield} = \left(\frac{\text{Actual mass of product obtained}}{\text{Theoretical mass of product obtainable}}\right) \times 100$$

## PROCEDURE

**Wear protective glasses.**

Accurately weigh 3.00 g of salicylic acid and transfer to a 125 mL **dry** Erlenmeyer flask. Next add 6 mL of acetic anhydride and 5 to 8 drops of 85% phosphoric acid to the flask. Swirl the flask gently to mix the reagents and place it in a beaker of warm (70–80°C) water for 15 minutes.

Then, while the reaction mixture in the flask is still warm, **carefully** add, drop by drop, 20 drops of cold water from a medicine dropper to destroy the excess acetic anhydride.

⚠ CAUTION: The vigorous reaction of the excess acetic anhydride with water may cause spattering.

Add 20 mL of water to the flask; then put the flask in an ice bath to cool the reaction mixture and speed the crystallization of aspirin. When the crystallization seems to be complete, collect the crystals by suction filtration using a small Buchner funnel (Fig. 31.1) or an alternate filtering device. Wash the crystals on the filter once with a few milliliters of cold, preferably iced, water.

Recrystallization: Transfer the crystals to a 100 mL beaker, add 10 mL of ethyl alcohol and stir to dissolve. If necessary warm the beaker in a 250 mL beaker containing warm water to complete the solution formation. When all of the crystals have dissolved, pour 25 mL of warm (60-70° C) water into the alcohol solution. Cover the beaker with a watch glass and set aside to cool until crystals begin to form. Then place the beaker in an ice bath for about 10 minutes to complete crystallization.

After the crystallization is complete, filter the crystals out using a suction filter. Spread the crystals out on the filter with a spatula and press them with dry filter paper to aid in drying. Transfer the crystals to a watch glass, cover with filter paper and finish drying by storing in your locker for at least one day. Transfer the dried aspirin to a weighed vial or test tube. Then re-weigh and determine the weight of aspirin.

Determine the melting point of your dried aspirin crystals using the method given in Experiment 27, Pages 209-214. Look up the melting point given for aspirin in the chemical literature. The *Handbook of Chemistry and Physics* or Lange's *Handbook of Chemistry* are suitable sources.

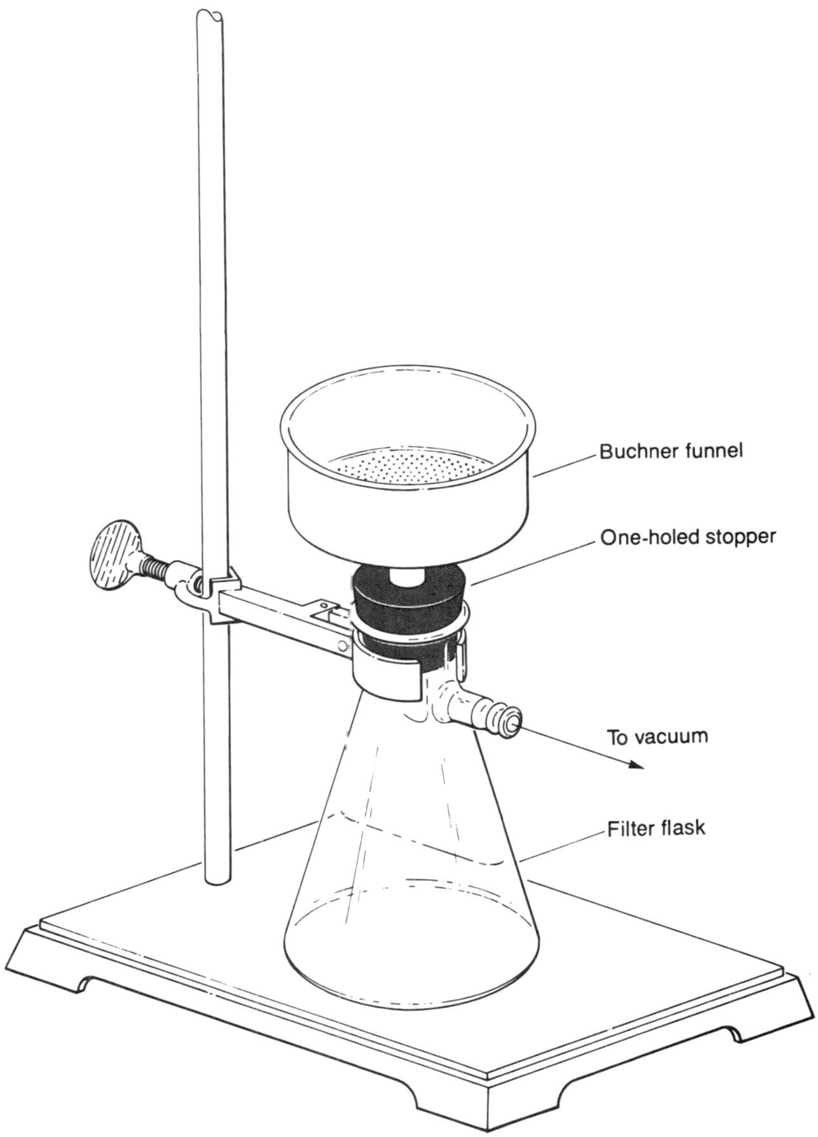


**Figure 31.1. Buchner Funnel Setup**

NAME ____________________

SECTION __________ DATE __________

INSTRUCTOR ____________________

# *REPORT FOR EXPERIMENT 31*

## Synthesis of Aspirin

Percentage yield of aspirin:

1. Mass of salicylic acid used ____________
2. Mass of aspirin crystals obtained ____________
3. Mass of aspirin theoretically obtainable from salicylic acid used. Show calculation setup. ____________

4. Percentage yield of aspirin on theoretical mass from 3. Show calculation setup. ____________

Melting point:

1. Melting point range of your aspirin ____________
2. Melting point of aspirin from literature ____________

### QUESTIONS AND PROBLEMS

1. Write the chemical equation representing the synthesis of aspirin from acetyl chloride.

2. Could aspirin be prepared by the method used in this experiment from these compounds? Explain your answer.

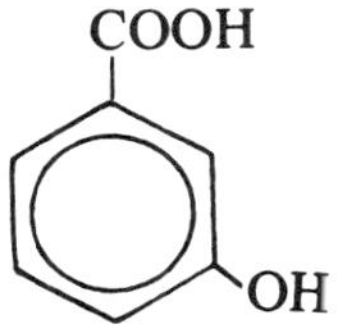


and $CH_3\overset{\underset{\|}{O}}{C}-O-\overset{\underset{\|}{O}}{C}CH_3$

3. (a) How many moles of acetic anhydride were represented by the 6.00 mL (density = 1.08 g/mL) used in the synthesis? (b) How many excess moles of acetic anhydride were used—over and above the amount needed to react with 3.00 g of salicylic acid? Show calculations.

(a) ______________________

(b) ______________________

4. Aspirin is insoluble in water. Explain why it is soluble in sodium hydroxide solution.

5. If the annual production of acetyl salicylic acid in the U.S.A. were converted to 5-grain aspirin tablets, approximately how many tablets would there be for each person? Assume that the population is about 250,000,000. Show calculation.

______________________

6. What is (a) an analgesic and (b) an antipyretic?

# *EXPERIMENT 32*

# Polymers—Macromolecules

## MATERIALS AND EQUIPMENT

Solids: Benzoyl peroxide $(C_6H_5COO)_2$. Liquids: methyl methacrylate $[CH_2{=}C(CH_3)COOCH_3]$. Solutions: 0.40 M adipoyl chloride in cyclohexane $(C_6H_8O_2Cl_2)$, 0.40 M hexamethylenediamine in 0.40 M NaOH $[(CH_2)_6(NH_2)_2]$, 5% sodium bicarbonate $(NaHCO_3)$. A 4 inch wire with a hook at one end.

## DISCUSSION

Polymers are high molecular weight compounds, many of which have molecular weights exceeding one million. Thus polymers are often referred to as macromolecules (very large molecules).

The process of forming very large, high molecular weight molecules from smaller units is called polymerization. The large molecule is called the polymer, and the small unit, the monomer. The monomers may be alike or they may be different. When there are two or more monomers, the macromolecule is known as a copolymer.

Starch, glycogen, cellulose, and proteins are examples of naturally occurring polymers. Man-made polymers touch every phase of modern living. Examples of these are polyethylene, Nylon, Dacron, Bakelite, Lucite, polyvinyl chloride, etc.

An example of a naturally occurring polymer is starch, a sugar storage molecule found in plants. The monomer of starch is glucose and a portion of the starch molecule is represented as follows:

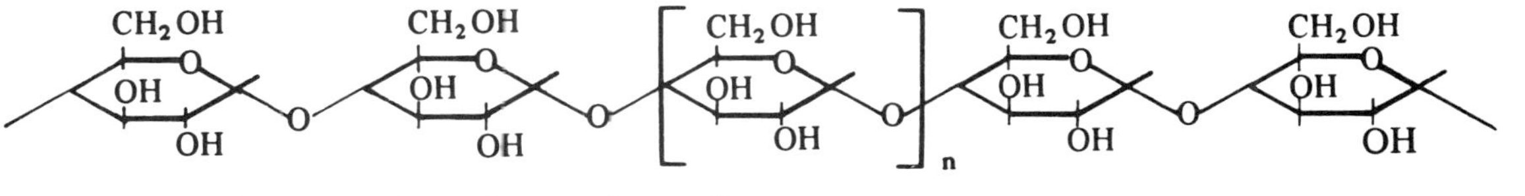

where n represents a large number of repeating units.

Alkenes, or substituted alkenes, are among the most common monomers for making synthetic polymers. For example, ethylene can be polymerized to polyethylene.

$$n\,CH_2{=}CH_2 \longrightarrow -CH_2CH_2{+}CH_2CH_2{+}_nCH_2CH_2-$$

Ethylene Polyethylene

Polymers that soften on heating, and therefore can be changed into different usable shapes, are known as thermoplastic polymers. Polymers that set to infusible solids and do not soften on reheating are known as thermosetting polymers.

All polymers may be classified as either addition polymers or condensation polymers. An addition polymer is one that is formed by the successive addition of repeating monomer molecules. A condensation polymer is one that is formed from monomers with the elimination of water or some other simple substance.

In this experiment, you will prepare a condensation polymer, Nylon (a polyamide) and an addition polymer, Lucite (polymethyl methacrylate).

Nylon 6-6 is a condensation polymer made from two different monomers, adipic acid and hexamethylenediamine.

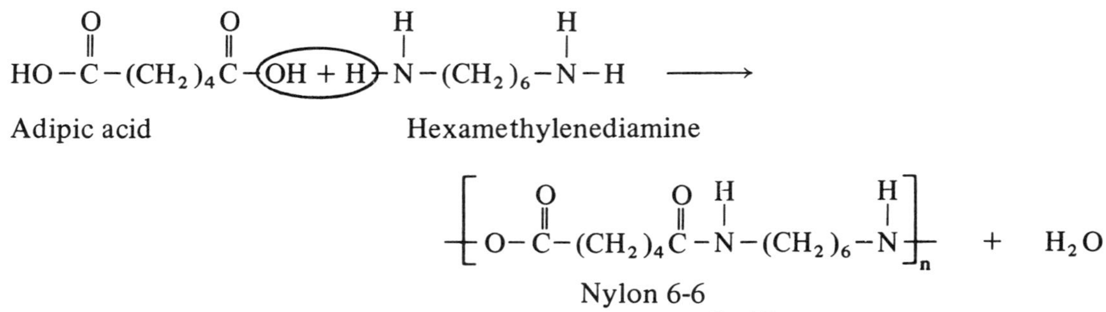


Since the bonds linking the monomers are amide bonds $(-\overset{O}{\overset{\|}{C}}-\overset{H}{\overset{|}{N}}-)$, Nylon is known as a polyamide. Polyamides can be made from diacids and diamines as shown above or from diacid chlorides and diamines. In this experiment, the diacid chloride is used, eliminating HCl to form the amide linkage.

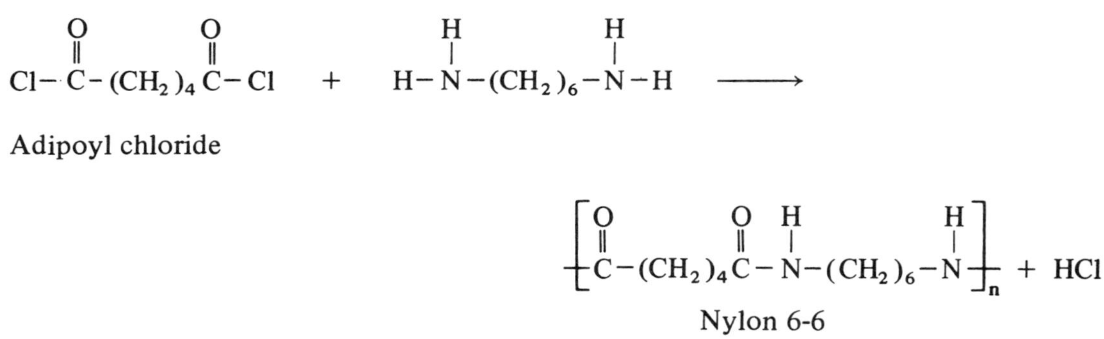


Nylon has good tensile strength, elasticity, and resistance to abrasion. It is used in fabrics, surgical sutures, thread, gears and bearings, rope, hosiery and clothing, tire cords, and so on.

Lucite, also known as Plexiglas, is an addition polymer made from the monomer methyl methacrylate. Lucite is one of the polyacrylics, the monomers of which are derivatives of acrylic acid.

The polymerization of methyl methacrylate is a chain reaction catalyzed by benzoyl peroxide, which forms the chain-initiating species. The overall reaction is represented by this equation.

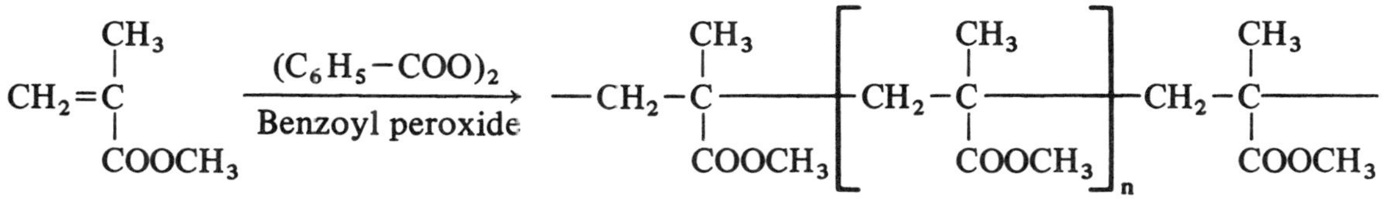

Methyl methacrylate — Polymethyl methacrylate (Lucite)

Polyacrylics are clear, colorless polymers. They take a high polish, are transparent to visible and ultraviolet light, and have excellent optical properties. Lucite is used for windshields in airplanes, for contact lenses, automobile finishes, molded ornamental objects, etc.

## PROCEDURE

**Wear protective glasses.**

### A. Synthesis of Nylon

CAUTION: The monomers used in this experiment are skin irritants, so be careful not to get them on your skin.

Place 5 mL of aqueous 0.40 M hexamethylenediamine solution into a test tube. Obtain 5 mL of 0.40 M adipoyl chloride solution in cyclohexane and carefully pour this solution down the wall of the slightly tilted test tube containing the diamine. Do not allow the solutions to mix. Since the solution in cyclohexane is less dense it will float on the aqueous solution.

Observe that a thin film forms at the interface of the two solutions. This film is Nylon 6-6. Record your observations.

Using a piece of wire with a hook bent into one end, hook the polymer mass at the solution interface and slowly raise the wire so that a continuous "rope" of polymer is obtained. Wind the rope around another test tube as you pull it from the reaction mixture. If the Nylon rope breaks, start a new strand with the hooked wire. Do not touch the polymer with your hands since it may contain unreacted reagents.

Continue to wind the polymer around the test tube. When no more polymer is obtained transfer the Nylon to a small beaker, add 25 mL of 5% $NaHCO_3$ solution, stir, and decant the bicarbonate wash. Finally, wash the Nylon several times with water. Once the Nylon is washed it is safe to touch. Dry the polymer by pressing it between pieces of paper towel.

Stir the remaining solutions to form additional Nylon and dispose of this mixture according to the directions of your instructor.

### B. Synthesis of Polymethyl methacrylate (Lucite)

Measure 5 mL of water in a graduated cylinder and pour it into a test tube. Now pour an equal volume of methyl methacrylate from the reagent bottle into a second clean, dry test tube. Discard the 5 mL of water. Have your instructor add about 25 mg of the catalyst, benzoyl peroxide, to the methyl methacrylate. Dissolve the benzoyl peroxide by swirling the mixture.

Place the test tube in a beaker of gently boiling water. The volume of water in the beaker should be such that about one-half of the content of the test tube is submerged in the water. Watch the tube closely, and when gas bubbles start to form (5 to 10 minutes) remove the tube from the water and place it in the test tube rack. Continue to observe the tube as the polymerization proceeds to form a hard, glass-like polymer. While the polymerization is occurring, carefully feel the test tube to determine if the reaction is exothermic or endothermic.

NAME ______________________

SECTION __________ DATE __________

INSTRUCTOR ______________________

## *REPORT FOR EXPERIMENT 32*

# Polymers—Macromolecules

### A. Synthesis of Nylon

1. Describe what you observed at the interface of the two solutions.

2. Why did Nylon 6-6 polymer form only at the interface of the two liquid phases?

3. Nylon 6-8 has the formula

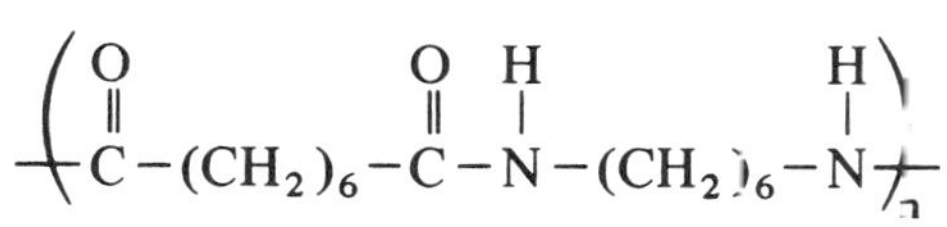

In the names, Nylon 6-6 and Nylon 6-8, to what do the numbers refer?

4. What would happen if you used $CH_3(CH_2)_4CH_2NH_2$ instead of hexamethylenediamine in the synthesis of Nylon 6-6?

5. Suppose you used citroyl chloride instead of adipoyl chloride. Draw enough of the polymer formula to show how the monomers would react with one another.

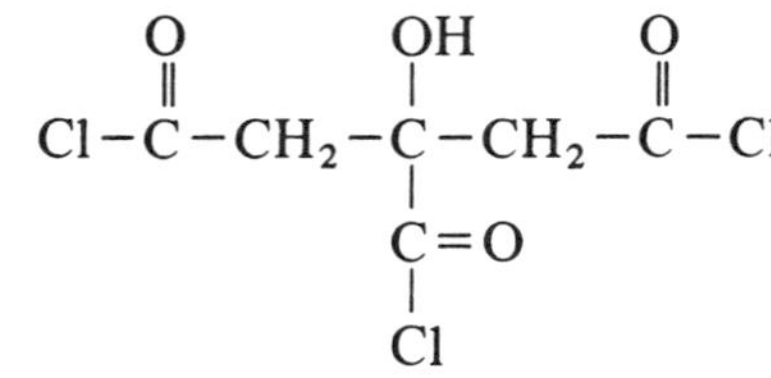

citroyl chloride

6. Another type of Nylon, Nylon 6, is made from caprolactam. In the reaction the ring opens between the $-C=O$ and the $-NH$ groups to give the monomer that polymerizes to the polyamide Nylon 6. Write the structure showing 3 units in Nylon 6.

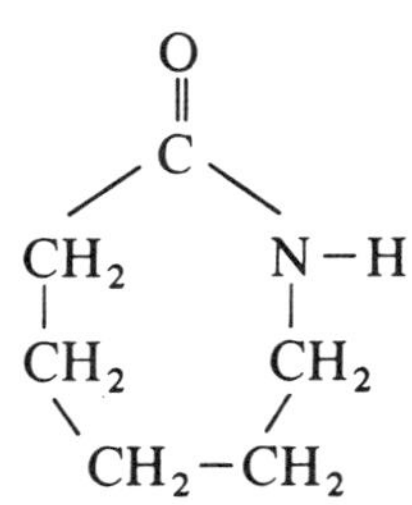

**B. Synthesis of Lucite**

1. Is this polymerization an exothermic or an endothermic reaction? What was your evidence?

2. Describe how the physical appearance of the polymer compares with that of the monomer.

3. What change in the carbon-carbon bonding occurs in this polymerization?

4. Why is this polymerization reaction known as addition polymerization?

5. The polymerization of styrene to polystyrene can also be catalyzed by benzoyl peroxide. Write an equation for this reaction showing at least three units of the polymer.

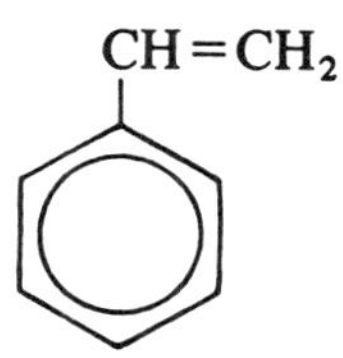


Styrene

# EXPERIMENT 33

# Carbohydrates

## MATERIALS AND EQUIPMENT

Solids: Glucose, sucrose. Liquids: Concentrated sulfuric acid. Solutions: 1% solutions of arabinose, fructose, glucose, maltose, starch, sucrose (freshly prepared), xylose; concentrated hydrochloric acid, 1% iodine in 2% potassium iodide; 10% sodium hydroxide; reagent solutions for Barfoed's, Benedict's, Bial's, Molisch's, and Seliwanoff's tests; fruit juices such as orange, lemon, lime, grapefruit, apple, etc.

## DISCUSSION

Carbohydrates are one of the three principal classes of foods. They are major constituents of plants and are also found in animal tissues. They were so named because the carbon, hydrogen, and oxygen atom ratio in most carbohydrates approximates that of $C \cdot H_2O$. Carbohydrates, of course, are not hydrates of carbon but are relatively complex substances such as sugars, starches, and cellulose. Chemically, **carbohydrates** are polyhydroxy aldehydes or ketones or substances that when hydrolyzed, yield polyhydroxy aldehydes or ketones.

Carbohydrates are classified as monosaccharides, disaccharides, oligosaccharides, or polysaccharides based on the number of monosaccharide units present in the molecule. A **monosaccharide** is a carbohydrate that cannot be hydrolyzed to simpler carbohydrate molecules. A **disaccharide** yields 2 monosaccharide molecules, alike or different, when hydrolyzed. **Oligosaccharides**, on hydrolysis, yield 2 to 10 monosaccharide molecules. The monosaccharide molecules may be of only one kind, or they may be of two or more different kinds. **Polysaccharides**, when hydrolyzed, yield many monosaccharide molecules. These monosaccharide molecules are typically of only one kind.

The monosaccharides are futher classified according to the length of the carbon chain, such as trioses ($C_3H_6O_3$), tetroses ($C_4H_8O_4$), pentoses ($C_5H_{10}O_5$), and hexoses ($C_6H_{12}O_6$). If they are aldehydes, they are called aldoses; if ketones, they are called ketoses. The most common monosaccharides are glucose, galactose, and fructose. Glucose and galactose are aldohexoses; fructose is a ketohexose. The three common disaccharides are sucrose (glucose and fructose), maltose (glucose and glucose), and lactose (galactose and glucose). All three have the formula $C_{12}H_{22}O_{11}$ and can be hydrolyzed to yield their respective monosaccharides by heating in a water solution containing a small amount of HCl or $H_2SO_4$. The three most common polysaccharides, starch, glycogen, and cellulose, have the formula $(C_6H_{10}O_5)_x$, where x ranges from about 200 to several thousand. All three are polymers of glucose.

Numerous tests have been devised for determination of the properties and for the differentiation of carbohydrates. A brief description of some of these tests follows.

### Molisch's Test

This is a very general test for carbohydrates. The test is based on the formation of furfural, or hydroxyfurfural when a carbohydrate reacts with concentrated sulfuric acid. The furfural reacts with the Molisch reagent, α-naphthol, to yield colored condensation products.

### Seliwanoff's Test

This test distinguishes fructose, a ketohexose, from aldohexoses and disaccharides. The reaction between fructose and the reagent (resorcinol in dilute HCl) occurs within one minute in boiling water. A reddish-colored product is formed, the color intensifies with further heating. Other carbohydrates subjected to this test will produce a faint red color if heating is prolonged. The color formation is attributable to transformation of glucose to fructose by the catalytic action of hydrochloric acid or by hydrolysis of sucrose to yield fructose.

The Benedict's test and Barfoed's test are reduction tests. Certain carbohydrates have a reducing ability because they have, or are able to form, a free aldehyde or ketone group in solution. In alkaline solution, copper(II) or silver ions are reduced to characteristic precipitates of $Cu_2O$ or free Ag. All the common monosaccharides are reducing sugars. Some disaccharides and polysaccharides may initially be nonreducing but show reducing properties after heating in an acidic solution. During the heating, hydrolysis to monosaccharides occurs. Some widely used tests for sugars are based on this reducing ability of carbohydrates. An example is Benedict's test, which is used to detect sugar (glucose) in urine.

### Benedict's Test

In this test $Cu^{2+}$ is reduced to $Cu^+$, forming $Cu_2O$, which is a brick-red colored precipitate. However, the color in a positive Benedict's test may appear as green, yellow, orange, or red depending upon the amount of $Cu_2O$ suspended in the dark blue reagent. The amount of $Cu_2O$ formed depends on the concentration of sugar in the solution. Reducing sugars give a positive test.

### Barfoed's Test

The Barfoed reagent is used to distinguish between mono- and disaccharides. This test is also a copper reduction reaction but differs from Benedict's in that the reagent is made in an acidic medium [copper(II) acetate and acetic acid]. Within the stated time interval, only monosaccharides will reduce the $Cu^{2+}$ ions. If heated long enough, disaccharides will be hydrolyzed by the acid present and give a positive test. Barfoed's reagent is not suitable for detection of sugar in urine.

### Bial's Test

This test is used to distinguish pentoses from hexoses. Pentoses occur in both plants and animals. The pentoses ribose and deoxyribose are universally found in the nucleic acid portion of nucleoproteins of the cells. Bial's reagent contains orcinol (5-methylresorcinol) dissolved in concentrated HCl plus a small amount of $FeCl_3$. When mixed with the reagent, pentoses are converted to furfural, which reacts to yield a blue-green colored compound.

### Iodine Test

Iodine reacts with starch to form a deep blue complex. When an acidified starch solution is boiled, it hydrolyzes to yield glucose ultimately. The iodine test can be used to follow the course of this hydrolysis.

## PROCEDURE

**Wear protective glasses.**

### A. Molisch Test

Run this test on each of the following 1% carbohydrate solutions and on water as a reference blank: (1) arabinose, (2) glucose, (3) fructose, (4) maltose, (5) sucrose, (6) starch, and (7) water (blank). To 5 mL of the carbohydrate solution in a test tube, add 3 drops of Molisch reagent and mix well. Now tilt the test tube at an angle of about 45 degrees and very carefully and slowly pour 2-3 mL of concentrated sulfuric acid from a 10 mL graduated cylinder down the side of the test tube so that the sulfuric acid forms a layer underneath the solution being tested.

> NOTE: It is very important that the lip of the graduated cylinder be touching the inner top of the test tube containing the carbohydrate and that the **acid be poured slowly**.

Set the test tubes in the rack and observe for evidence of reaction at the interface of the two liquid layers. Some reactions may take as long as 15 to 20 minutes, so proceed with the next part of the experiment.

### B. Seliwanoff's Test

Run this test on each of the following solutions: (1) arabinose, (2) glucose, (3) fructose, (4) maltose, (5) sucrose, and (6) water (blank). Mix together in a test tube 1 mL of the carbohydrate solution and 4 mL of Seliwanoff's reagent. Place all six tubes in a beaker of boiling water for 2 minutes. (Do not overheat.) Observe and record the results.

### C. Benedict's Test

Run this test on each of the following solutions: (1) arabinose, (2) glucose, (3) fructose, (4) maltose, (5) sucrose, (6) starch, and (7) water (blank). Mix together in a test tube 1 mL of the carbohydrate solution and 5 mL of Benedict's reagent. Place all seven tubes in a beaker of boiling water for 5 minutes. Observe and record the results.

### D. Barfoed's Test

Run this test on the following solutions: (1) arabinose, (2) glucose, (3) maltose, (4) sucrose, and (5) water (blank). Mix together in a test tube 1 mL of the carbohydrate solution and 5 mL of Barfoed's reagent. Place the tubes in a beaker of boiling water for 5 minutes. Observe and record the results.

### E. Bial's Test

Run this test on each of the following solutions: (1) arabinose, (2) xylose, (3) glucose, (4) fructose, and (5) water (blank). Mix together in a test tube 2 mL of the carbohydrate solution and 3 mL of Bial's reagent. Carefully heat (with agitation) each tube over a burner flame until the mixture just begins to boil. Observe and record the results.

### F. Dehydration

1. Place about 2 grams (no more) of sucrose in a test tube; add 1 mL of concentrated sulfuric acid. After about 30 seconds very carefully touch the test tube to feel the heat evolved. Observe and record the results. Allow the tube to cool for about ten minutes. Successively add and pour out small amounts of water to loosen the residue in the tube. Try to avoid letting the residue go down the sink drain.

2. Repeat this experiment using 2 grams of glucose and 1 mL concentrated sulfuric acid.

### G. Hydrolysis of Disaccharides

Mix together 10 mL of sucrose solution and 5 drops of concentrated hydrochloric acid in one test tube and, in a second test tube, 10 mL of maltose solution and 5 drops of concentrated hydrochloric acid. Place both tubes in a beaker of boiling water for about 10 minutes. Cool the solutions and neutralize the acid with 10% NaOH solution (requires 18-20 drops of base). Use red litmus paper as an indicator. Now run Benedict's and Seliwanoff's tests using 2 mL samples of the hydrolyzed sugar solutions and 5 mL of each test reagent. Record your results.

### H. Hydrolysis of Starch

Mix together in a 50 mL beaker 10 mL of 1% starch solution, 20 mL of water, and 10 drops of concentrated hydrochloric acid. Withdraw a medicine dropper full of this solution, and transfer the solution to a test tube. Prepare a blank by adding a medicine dropper full of distilled water to a test tube. Save these two samples for later reference.

Label three clean test tubes, 5 min., 10 min., and 15 min., Gently boil the starch solution for 15 minutes, stirring occasionally. During this period, withdraw a medicine dropper full of the hot solution every 5 minutes and transfer the solution to the appropriate tube. Now add 1 drop of the $I_2-KI$ solution to each of the five tubes and mix. Record your observations.

After the heating is completed, withdraw another medicine dropper full of the solution and transfer to a test tube. Neutralize the acid present with 10% NaOH solution and test for the presence of a reducing sugar with Benedict's reagent.

### I. Fruit Juices

Test available fruit juices (preferably fresh), e.g., orange, lemon, lime, grapefruit, apple, etc., for the presence of reducing sugars and fructose using Benedict's and Seliwanoff's reagents. Use about 1 mL samples of juice for each test.

## INSTRUCTOR DEMONSTRATION (optional)

### Colloidal Nature of Starch

Mix 4 mL of 1% starch with 200 mL distilled water, divide into two equal portions and place in 100 mL beakers. Fill a third 100 mL beaker with distilled water, then line up the three beakers with water in the center and pass a narrow beam of light from a microscope illuminator (or other high intensity source) through the beakers. Note the marked Tyndall effect apparent in even this very dilute starch. Observe the suspension from both the side and the top views.

NAME ____________________

SECTION __________ DATE __________

## *REPORT FOR EXPERIMENT 33*

INSTRUCTOR ____________________

# Carbohydrates

### A. Molisch Test

1. Describe the evidence for a positive test.

2. What is the chemical basis for a positive test?

3. Circle the names of the substances that gave a positive test.

   arabinose glucose fructose maltose sucrose starch water

### B. Selinwanoff's test

1. Describe the evidence for a positive test.

2. Circle the name(s) of the substances tested that gave a positive test.

   arabinose glucose fructose maltose sucrose water

3. Upon prolonged heating, would you expect sucrose to give a positive Seliwanoff's test? Explain.

### C. Benedict's Test

1. Describe the evidence for a positive test.

2. Circle the name(s) of the substances tested that gave a positive test.

   arabinose  glucose  fructose  maltose  sucrose  starch  water

3. Circle the name(s) of the substances that are reducing carbohydrates.

   arabinose  glucose  fructose  maltose  sucrose  starch

4. If the $Cu^{2+}$ ions in Benedict's reagent are reduced, what must have happened to the sugar molecules?

5. Complete and balance:

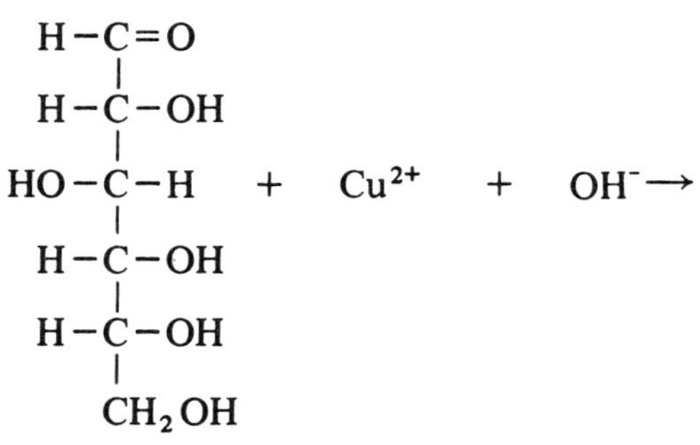

### D. Barfoed's Test

1. What is the evidence for a positive test?

2. Circle the name(s) of the substances that gave a positive test.

   arabinose  glucose  maltose  sucrose  water

3. How is Barfoed's reagent able to distinguish a reducing monosaccharide from a reducing disaccharide?

4. Circle the name(s) of any of the following carbohydrates that will give a positive Barfoed's test.

   ribose  lactose  mannose  starch

### E. Bial's Test

1. What is the evidence for a positive test?

2. Circle the name(s) of the substances that gave a positive test.

arabinose xylose glucose fructose water

3. Write the structural formulas of two carbohydrates (other than those tested) that will give a positive Bial's test.

## F. Dehydration

1. Describe the effects that you noted after the addition of concentrated sulfuric acid to sucrose.

2. Write an equation for the reaction of sucrose and sulfuric acid. Assume complete dehydration and the formation of $H_2SO_4 \cdot H_2O$ as one of the products.

## G. Hydrolysis of Disaccharides

1. Results after hydrolysis. Use + or − signs to indicate whether test was positive or negative.

| | Sucrose | Maltose |
|---|---|---|
| Benedict's Test | ______ | ______ |
| Seliwanoff's Test | ______ | ______ |

2. What do the results of these tests indicate about the composition of sucrose and maltose?

## H. Hydrolysis of Starch

1. What is the evidence for a positive iodine test for starch?

2. What evidence did you observe to indicate that the starch was hydrolyzing while being heated?

3. Did the Benedict's tests prove that the starch actually was hydrolyzed? Explain.

## I. Fruit Juices

1. Tabulate your results using + or − signs to indicate whether a positive or a negative result was obtained with each fruit juice tested.

| Fruit Juice | Benedict's Test | Seliwanoff's Test |
|---|---|---|
| | | |
| | | |
| | | |
| | | |

2. Which carbohydrate(s) is proven to be present in each fruit juice tested?

## QUESTIONS

1. The possible carbohydrates in the following problems are limited to these: arabinose, glucose, fructose, maltose, sucrose, and starch.

   (a) A carbohydrate solution gave a positive Molisch test and negative Benedict's, Barfoed's, Bial's, and Seliwanoff's tests. When treated with hydrochloric acid and boiled for several minutes, the solution showed positive Benedict's, Barfoed's, and Seliwanoff's tests and a negative Bial's test. What carbohydrate was in the original solution?

   (b) A solution containing only one carbohydrate gave a $Cu_2O$ precipitate with Benedict's reagent. Which possible carbohydrates are present in the solution?

   (c) Another sample of the solution from (b) failed to give a $Cu_2O$ precipitate with Barfoed's reagent. Which carbohydrate is present in the solution?

2. Lactose is a reducing disaccharide. Describe how a solution of lactose would react toward these reagents.

   (a) Benedict's

   (b) Barfoed's

   (c) Seliwanoff's

# EXPERIMENT 34

# Glucose Levels in Whole Blood

## MATERIALS AND EQUIPMENT

Solutions: Whole blood (**certified AIDS virus free**), 0.2 M zinc sulfate ($ZnSO_4 \cdot 7H_2O$), 0.2 M barium hydroxide [$Ba(OH)_2 \cdot 8H_2O$], arsenomolybdate reagent, copper reagent, standard glucose solutions ($C_6H_{12}O_6$). Marbles (~ 20 mm diameter), 10 mL graduated pipet, protective gloves, spectrophotometer.

## DISCUSSION

Blood is a metabolic ocean of life-giving molecules. One of the most critical constituents is glucose. Glucose is a source of energy and a building block for other cellular components. Normal adult levels of glucose fall between 80 mg and 120 mg per 100 mL of whole blood. Glucose is a reducing sugar capable of reducing $Cu^{2+}$ to $Cu^{+}$. The $Cu^{+}$ ion is also a reducing agent. The reducing ability of $Cu^{+}$ enables it to form blue complexes with arsenomolybdate reagent. The color intensity of these blue complexes is proportional to the concentration of the original glucose.

To quantitatively determine the concentration of glucose in whole blood, you will use an instrument called a spectrophotometer to measure the amount of light absorbed by the blue complexes. The absorption depends primarily upon the concentration of the colored species. However, the distance the light travels through the solution (pathlength) and the wavelength of the light source also affect the amount of absorption. A spectrophotometer produces various wavelengths of light and can be set at a predetermined wavelength. In order to determine the concentration of an unknown sample, its absorption is compared to the absorption of standard solutions whose concentrations are known. In this comparison process, identical sample cuvettes (cells) are used. Since all samples have the same pathlength, pathlength becomes a constant. Using a pre-set wavelength of 495 nanometers (nm) for the blue complexes and fixing the pathlength, the absorption of light varies directly with the concentration of glucose.

To determine the concentration of glucose, a calibration curve is used. This curve is created by plotting the absorbance of each standard glucose sample versus its concentration. Figure 34.1 shows a sample calibration curve.

This curve can then be used as a comparison tool. The solution derived from the whole blood is placed in a cuvette, and its absorbance measured. The corresponding concentration of glucose can then be determined from the calibration curve.

Glucose is only one of many components found in whole blood. Since we want to measure only the concentration of the blue complexes formed by the reducing sugar glucose, other substances that interfere, such as proteins, must be removed. Otherwise, values of glucose will be obtained that are higher than the true values. A method, known as the Somogyi-Nelson procedure, is used to remove both proteins and nonsugar-reducing substances from whole blood.

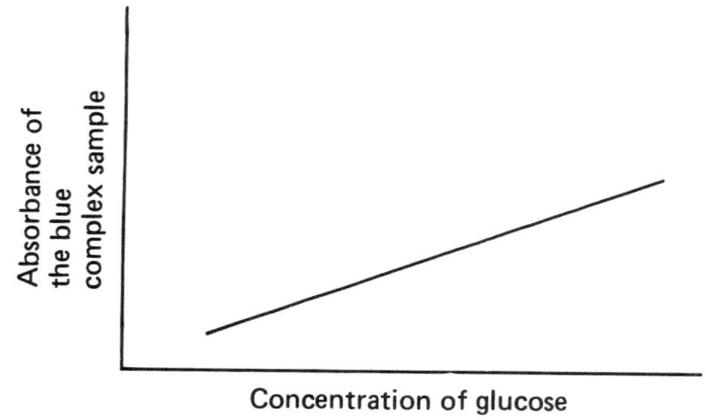


**Figure 34.1. Calibration Curve for Glucose**

In the Somogyi-Nelson procedure, barium hydroxide and zinc sulfate are added to precipitate both proteins and nonsugar-reducing substances. Proteins form insoluble zinc proteinates while the various reducing species are removed as either zinc or barium salts. These precipitates must be filtered or centrifuged from the mixture to obtain a clear solution for glucose analysis.

## PROCEDURE

**Wear protective glasses.**

**Wear protective gloves.**

⚠ CAUTION: Use rubber suction bulb or water aspirator pump to draw liquids into your pipet. **Do not pipet by mouth.**

### A. Removal of Proteins and Nonsugar-reducing Substances

1. Pipet 2.0 mL of whole blood into a small Erlenmeyer flask.
2. Using a pipet, add 14.0 mL of water and mix. This ruptures the red blood cells.
3. Using a pipet, add 2.0 mL barium hydroxide solution and mix.
4. After at least one minute, pipet 2.0 mL zinc sulfate solution into the mixture. Mix thoroughly and let stand for five minutes.
5. Filter the mixture and collect the filtrate. The blood filtrate should be a water clear solution. Do not wet the filter paper before filtering.

### B. Preparation of Samples

1. Thoroughly clean and dry nine test tubes and label them 1 to 9.
2. Fill one test tube with 24.0 mLof water and mark a line on the side to indicate this volume. Use this as a guide and mark all of the other tubes in a similar manner.
3. Pipet 3.0 mL of the appropriate liquid into each tube according to Table 34.1.

### C. Production of Blue Complexes

1. Pipet 1.0 mL of the copper reagent into each test tube and swirl to mix.
2. Transfer all 9 test tubes (simultaneously, if possible) to a beaker containing boiling water. To prevent evaporation, place a marble on top of each test tube. Let the test tubes stand in this bath for at least 10 minutes.

Table 34.1

| Tube No. | Material |
|---|---|
| 1 | Distilled water |
| 2 | 2.0 mg/100 mL glucose standard |
| 3 | 5.0 mg/100 mL glucose standard |
| 4 | 8.0 mg/100 mL glucose standard |
| 5 | 12.0 mg/100 mL glucose standard |
| 6 | 15.0 mg/100 mL glucose standard |
| 7 | 18.0 mg/100 mL glucose standard |
| 8 | Blood filtrate |
| 9 | Blood filtrate |

3. Remove the marbles and place the test tubes in an ice-water bath for 5 minutes. Remove the test tubes from the ice-water bath and place them in the test tube rack.

4. Pipet 1.0 mL of the arsenomolybdate reagent into each test tube and gently mix. Let these solutions stand at room temperature for 5 minutes.

5. Dilute each solution with distilled water to the 24 mL mark. Using a rubber stopper to close the test tube, thoroughly mix the contents by repeated inversions.

### D. Spectrophotometric Analysis

1. Set the wavelength of your instrument to 495 nm. Since there are different types of spectrophotometers, your instructor will demonstrate the necessary procedures for adjusting your instrument.

2. Pour approximately 2 mL of Solution #1 into a cuvette. This solution is used to zero the instrument and is called the "blank." No glucose is contained in this sample; therefore, set the spectrophotometer at 0.0 Å (no absorbance).

3. Remove the cuvette and thoroughly rinse it with the next solution to be tested (Tube No. 2). Measure the absorbance of this solution and record your data. Repeat this procedure with Tubes 3-9. Note: rinse the cuvette twice with the solutions in Tubes 8 and 9.

NAME ______________________

SECTION __________ DATE __________

## *REPORT FOR EXPERIMENT 34*

INSTRUCTOR ______________________

# Glucose Levels in Whole Blood

**Data Table**

| Sample No. | Absorbance | Concentration (mg glucose/100 mL) |
|---|---|---|
| 1 | | 0.0 |
| 2 | | 2.0 |
| 3 | | 5.0 |
| 4 | | 8.0 |
| 5 | | 12.0 |
| 6 | | 15.0 |
| 7 | | 18.0 |
| 8 | | |
| 9 | | |

Plot the data obtained from Samples 2 through 7 on the graph paper provided. Draw the best straight line through the data points. Read the concentration values for Samples 8 and 9, and record them in the data table.

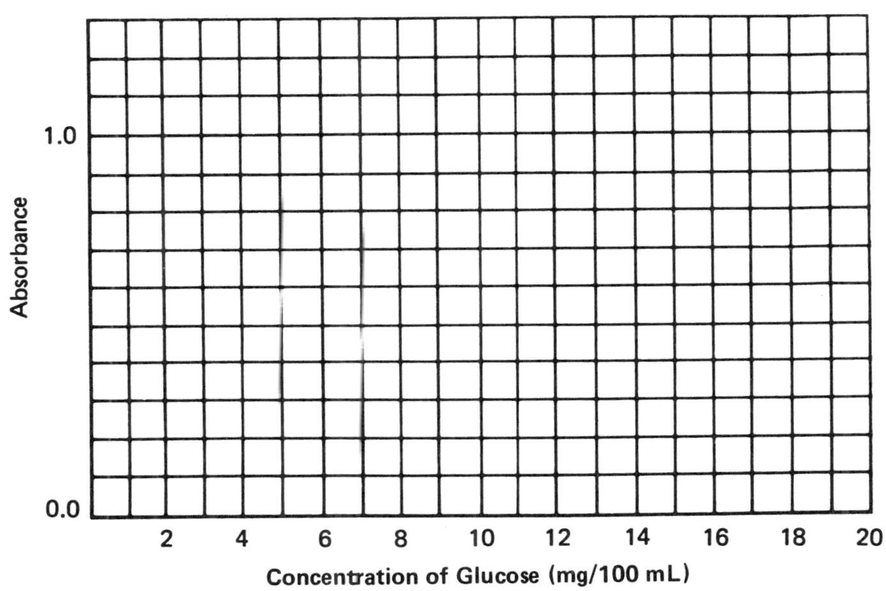

## QUESTIONS AND PROBLEMS

1. Average the concentration values for Samples 8 and 9. Calculate the concentration of glucose in the whole blood. (In your calculation consider that the whole blood has been diluted.)

2. The concentrations you have been working with have units of mg/100 mL. Calculate the molar concentration of glucose in a blood sample containing 95 mg/100 mL.

3. Did your blood glucose samples have exactly the same absorbance? List some reasons why they could vary in value.

# EXPERIMENT 35

# Amino Acids and Proteins

## MATERIALS AND EQUIPMENT

Solids: Tyrosine [p-$HOC_6H_4$–$CH_2CH(NH_2)COOH$], urea [$CO(NH_2)_2$]. Liquids: Glacial acetic acid ($HC_2H_3O_2$), nonfat (skim) milk. Solutions: 2% albumin, 1% copper(II) sulfate ($CuSO_4$), 2% gelatin, 1% glycine ($H_2NCH_2COOH$), dilute (6 M) hydrochloric acid (HCl), 0.1 M lead(II) acetate [$Pb(C_2H_3O_2)_2$], Millon's reagent (must be freshly prepared), 0.3% ninhydrin in acetone, concentrated nitric acid ($HNO_3$), 1% phenol ($C_5H_5OH$), 10% sodium hydroxide (NaOH).

## DISCUSSION

Proteins are present in each cell in every living thing on earth. They function as structural materials and as enzymes (catalysts) which regulate the multitude of chemical reactions that are necessary for life. Chemically proteins are polymers of α-amino acids (see your text for a list of the common α-amino acids and their structures).

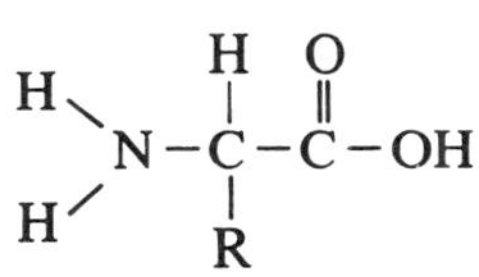

An α-amino acid

Individual amino acids are joined together by bonds called **peptide linkages**. A peptide linkage is an amide structure formed by splitting out a molecule of water between the carboxyl group of one amino acid and the amino group of another. In a typical protein, hundreds—sometimes thousands—of amino acids are linked together to form a polypeptide chain (Figure 35.1).

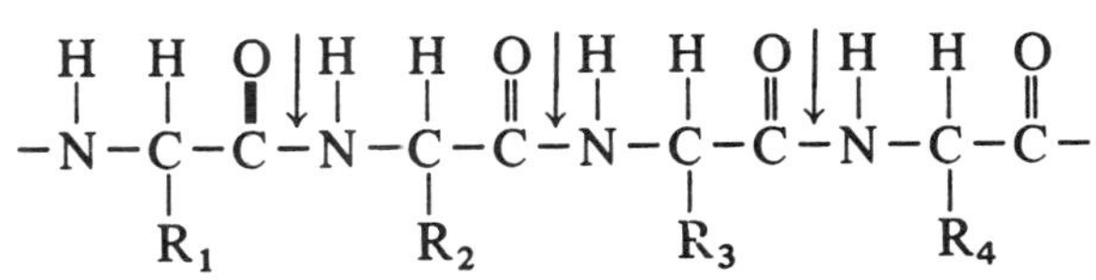

**Figure 35.1. Segment of a polypeptide chain formed from α-amino acids. The R's represent the remainders of the amino acids and the arrows point out the peptide linkages in the polypeptide chain.**

The sequence of amino acids along the polypeptide chain is the primary structure of a protein. The polypeptide chains of proteins also have a secondary structure or configuration: the polypeptide chains are coiled into alpha helices or arranged side by side as pleated sheets.

Proteins are one of the three major classes of foods (proteins, carbohydrates, and fats). Proteins differ from carbohydrates and fats not only in their functions in the living organism, but also in elemental composition. In addition to carbon, hydrogen, and oxygen, which are present in carbohydrates and fats, proteins contain nitrogen. Most proteins also contain sulfur; sulfur is present in those that contain the amino acids cystine, cysteine, and methionine. Additional elements such as phosphorus, iron, copper, zinc, and iodine, also occur in certain complex proteins. These elements are not part of the primary protein structure but are generally constituents of nonprotein substances combined with proteins.

**Color Reactions of Proteins:** Characteristic colors are produced when certain reagents react with one or more of the constituent groups in a protein molecule. The color produced with a given reagent will vary in intensity with different proteins. This is because all proteins either do not contain the same amino acids or do not contain the same amounts of a color-producing group. In this experiment, color-producing reactions will be used to obtain qualitative information about the composition of selected proteins. Various substances other than proteins or amino acids also give colors with some of the reagents. The colors produced with these substances will be compared to those produced with proteins and amino acids.

### A. Biuret Test

This test derives its name from biuret, a compound formed by heating urea to 180°C.

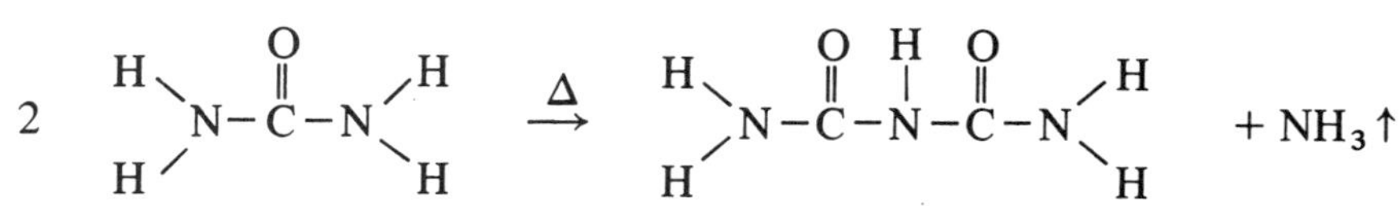

Urea Biuret

The biuret test involves heating a strongly alkaline solution of the test material with a little copper(II) sulfate and is very sensitive for proteins and polypeptides containing at least three peptide units. Biuret produces a violet color, but dipeptides do not give a positive biuret test. Colors ranging from pink to blue are produced both by polypeptides and proteins containing a structure with at least two peptide linkages (indicated by the arrows) thus

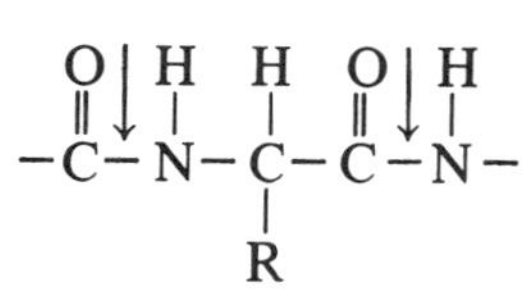

### B. Millon's Test

This test is a very sensitive method for detecting the hydroxyphenyl group, $-C_6H_4OH$, in proteins or in any phenolic compound that is unsubstituted in the 3, 5-position.

### C. Ninhydrin Test

Alpha-amino acids react with ninhydrin (triketohydrindene hydrate) to form a blue colored complex. This reaction is the basis for identification of amino acids by chromatography and also for the quantitative colorimetric determination of amino acids present in solution. The amino acid or protein must have a free amino ($-NH_2$) group and a free carboxylic acid ($-COOH$) group which, during the reaction, liberate ammonia and carbon dioxide. The ammonia reacts with ninhydrin to give the blue colored complex.

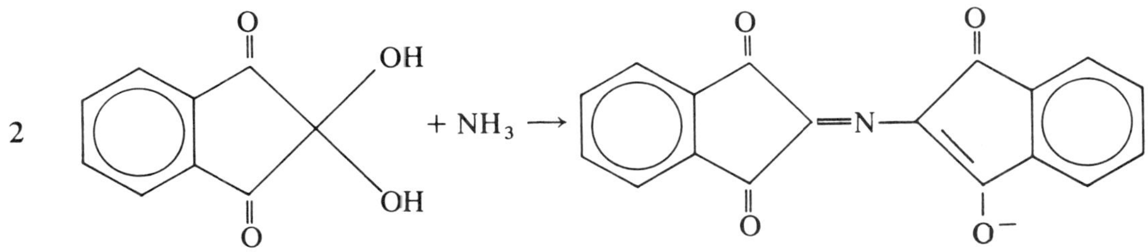

Ninhydrin                Colored Complex

## D. Xanthoproteic Test

The xanthoproteic test (pronounced zan-tho-pro-teyic) is shown by most proteins and is due to the presence of the phenyl group, $-C_6H_5$, which reacts with nitric acid to form colored nitro compounds. (Phenylalanine generally does not give a positive test.)

## E. Test for Cystine and Cysteine Sulfur in Proteins

When proteins containing cysteine are heated with a NaOH solution, the sulfur is converted into sulfide ions ($S^{2-}$). (Methionine usually does not react.) The presence of sulfide ion in solution is detected by reacting with lead(II) acetate to form a black precipitate of lead(II) sulfide.

## F. Separation of Casein from Milk

Nutritionally milk is an almost complete food containing proteins, fats, carbohydrates, many minerals, and a number of important vitamins. The proteins include casein, lactalbumin, and lactoglobulin. Casein makes up about 80% of the protein in cow's milk and about 40% of that in human milk. The casein in milk is a phosphoprotein containing about 0.7% phosphorus. It is probably present as calcium caseinate in cow's milk and as potassium caseinate in human milk.

Casein is released from its salts and precipitated from nonfat milk by treating with dilute acetic or hydrochloric acids. However, care must be taken not to use too much acid because free casein acts as a base and redissolves in excess acid. In addition to being a food, casein is used industrially in the manufacture of adhesives, paints, paper coatings, and in printing textiles and wallpaper.

# PROCEDURE

**Wear protective glasses.**

## A. Biuret Test

1. Place 0.3 to 0.5 gram of urea in a dry test tube and carefully heat over a Bunsen flame. After the urea has melted, continue heating for about 30 seconds. Stop heating, and immediately note the odor (CAUTION) of the gas being emitted. Allow the tube to cool, add about 5 mL of water, and mix to dissolve some of the residue. Set aside for the Biuret test procedure.

   **Biuret Test Procedure:** To a clean test tube add 4 mL of the solution to be tested, 1.0 mL 10% NaOH solution, and 4 drops 1% $CuSO_4$ solution. Mix and note the color that develops.

2. Do biuret tests on these solutions. Record your observations in the data table.

   (a) Solution derived from heating urea in 1.

   (b) 2% albumin

   (c) 1% glycine

   (d) Water (blank)

**B. Millon's Test**

**Millon's Test Procedure:** To a clean test tube add 5 mLof the solution to be tested and 5 drops of Millon's reagent and mix. Heat the tube for 2 to 3 minutes in a boiling water bath. Note the color that develops.

Do Millon's test on these solutions. Record your observations.

(a) 2% albumin

(b) 1% phenol

(c) 1% glycine

(d) 5 mL water to which a very small amount of tyrosine has been added (from the tip of a spatula)

(e) 2% gelatin

(f) Water (blank)

**C. Ninhydrin Test**

**Ninhydrin Test Procedure:** To a clean test tube add 5 mL of the solution to be tested and 10 drops of ninhydrin solution. Mix and heat the mixture in a boiling water bath for about 2 minutes. Note the color that develops.

Do the ninhydrin test on these solutions. Record your observations.

(a) 2% albumin

(b) 1% glycine

(c) 1% phenol

(d) Water (blank)

**D. Xanthoproteic Test**

**Xanthoproteic Test Procedure:** To a clean test tube add 3 mL of the solution to be tested and 10 drops of concentrated nitric acid. Mix and note any changes. Heat the tube 3 to 4 minutes in beaker of boiling water and note any color changes. Cool and add 10% NaOH solution (3-4 mL) until the solution is alkaline (test with red litmus). Note the color change.

Do the xanthoproteic test on the following solutions. Record your observations.

(a) 2% albumin

(b) 1% glycine

(c) 1% phenol

(d) 2% gelatin

(e) Water (blank)

## E. Test for Cysteine Sulfur in Proteins

**Test Procedure:** To a clean test tube add 3 mL of the solution to be tested and 2 mL of 10% NaOH solution. Heat the tube in a boiling water bath for 1 to 2 minutes. Remove the tube from the water bath and add 4 drops of lead(II) acetate solution and note the results. Add dilute (6M) HCl (2 to 3 mL) with slight warming until the dark color disappears. Carefully smell and note the odor of the solution.

Do the sulfur test on these solutions.

(a) 2% albumin

(b) 2% gelatin

(c) Water (blank)

## F. Separation and Testing of Casein

1. Separation of Casein: Mix 50 mL of nonfat milk with 50 mL of water in a 250 mL beaker. Add dilute acetic acid (1 vol. glacial acetic acid + 9 vol. $H_2O$) dropwise from a pipet or a medicine dropper, with vigorous stirring, until a flocculent precipitate forms. From 2 to 3 mL of the dilute acid is usually required. Avoid excess acid. Allow the precipitated casein to settle for a few minutes. Decant the supernatant liquid through filter paper in a funnel. Then pour all of the precipitate into the funnel. Finally remove excess moisture from the casein by pressing the precipitate between absorbent paper.

2. Tests On Casein: For each of the following tests, mix a quantity of casein about the size of a small pea with the appropriate amount of water (3-5 mL). Perform the test and record whether the result is positive or negative.

(a) Biuret

(b) Millon's

(c) Ninhydrin

(d) Xanthoproteic

NAME ____________________

SECTION __________ DATE __________

INSTRUCTOR ____________________

## *REPORT FOR EXPERIMENT 35*

# Amino Acids and Proteins

### A. Biuret Test

**Data Table**

| Solution | Color observations |
|---|---|
| (a) Heated urea | |
| (b) 2% albumin | |
| (c) 1% glycine | |
| (d) Water (blank) | |

1. What is the odor of the gas emitted by the heated urea? ____________________

2. How could you test for this gas other than by smelling it?

3. What is the evidence for a positive biuret test?

4. Write the formula of the organic substance responsible for the color reaction noted in tube (a).

5. Does albumin give a positive biuret test? ____________________

6. Does glycine give a positive biuret test? ____________________

7. Write the structure that must be present in a protein for a positive biuret test.

## B. Millon's Test

**Data Table**

| Solution | Observations |
|---|---|
| (a) albumin | |
| (b) 1% phenol | |
| (c) 1% glycine | |
| (d) Water + tyrosine | |
| (e) 2% gelatin | |
| (f) Water (blank) | |

1. What is the evidence for a positive Millon's test?

____________

2. Write the structural formula(s) of any amino acid(s) that will give a positive Millon's test.

3. Which of the proteins tested contains this/these amino acids?

____________

4. Write the structural formula for a tripeptide that will give a positive Millon's test.

## C. Ninhydrin Test

**Data Table**

| Solution | Observations |
|---|---|
| (a) 2% albumin | |
| (b) 1% glycine | |
| (c) 1% phenol | |
| (d) Water (blank) | |

1. What is the evidence for a positive ninhydrin test?

____________________

2. What groups in the amino acid or protein is responsible for a positive ninhydrin test?

____________________

3. Write the name and structural formula of an amino acid (among those listed in your text) that will not give a positive ninhydrin test.

____________________

4. Is it likely that there are proteins that will not give positive ninhydrin tests? Explain your answer.

**D. Xanthoproteic Test**

**Data Table**

| Solution | | Observations | |
|---|---|---|---|
| | After adding $HNO_3$ | After heating | After adding NaOH |
| (a) 2% albumin | | | |
| (b) 1% glycine | | | |
| (c) 1% phenol | | | |
| (d) 2% gelatin | | | |
| (e) Water (blank) | | | |

1. Which amino acids may be present in a protein showing a positive xanthoproteic test?

2. Why is the skin stained yellow when it comes in contact with nitric acid?

## E. Test for Cysteine Sulfur in Proteins

**Data Table**

| Solution | Observations | |
|---|---|---|
| | After adding lead(II) acetate | After adding hydrochloric acid |
| (a) 2% albumin | | |
| (b) 2% gelatin | | |
| (c) Water (blank) | | |

1. What is the evidence that sulfur is present?

2. Write the formula of the amino acid that may cause a protein to show a positive sulfur test.

3. What is the dark-colored substance that is formed when lead(II) acetate is added to the test solution? Write the equation for its formation.

4. What is the compound that you can smell after HCl is added to the dark-colored test mixture? Write the equation for its formation.

## F. Separation of Casein from Milk

Use (+) and (−) signs to indicate whether each test on casein was positive or negative.

| Biuret | Millon's | Ninhydrin | Xanthoproteic |
|---|---|---|---|
| | | | |

# EXPERIMENT 36

# Paper Chromatography

## MATERIALS AND EQUIPMENT

Solutions: 0.2 M aspartic acid, 0.2 M alanine, 0.2 M leucine, 0.2 M lysine, a mixture of aspartic acid, alanine, leucine, and lysine all at 0.2 M; 0.3% ninhydrin in acetone; yellow, green, and blue food colors, isopropyl alcohol ($C_3H_7OH$)-water (2:1 by volume); and acetic acid ($CH_3COOH$)-1-butanol ($C_4H_9OH$)-water (1:3:1 by volume). A millimeter ruler, two 500 mL Erlenmeyer flasks, Whatman #1 filter paper (14 X 14 cm), micropipets, Pasteur pipets, aluminum foil (7 X 7 cm), hair dryer, and a spray applicator for the ninhydrin.

## DISCUSSION

Suppose that you have a solution containing two different types of molecules and you need to separate these two components from one another. How would you go about this task? Chemists face this problem often–if not daily. One possible separation technique is known as chromatography. In this experiment you will use one form of chromatography called paper chromatography. Mixtures can be separated because individual compounds (or components) are carried along the surface of paper at different rates by a solvent. This difference in mobility is due to the existence of two phases, stationary and mobile. Water absorbed on the paper fibers is the stationary phase and the free solvent is the mobile phase. The components which are the most water soluble are attracted to the stationary water phase and, therefore, will move along the paper surface more slowly, thus effecting a separation of the components in the mixture.

The ratio of the distance traveled by the component to that traveled by the solvent is the $R_f$ value.

$$R_f = \frac{\text{The distance traveled by the component}}{\text{The distance traveled by the solvent front}}$$

This $R_f$ value is a function of the characteristics of both the solvent and the component. The closer the $R_f$ value is to 1.0, the more the component stayed in the mobile phase. Unknown components can be identified by comparing their $R_f$ values to those of standard known compounds. The components must be compared with the standards under identical conditions.

You will investigate the separation of the components in (a) food colors and in (b) amino acid mixtures. To obtain a desired shade, most food colors contain more than one dye. In the first part of this experiment, three food colors are separated into their components. The second part of this experiment points out how chromatography can be used to separate and identify amino acids. By comparing the $R_f$ value of an unknown amino acid with the values for those of known amino acids, you can determine the identity of the unknown.

Literally chromatography means "writing with colors." However, unlike the food colors, amino acids have no color. Therefore, to identify the positions of the amino acids, the dried paper is sprayed with ninhydrin, which reacts with amines to yield a purple to reddish-brown spot at each amino acid location.

These are the structures for the amino acids that will be used in this experiment.

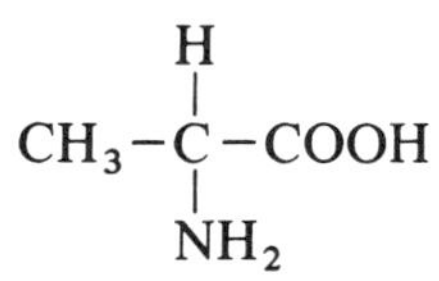

Alanine (ala)

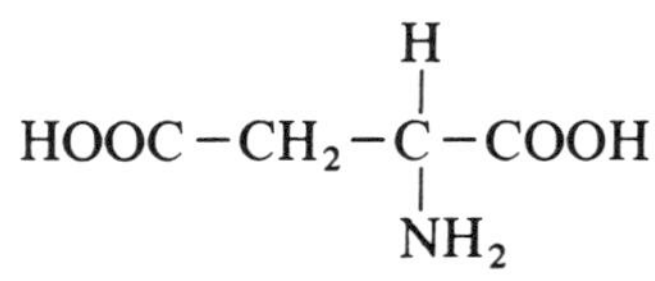

Aspartic acid (asp)

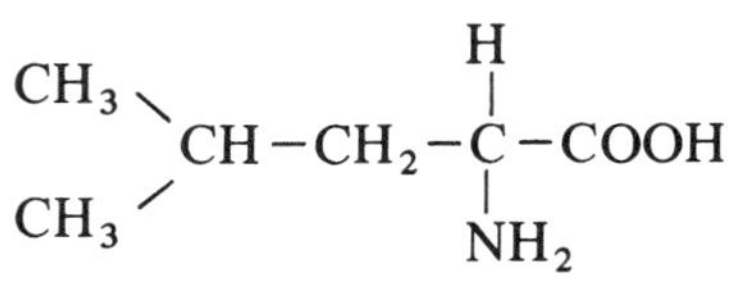

Leucine (leu)

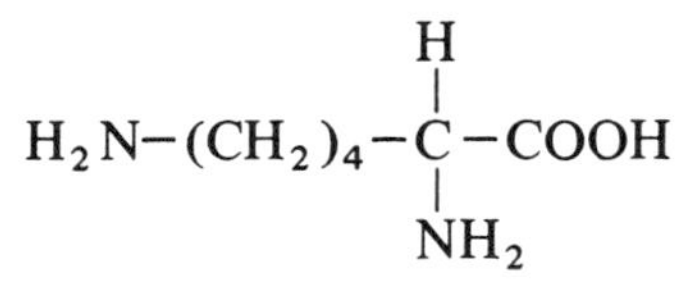

Lysine (lys)

## PROCEDURE

**Wear protective glasses.**

NOTE: After starting Part A, begin Part B as soon as possible. The two parts should overlap.

### A. Food Color Separation

1. Cut a piece of Whatman #1 filter paper into a 14 cm square. Using a pencil, not a pen, draw a thin line across the sheet 2 cm from one edge. Measure in 3 cm, 7 cm, and 11 cm along this line and make a small cross at each point. Label them 1, 2, and 3 as shown in Figure 36.1.

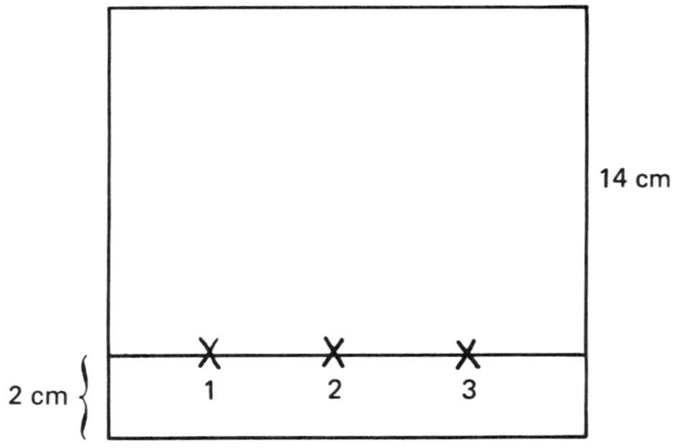


**Figure 36.1.**

2. At the first cross apply a small drop of yellow food color, at the second cross, a drop of green, and at the third cross a drop of blue. Each spot should be about 0.5 to 1 cm in diameter. Allow all the spots to dry for 5 minutes.

3. Pour 50 ml of isopropyl alcohol-water solvent into a 500 ml Erlenmeyer flask.

4. Roll the sheet of filter paper into a cylinder small enough to pass through the mouth of the flask. The spots should be on the outside surface at the end of the cylinder. The cylinder is then carefully inserted into the flask so that the spots are about 1 cm from the surface of the solvent. Just the bottom edge of the paper should touch the solvent. Tightly cover the mouth of the flask with a piece of aluminum foil. **Be careful not to splash the solvent on the paper.**

5. The solvent will begin to move up the paper. Allow the solvent front to travel up the paper about 7 cm above the pencil line.

NOTE: Start Part B while the solvent in Part A is moving up the paper.

6. Remove the paper and **immediately mark** the solvent front with a light pencil line. Place a cross at the leading edge of each colored spot or smear.

7. Measure the distance, to the nearest 0.1 cm, between the original line at the bottom of the paper and the solvent front. Also, measure the distances from the bottom line to each cross. These are the measurements used to calculate the $R_f$ values. Record your data.

**B. Amino Acid Separation and Unknown Identification**

In this part of the experiment, the $R_f$ values for four known amino acids are determined. By using these $R_f$ values, the amino acids in an unknown sample may be determined.

1. Cut a piece of Whatman #1 filter paper into a 14 cm square. Using a pencil, not a pen, mark a line across the sheet 2 cm up from one side. At 2 cm intervals draw crosses on this line. Label the crosses ala, asp, leu, lys, mixture, and unknown as shown in Figure 36.2. **Be sure you handle the paper only at the extreme edges.**

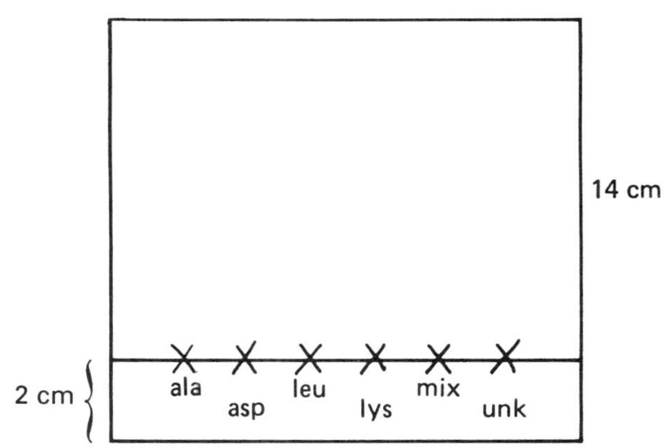


**Figure 36.2.**

2. Fill a micropipet with one of the samples (the four amino acids, the known mixture of the four amino acids, or the unknown) and touch it to the appropriate cross. Use a different micropipet for each sample. Repeat the application process to each of the six spots twice more to increase the amount of sample on each spot.

3. Pour 50 ml of the acetic acid-butanol-water solvent into a 500 ml Erlenmeyer flask.

4. Roll the sheet of filter paper into a cylinder small enough to pass through the mouth of the flask. The spots should be on the outside surface at the end of the cylinder. The cylin-

der is then carefully inserted into the flask so that the spots are about 1 cm from the surface of the solvent. Just the bottom edge of the paper should touch the solvent. Tightly cover the mouth of the flask with a piece of aluminum foil. **Be careful not to splash the solvent on the paper.**

5. The solvent will begin to move up the paper. Allow the solvent front to travel up the paper about 7 cm above the pencil line.

6. Remove the paper and **immediately mark** the solvent front with a light pencil line. Allow the paper to dry thoroughly. If available, use a hair dryer to speed the drying process.

7. Do this in a hood.

   Spray the paper with a fine mist of ninhydrin solution. Colored spots should appear within 15 minutes. The application of a small amount of heat will make these spots appear more quickly. Now mark each spot at its leading edge. Measure the distance, to the nearest 0.1 cm, between the original line at the bottom of the paper and the solvent front. Also, measure the distances from the bottom line to each of the spots. These are the measurements used to calculate the $R_f$ values. Record your data.

NAME ______________________

SECTION __________ DATE __________

## *REPORT FOR EXPERIMENT 36*

INSTRUCTOR ______________________

# Paper Chromatography

Attach both chromatograms here.

 NAME ____________

**A. Data from your food color chromatogram.**

Solvent front distance ____________

Yellow spot(s) distance(s) ____________

Yellow $R_f$ value(s) ____________

Green spot(s) distance(s) ____________

Green $R_f$ value(s) ____________

Blue spot(s) distance(s) ____________

Blue $R_f$ value(s) ____________

**B. Data from your amino acid chromatogram.**

Solvent front distance ____________

Alanine distance ____________

Alanine $R_f$ ____________

Aspartic acid distance ____________

Aspartic acid $R_f$ ____________

Leucine distance ____________

Leucine $R_f$ ____________

Lysine distance ____________

Lysine $R_f$ ____________

Unknown distance(s) ____________

Unknown $R_f$ value(s) ____________

Unknown number is ____________

Which amino acid(s) is in your unknown? ____________

____________

____________

____________

## QUESTIONS AND PROBLEMS

1. Why did you use pencil to mark the paper and not a pen?

2. In each part, A and B, what was the mobile phase and what was the stationary phase?

3. If two samples have identical $R_f$ values does this mean that they are necessarily identical molecules? Explain.

4. For the green food color, how would it be possible to show that the component spots you separated do produce a green food color?

5. Why did you chromatograph a mixture containing all four of the amino acids?

6. Why is it very important in the amino acid chromatogram not to touch the paper with your fingers or hands? Does this matter in the food color experiment?

# EXPERIMENT 37

# Ion-Exchange Chromatography of Amino Acids

## MATERIALS AND EQUIPMENT

Solutions: Millon's reagent (freshly prepared), 0.2% ninhydrin in 1-butanol, 0.2 M phosphate buffer, 0.1% tyrosine-0.1% arginine mixture, Dowex-50 slurry. A screw clamp, rubber tubing (about 3 cm long), glass wool (Pyrex), 30 test tubes, marker capable of writing on glass, a wire (about 20 cm long), a 600 mL beaker, and either a chromatography column (polypropylene column with funnel from Kontes) or a 10 mL graduated pipet.

### Discussion

In many areas of chemistry the separation of components in a mixture is a major challenge. One very powerful separation technique is chromatography. Many types of chromatographic techniques are used in the modern laboratory. This experiment involves ion-exchange chromatography.

Ions present in a mixture can be separated by using ion-exchange chromatography. Separation occurs because the ions that are being selectively removed bind electrostatically to ion-attracting groups having an opposite charge. The ion-attracting groups are part of an insoluble resin. Positively charged ions will bind to this resin, hence the name cation-exchange resin. Anion-exchange resins have positively charged groups capable of binding to negative ions (anions).

Amino acids exist in ionized forms (see the Discussion in Experiment 38). Amino acids which have more than one positive group will bind to the cation-exchange resin more tightly than one with only one positive charge. In this experiment you will use the amino acids tyrosine and arginine.

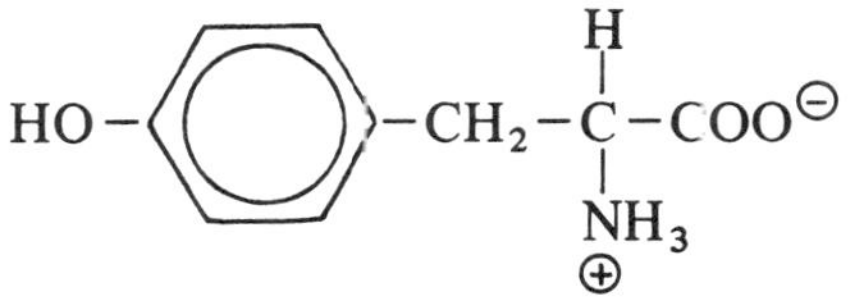

Tyrosine at pH 6

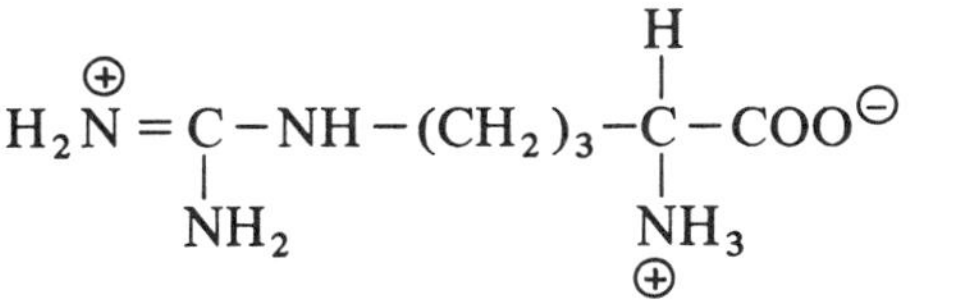

Arginine at pH 6

At a pH of 6 arginine has one more positive charge than tyrosine and will bind more tightly (with greater affinity) to Dowex-50 than tyrosine. This equilibrium is involved:

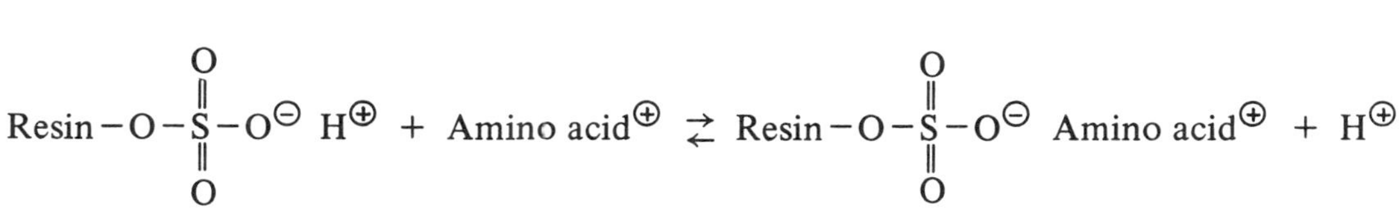

In this experiment, a column filled with Dowex-50 is used. The Dowex-50 is suspended in phosphate buffer at pH 6. The buffer keeps the pH constant and thus maintains the charged states of the amino acids. A mixture of 0.1% tyrosine and 0.1% arginine in phosphate buffer at pH 6 is added to the column. Once the tyrosine-arginine sample is on the resin, additional buffer is allowed to flow through the resin, and fifteen 5 mL fractions are collected. Separation is achieved since both amino acids are continuously leaving and returning to the resin (the equilibrium process). Since the arginine has a more positive charge, it has a higher affinity for the Dowex-50 and stays on the column longer than does the tyrosine. The early fractions will contain the tyrosine, and the arginine will be found in the later fractons.

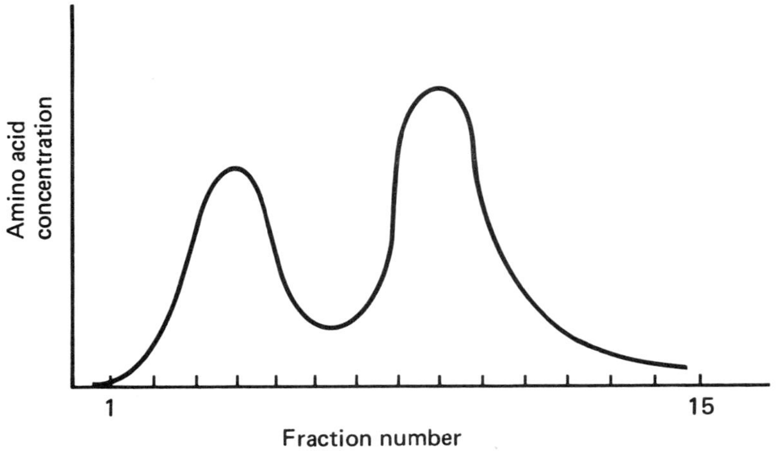


**Figure 37.1. Elution profile for tyrosine and arginine.**

A typical "elution" profile is shown in Figure 37.1. In this particular example, Fraction 4 would contain the highest concentration of one amino acid and Fraction 9 the highest concentration of the other amino acid.

Since amino acids are not colored there is no visible evidence of separation when the fractions are collected. We can detect the amino acids by using ninhydrin, which reacts to give a blue colored complex. Tyrosine can be specifically identified by using Millon's reagent. A positive test is indicated by a red-colored complex. With each of these reagents, the intensity of color reflects the relative amount of amino acid present in any fraction.

## PROCEDURE

**Wear protective glasses.**

1. Label 15 test tubes with numbers from 1 to 15. Fill one test tube with 5 mL of water and mark a line on the side to indicate this volume. Use this as a guide and mark all of the others in a similar manner. This will allow you to know immediately when you have collected a 5 mL fraction from the column.

2. Mount a chromatography column (or 10 mL graduated pipet) on a ringstand with the tip down (see Figure 37.2). Force a piece of rubber tubing, about 3 cm long, onto the tip and close the tubing with a screw clamp. Place a small amount of Pyrex glass wool inside the column. Using a wire, push the wool down the column to form a loose mat that covers the hole. When the clamp is open, this mat of glass wool prevents the Dowex-50 from running out the opening but allows the buffer to flow freely. If the glass wool is packed too tightly it will prevent or significantly impede the flow of liquid through the column.

3. Obtain 25 mL of Dowex-50 slurry in a 50 mL beaker. Be sure to stir the slurry while pouring it into the beaker. Pour the slurry into the column until the resin level is about 1 cm from

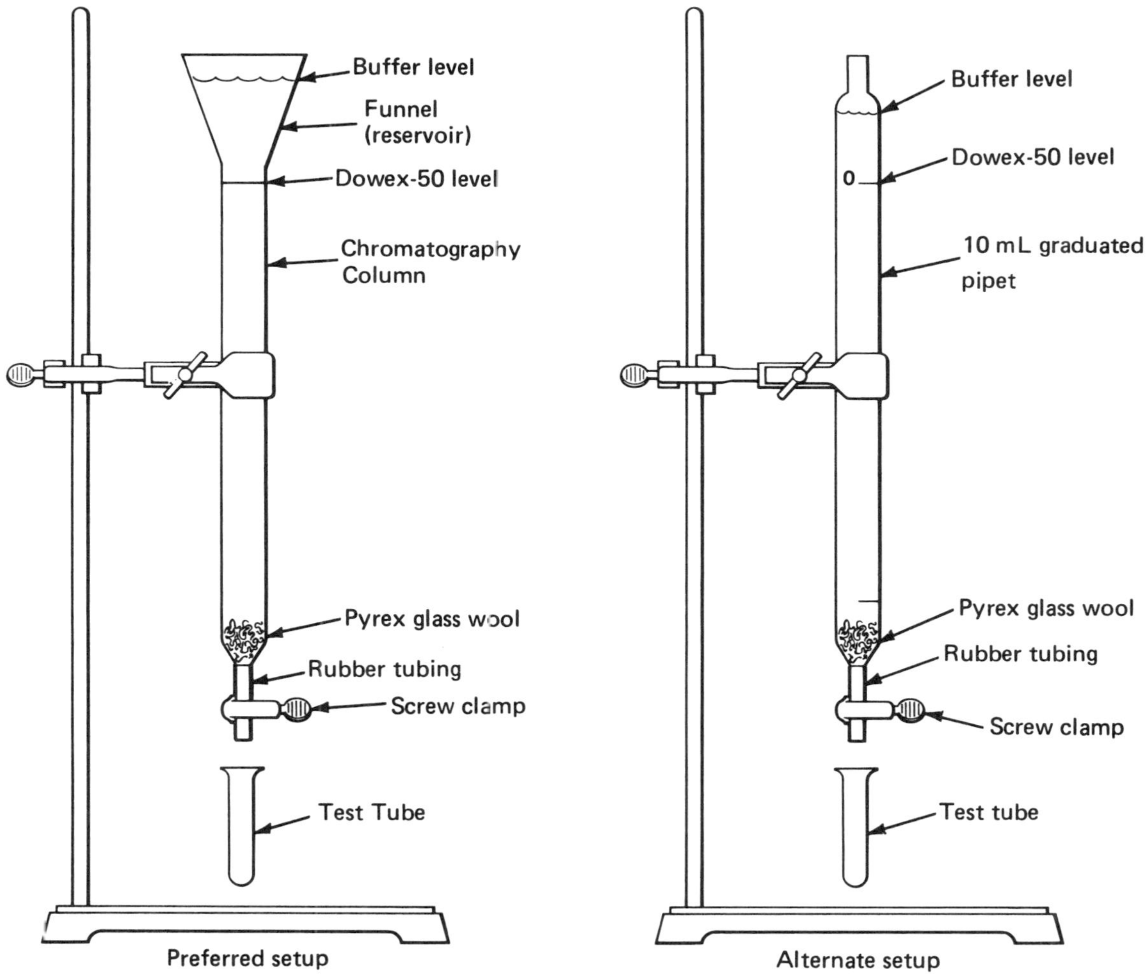


**Figure 37.2. Setup of chromatographic column**

the top of the column (not to the top of the funnel). You will need to drain the column slowly to pack the resin into place. **It is very important to never allow the buffer to drop below the resin level.** There must always be a layer of liquid above the resin.

4. Drain the buffer until only 2 mm of liquid remain above the resin, and close the clamp.

5. Using a thin dropper, carefully "load" the column by adding 1 mL of a solution containing 0.1% tyrosine and 0.1% arginine. Do not disturb the surface of the resin.

6. Open the clamp and slowly (about 1 mL/min) drip the solution into the first test tube. Stop just before the surface of the sample touches the resin. Carefully fill the funnel (reservoir) with buffer solution. Finish collecting the first 5 mL fraction. As soon as you collect the first fraction, switch to the second test tube, and so on, until you have collected all fifteen fractions. **Fill the reservoir with buffer solution after you collect each fraction.**

7. Number a second set of fifteen test tubes. Pour approximately half of each 5 mL fraction into the empty test tube with the corresponding number.

8. Test the contents of the first set of tubes for the presence of an amino acid by adding 10 drops of the ninhydrin solution to each test tube with mixing. Simultaneously place all of the test tubes in a 600 mL beaker containing 100 mL of boiling water, and heat the solutions for about three minutes. Remove the test tubes from the beaker, set aside, and record your observations.

9. Test the contents of the second set of tubes for the presence of tyrosine by adding 5 drops of Millon's reagent to each test tube with mixing. Simultaneously place all of the test tubes in a 600 mL beaker containing 100 mL of boiling water and heat the solutions for about three minutes. Remove the tubes from the beaker, and record your observations.

NAME ____________________

SECTION __________ DATE __________

*REPORT FOR EXPERIMENT 37* INSTRUCTOR ____________________

# Ion-Exchange Chromatography of Amino Acids

**Data Table:** In recording your observations estimate the color intensity of each test tube. Place a 0 in the result column if no color is detected. Use one + to indicate a light color; two +'s for medium color; and three +'s for the most intense color.

**Data Table**

| Tube No. | Ninhydrin Results | Millon's Reagent Results |
|---|---|---|
| 1 | | |
| 2 | | |
| 3 | | |
| 4 | | |
| 5 | | |
| 6 | | |
| 7 | | |
| 8 | | |
| 9 | | |
| 10 | | |
| 11 | | |
| 12 | | |
| 13 | | |
| 14 | | |
| 15 | | |

Plot all of the data on the graph provided. Draw a separate curve for each reagent.

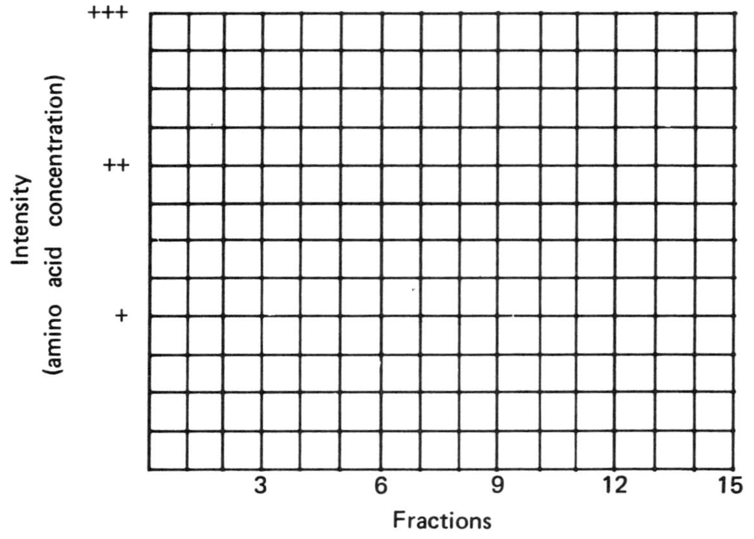


## QUESTIONS AND PROBLEMS

1. Could you use ion-exchange chromatography to separate aspartic acid from glutamic acid? Why?

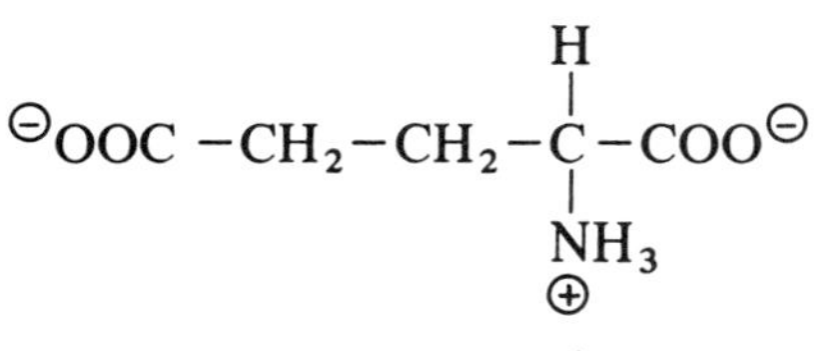

Glutamic acid at pH 6

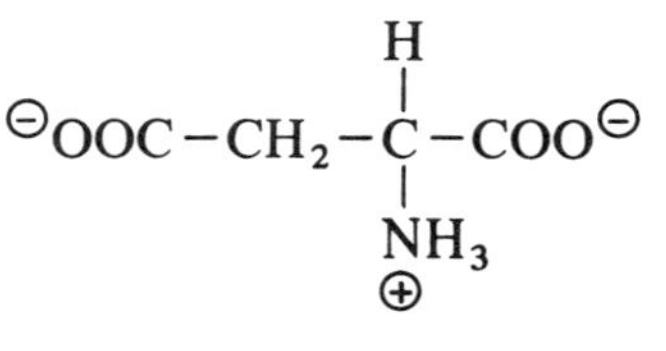

Aspartic acid at pH 6

2. How would you separate aspartic acid from tyrosine using ion-exchange chromatography at pH 6?

3. From the results of your experiment, are you guaranteed that arginine has been separated from the tyrosine? Explain.

4. Suppose you were careless for a moment and let the buffer drop 4 to 5 cm below the top of the resin during the collection of Fraction 2. What results do you imagine this would have on the separation process?

# EXPERIMENT 38

# Identification of an Unknown Amino Acid by Titration

## MATERIALS AND EQUIPMENT

Solutions: 0.10 M sodium hydroxide (NaOH), 0.10 M glycine•HCl, unknowns of 0.10 M glutamic acid•HCl, 0.10 M histidine•HCl, 0.10 M alanine•HCl, 0.10 M lysine•HCl, and/or 0.10 M cysteine•HCl. Standard pH 7.0 buffer. A pH meter, one buret, a magnetic stirrer with stirring bar, and a 10 mL pipet.

## DISCUSSION

Proteins are organic polymers formed from amino acid monomers. There are 20 common amino acids found in proteins. The identification of amino acids is an important part of biochemistry. Knowledge of the amino acid composition of proteins is of value in (1) planning adequate diets, (2) detecting certain diseases that are characterized by abnormal concentrations of amino acids, and (3) synthesizing proteins, such as human insulin. Many techniques have been developed to identify amino acids. The method used in this experiment is one of the first developed and is still an excellent technique.

The R group shown in Figure 38.1 differs for each of the 20 common amino acids.

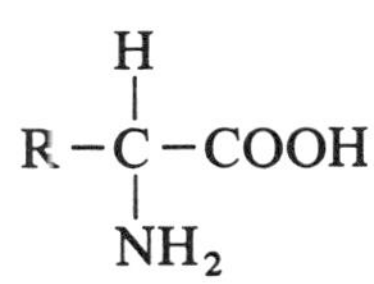

**Figure 38.1. General structure for an amino acid.**

Most organic molecules have relatively low melting points, but amino acids are exceptions. For example alanine has a melting point of 297°C and phenylalanine a melting point of 283°C. These unusually high values are attributable to the ionic nature of amino acids. Both the carboxyl group (−COOH) and the amino group ($-NH_2$) undergo ionization.

$$\underset{\text{Acid}}{-COOH} \rightleftarrows \underset{\text{Base}}{-COO^{\ominus}} + H^{\oplus} \qquad \text{(Equation 1)}$$

$$\underset{\text{Base}}{-NH_2} + H^{\oplus} \rightleftarrows \underset{\text{Acid}}{-NH_3^{\oplus}} \qquad \text{(Equation 2)}$$

Remember that an acid is a proton donor and a base is a proton acceptor. Equations 1 and 2 show that these functional groups in amino acids are either acids or bases, depending upon whether they donate or accept protons.

At the pH of blood, approximately 7.4, amino acids exist as di-ions or Zwitterions (Figure 38.2). These di-ion charges are the attractive forces that account for the high melting points of amino acids.

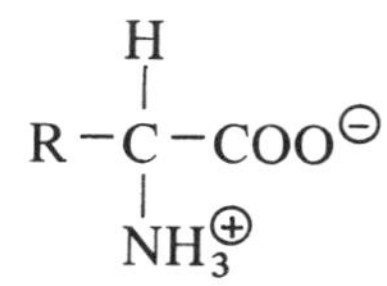

**Figure 38.2. An amino acid in Zwitterion form.**

Remember that pH is defined as the negative logarithm of the hydrogen ion concentration:

$$pH = -\log [H^{\oplus}] \qquad \text{(Equation 3)}$$

Every acid and every base has a characteristic ionization constant, $K_a$, that reflects the strength of that acid or base. The $pK_a$ is the negative logarithm of this ionization constant.

$$pK_a = -\log K_a \qquad \text{(Equation 4)}$$

Low $pK_a$ values (less than 4.5) indicate acids capable of giving up their protons easily, and high $pK_a$'s (greater than 8.0) indicate bases that accept protons readily.

In addition to the amino and carboxyl groups found in every amino acid, the R groups of some amino acids contain another ionizable group. For example the R group in lysine contains an amino group.

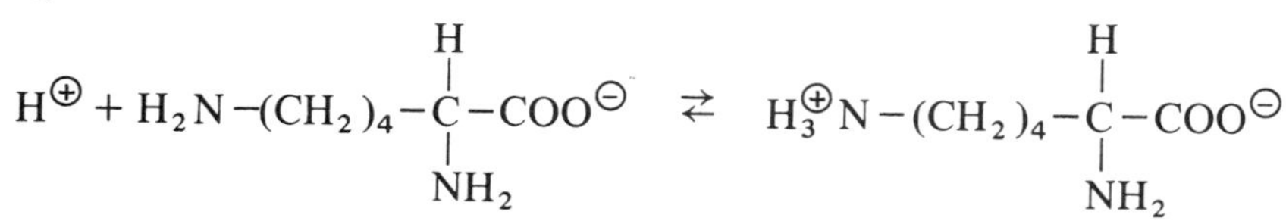

(Equation 5)

In this experiment, you will titrate the amino acid glycine and an unknown amino acid and plot their respective titration curves. The titration curve is a plot of the pH versus the volume of base added. An example of a typical titration curve is shown in Figure 38.3. This curve is for aspartic acid which has three ionizable groups. As the base (NaOH) concentration increases, each group will ionize at a specific and characteristic pH.

The mid-points of the flat portions of the curve labeled *1, 2,* and *3* in Figure 38.3, represent the $pK_a$'s for aspartic acid (i.e., 2.1, 3.9, and 9.8). Amino acids with only two ionizable groups (e.g., glycine) will have only two such points, one for the carboxyl group and one for the amino group.

The ionization equilibrium equation for the first ionizable group of aspartic acid is shown in Equation 6.

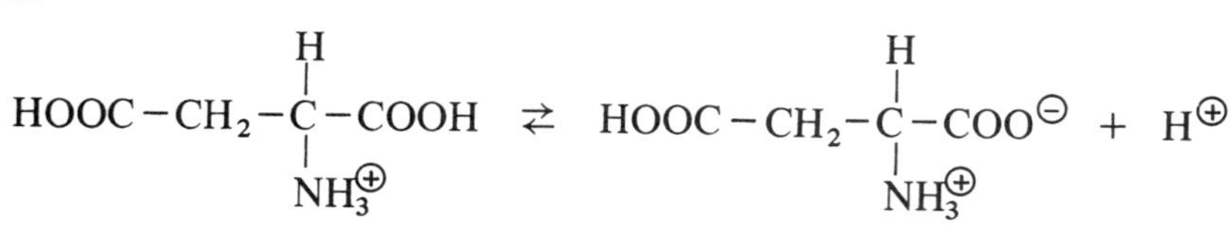

(Equation 6)

Equations similar to Equation 6 can be written for each of the other two ionizable groups of aspartic acid.

In this experiment, small volume increments of standard base are added from a buret to a glycine solution. The pH is determined during this titration by means of a pH meter. After the addi-

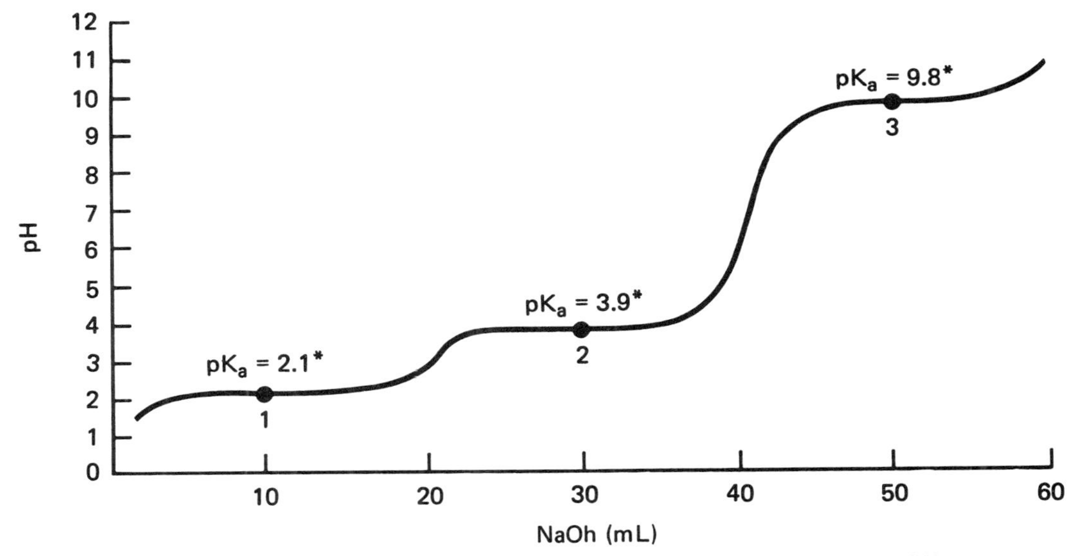


**Figure 38.3. Titration of 20 mL of 0.10 M aspartic acid with 0.10 M NaOH. (*Even though this represents pH at this point it is also equal to the pKa.)**

tion of each increment of base, the pH and total volume of base are recorded. This titration is repeated using an unknown amino acid. The data collected are then used to plot the titration curves for both amino acids. The $pK_a$'s are then determined from the titration curves. By comparing the $pK_a$'s of your unknown to the values in Table 38.1 you should be able to identify your unknown amino acid.

## PROCEDURE

**Wear protective glasses. Do not pipet by mouth.**

### A. Standardization of the pH Meter

Your laboratory instructor will demonstrate how to use and how to standardize a pH meter.

### B. Titration of Glycine

1. Pipet 10.0 mL of 0.10 M glycine into a 100 mL beaker, and add 30 mL of distilled water.
2. Continuous mixing is necessary. Place the beaker on a magnetic stirrer and put a stirring bar into the solution; lower the pH meter electrodes into the solution.

   CAUTION: Position the electrodes so that the bar will not strike them when it spins.

3. Start the magnetic stirrer and adjust it to a moderate speed.
4. Record the initial pH, and begin the titration by adding 1.0 mL of 0.10 M NaOH. Allow to mix for a few seconds, and record the pH and volume of added NaOH. Continue the addition of 1.0 mL increments of NaOH, recording the pH and total volume added after the addition of each increment. When the pH begins to change rapidly, use 0.5 mL increments of base during the rapid change. Continue the titration until a pH of 11.5 has been reached or 30 mL of base has been added.

### C. Titration of an Unknown Amino Acid

Pipet 10.0 mL of 0.10 M unknown into a second 100 mL beaker. Add 30 mL of water, and repeat the titration procedure.

**Table 38.1 Formulas, names and pK$_a$'s for selected amino acids.**

| Formula | Name | pK$_a$ −COOH | pK$_a$ $-NH_3^{\oplus}$ | pK$_a$ R-Group |
|---|---|---|---|---|
| $H-\underset{NH_2}{\overset{H}{C}}-COOH$ | Glycine | 2.3 | 9.6 | – |
| $CH_3-\underset{NH_2}{\overset{H}{C}}-COOH$ | Alanine | 2.3 | 9.7 | – |
| $HS-CH_2-\underset{NH_2}{\overset{H}{C}}-COOH$ | Cysteine | 1.7 | 10.8 | 8.3 |
| 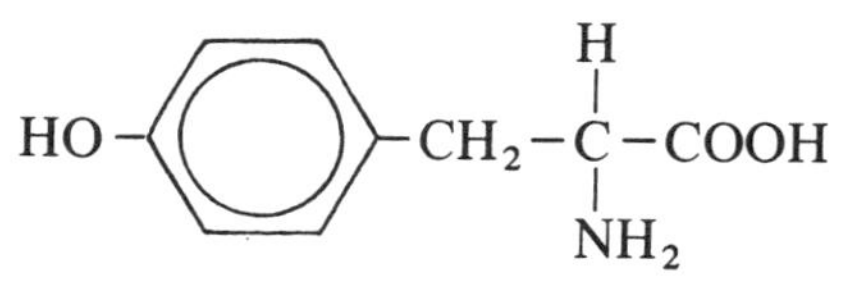 | Tyrosine | 2.2 | 9.1 | 10.1 |
| $HOOC-CH_2-\underset{NH_2}{\overset{H}{C}}-COOH$ | Aspartic acid | 2.1 | 9.8 | 3.9 |
| $HOOC-CH_2-CH_2-\underset{NH_2}{\overset{H}{C}}-COOH$ | Glutamic acid | 2.2 | 9.7 | 4.3 |
| 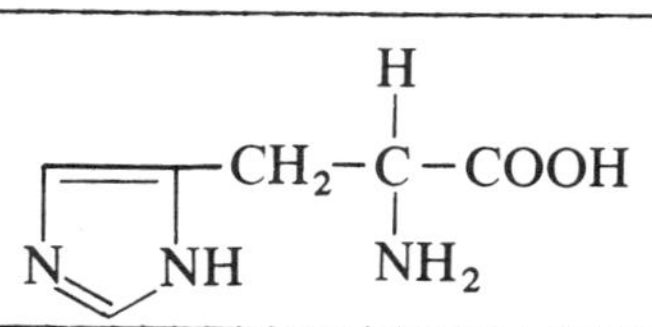 | Histidine | 1.8 | 9.2 | 6.0 |
| $H_2N-(CH_2)_4-\underset{NH_2}{\overset{H}{C}}-COOH$ | Lysine | 2.2 | 9.0 | 10.5 |

## D. Plotting and Reading the Data

1. Plot the data on the graphs provided.
2. Draw smooth curves through the plotted points to obtain the titration curves.
3. Read the pH corresponding to the middle of each horizontal plateau. These pH values are equal to the pK$_a$'s of the amino acids.

NAME ____________________

SECTION __________ DATE __________

INSTRUCTOR ____________________

## *REPORT FOR EXPERIMENT 38*

# Identification of an Unknown Amino Acid by Titration

### DATA TABLES

#### B. Titration of Glycine

| NaOH(mL) | pH | NaOH(mL) | pH |
|---|---|---|---|
| 0.00 | | | |
| | | | |
| | | | |
| | | | |
| | | | |
| | | | |
| | | | |
| | | | |
| | | | |
| | | | |
| | | | |
| | | | |
| | | | |
| | | | |
| | | | |
| | | | |
| | | | |
| | | | |
| | | | |
| | | | |
| | | | |
| | | | |
| | | | |

**Titration of Glycine** *table continued on page 300.*

#### C. Titration of Unknown

| NaOH(mL) | pH | NaOH(mL) | pH |
|---|---|---|---|
| 0.00 | | | |
| | | | |
| | | | |
| | | | |
| | | | |
| | | | |
| | | | |
| | | | |
| | | | |
| | | | |
| | | | |
| | | | |
| | | | |
| | | | |
| | | | |
| | | | |
| | | | |
| | | | |
| | | | |
| | | | |
| | | | |
| | | | |
| | | | |

**Titration of Unknown** *table continued on page 300.*

**B. Titration of Glycine** (continued)

| NaOH(mL) | pH | NaOH(mL) | pH |
|---|---|---|---|
| | | | |
| | | | |
| | | | |
| | | | |
| | | | |
| | | | |
| | | | |

**C. Titration of Unknown** (continued)

| NaOH(mL) | pH | NaOH(mL) | pH |
|---|---|---|---|
| | | | |
| | | | |
| | | | |
| | | | |
| | | | |
| | | | |
| | | | |

## TITRATION CURVES

### B. Titration Curve for Glycine

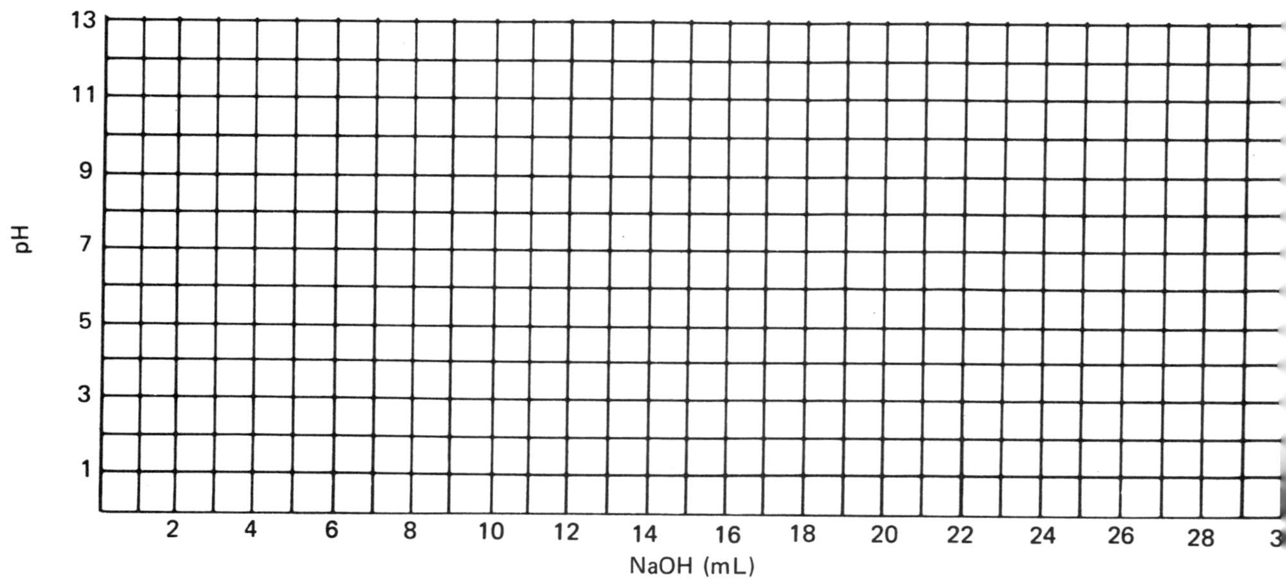

**C. Titration Curve for the Unknown**

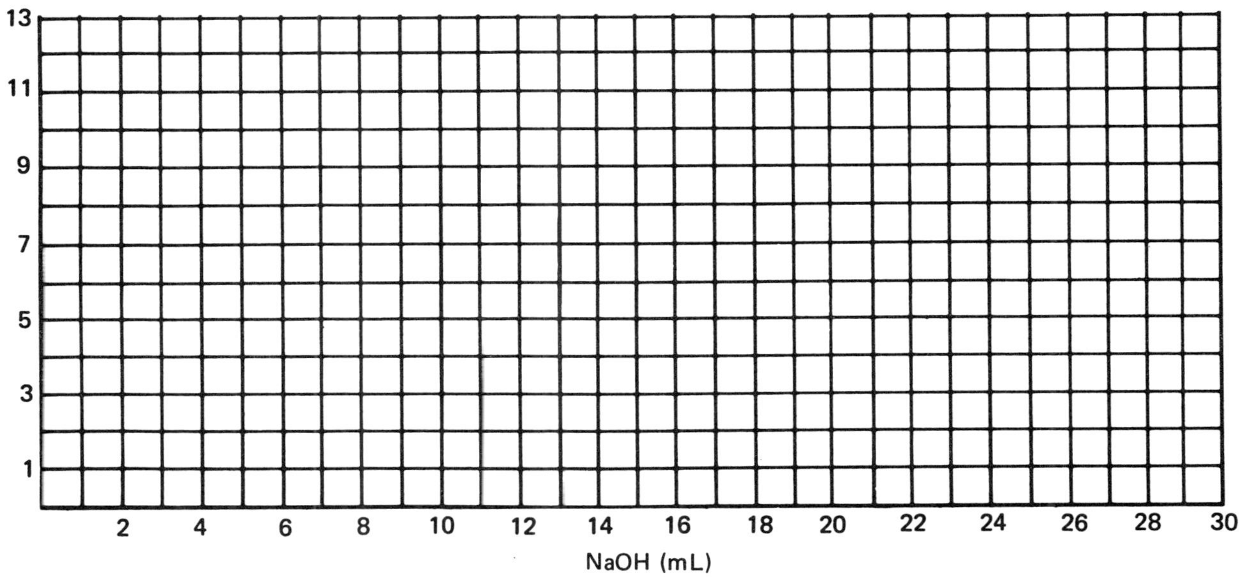


## QUESTIONS AND PROBLEMS

1. Based on your graph, what are the $pK_a$ values for glycine at the two plateaus? Explain why your values may differ from the reported ones.

2. (a) What are the $pK_a$'s for your unknown amino acid? Unknown No. ____________

   (b) What is your unknown? (Your choices are glutamic acid, histidine, alanine, lysine, or cysteine.)

3. On the basis of titration data, why might it be difficult to distinguish between lysine and tyrosine?

4. What is the pH of each of these solutions?

   (a) $10^{-3}$ M $HNO_3$

   (b) $6.21 \times 10^{-4}$ M HCl

5. Write the appropriate equilibria (see Equation 6) for each $pK_a$ of glycine and of your unknown.

 NAME ____________________

6. Suppose that you titrated a 0.10 M tyrosine solution with 0.10 M NaOH. Draw the titration curve that you would expect to obtain. (Assume that you have 10.0 mL of the tyrosine solution.)

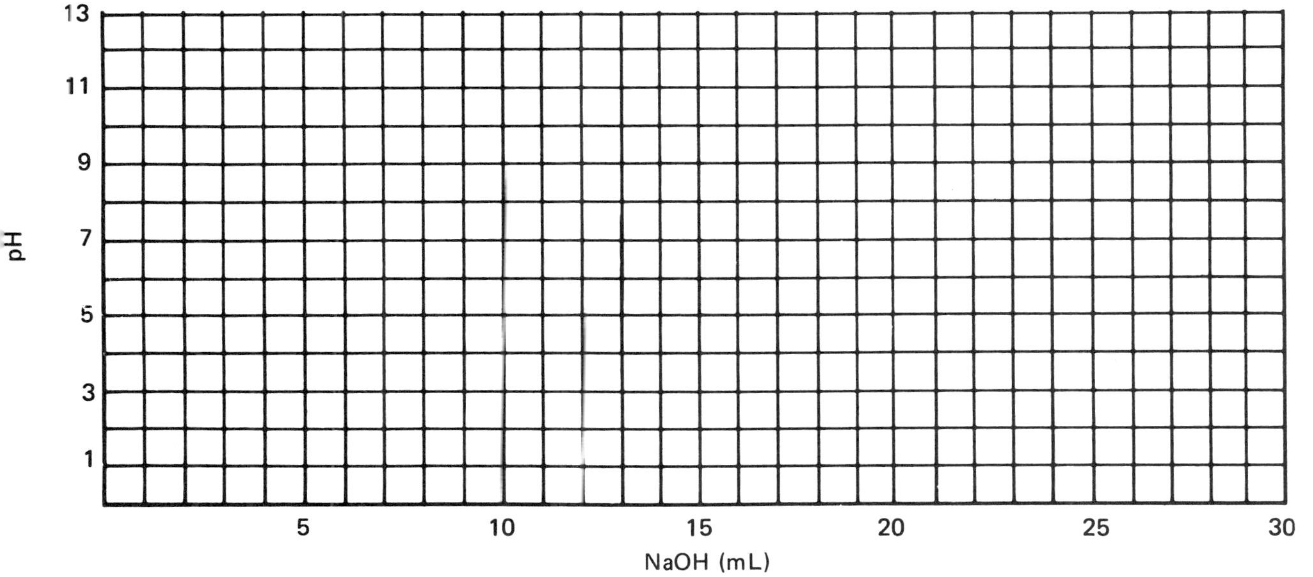

# EXPERIMENT 39

# Enzymatic Catalysis—Catalase

## MATERIALS AND EQUIPMENT

Solutions: 3% hydrogen peroxide ($H_2O_2$), 2 M copper(II) sulfate ($CuSO_4$), 10% sodium hydroxide (NaOH), 0.1 M hydrochloric acid (HCl) (gas collection apparatus, 2 mL and 3 mL volumetric pipets), a 10 mL graduated pipet, ice, and a potato.

## DISCUSSION

Almost all reactions in the body are controlled by enzymes. Many different enzymes are required to break down food (carbohydrates, fats, and proteins) into forms usable by the cells. Other enzymes are necessary for the cells to synthesize their own carbohydrates, fats, and proteins. Enzymes are special types of protein whose function is to control cellular reactions. These reactions occur very quickly. Speed is necessary to maintain constant cellular conditions. Enzymes are used over and over again by the cells to catalyze life-sustaining reactions.

Enzymes are very sensitive to their local surroundings. Changes in pH, temperature, ion concentration, and types of ions or molecules in solution drastically alter the enzyme's ability to catalyze a reaction. Enzymes are also very specific in catalyzing only one reaction out of the millions that are possible.

An enzyme is a protein and is built of amino acids. Each enzyme has a specific three-dimensional structure. This structure contains a pocket called the **active site** where the reaction occurs. The enzyme and substrate (reactant) combine to form the enzyme-substrate complex. The substrate, activated by the enzyme in the complex, reacts to form the products, regenerating the enzyme. (See Figure 39.1.)

Various forces (hydrogen-bonding, salt bridges, disulfide linkages, etc.) hold the enzyme in its specific shape. These forces can be overcome by denaturants, which inactivate the enzyme. One such denaturant is heat. A heated enzyme (boiled) loses its three-dimensional shape and becomes inactive or **denatured**.

Another method of inactivating an enzyme is to introduce a chemical **inhibitor** capable of specifically binding to the active site of the enzyme. These inhibitors prevent normal substrate-enzyme interactions. (See Figure 39.1.)

In this experiment we will investigate an enzyme called catalase. Catalase catalyzes the reaction:

$$2\,H_2O_2 \rightarrow 2\,H_2O + O_2\uparrow$$

Hydrogen peroxide, $H_2O_2$, is found in many cellular locations and is toxic above a certain concentration. Therefore, the removal of hydrogen peroxide is important to the cell's survival.

Catalase is found in high concentrations in red blood cells and in many plants. Our source of this enzyme will be the potato. We will study the effect of variations of pH and temperature on catalase activity. We will also see the effect of inhibition and denaturation in enzymatic catalysis.

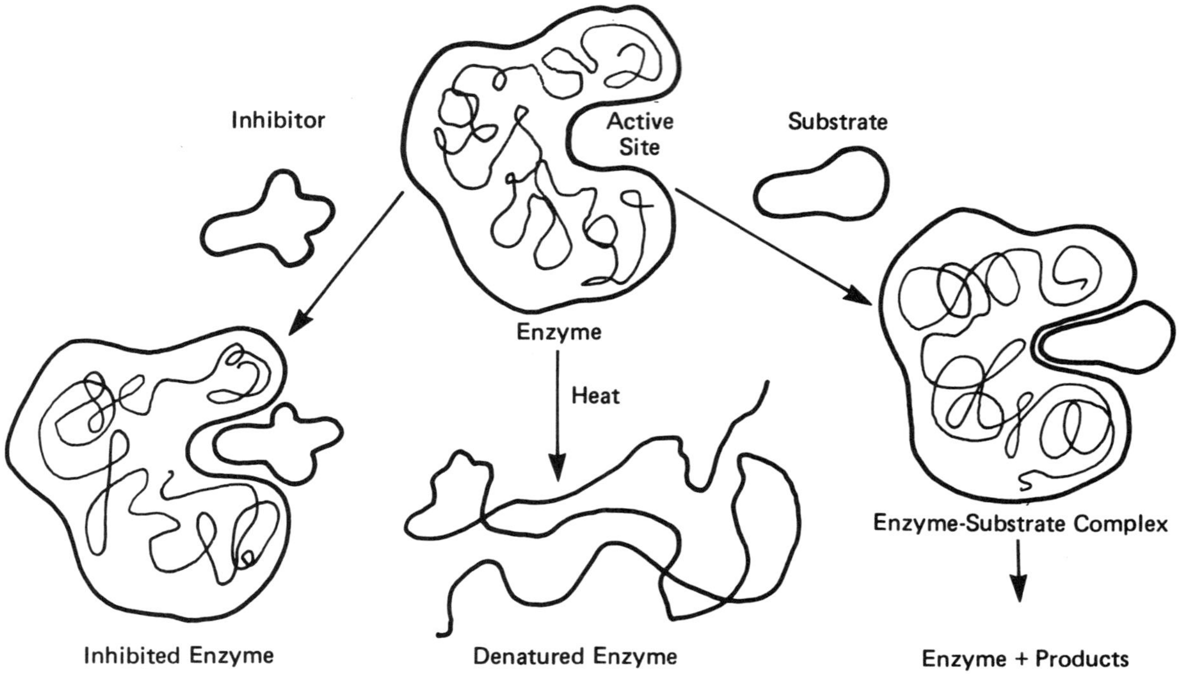


**Figure 39.1. Enzyme reactions.**

In order to determine the enzyme's activity (assay procedure), we must measure either the loss of a reactant or the increase of a product. In this experiment, we will measure the amount of oxygen produced during the reaction.

## PROCEDURE

**Wear protective glasses. Do not pipet by mouth.**

### A. Enzyme Preparation

Catalase can be extracted from a potato. Using **clean** equipment, cut approximately 30 grams of potato into small pieces and mash them to form a paste. Transfer the paste to an Erlenmeyer flask and add 20 mL of distilled water and swirl for 15 minutes. Filter this suspension to obtain a clear liquid containing the catalase. Filtration can be speeded up with a suction filter (see Experiment 31).

### B. Assays

The volume of oxygen produced from hydrogen peroxide can be determined by measuring the amount of water displaced (see Figure 39.2). The oxygen produced from the hydrogen peroxide decomposition is bubbled into the cylinder and displaces water. The volume of gas collected is measured by the graduations on the cylinder.

### 1. Temperature Effect

(a) Place a test tube in an ice-water bath and pipet 3.0 mL of 3% hydrogen peroxide into the test tube. Let this stay in the ice-water for at least 5 minutes. In the meantime, set up the

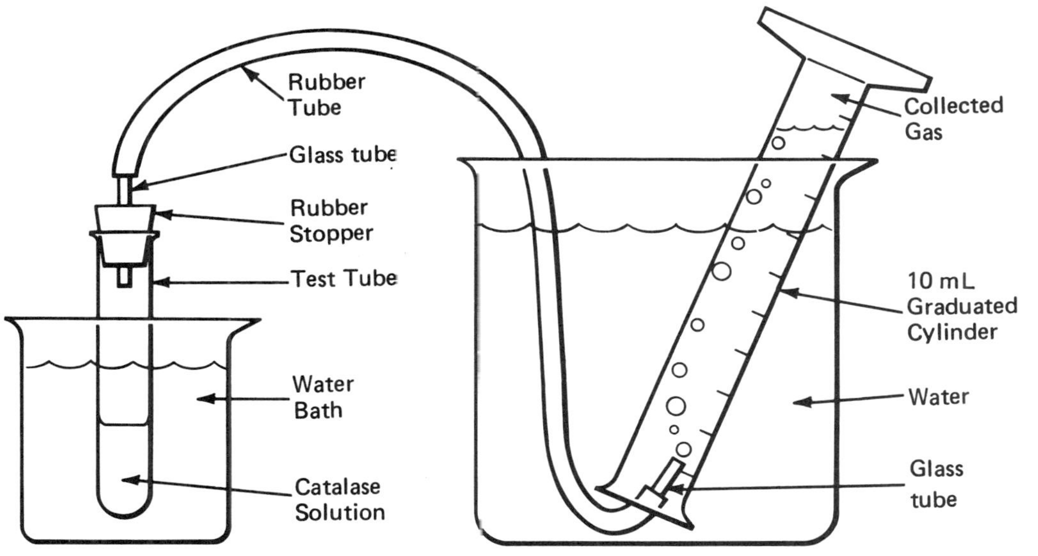


**Figure 39.2. Apparatus for measuring the volume of oxygen produced by decomposition of hydrogen peroxide.**

apparatus shown in Figure 39.2. Rinse pipet and pipet 2.0 mL of the catalase solution into the cooled hydrogen peroxide and mix quickly. With the test tube in the ice-water, replace the stopper so that the gas will bubble into the cylinder. After 3 minutes (more or less time may be needed depending on the activity of the enzyme) measure and record the volume of oxygen produced.

(b) Repeat the foregoing procedure using a room temperature water bath. Measure and record the volume of oxygen produced.

(c) Repeat, using a 37°C warm water bath. Measure and record the volume of oxygen produced.

### 2. Inhibitor Effect

Pipet 2.0 mL of the catalase solution into a test tube, add one drop of 2 M $CuSO_4$ solution, and mix. Place the test tube in a room temperature water bath. After 5 minutes, pipet 3.0 mL of 3% hydrogen peroxide into the test tube and quickly stopper. After 3 minutes, measure and record the volume of oxygen produced.

### 3. pH Effect

Label two test tubes A and B, and pipet 2.0 mL of the catalase solution into each tube. Add five drops of 0.1 M HCl to Tube A and five drops of 10% NaOH to Tube B. Place the two test tubes in a room temperature water bath. Pipet 3.0 mL of 3% hydrogen peroxide into Tube A and quickly stopper. Measure and record the volume of oxygen produced in 3 minutes. Repeat this process with Tube B.

### 4. Denaturation by Heat

Pipet 2.0 mL of the catalase solution into a test tube and place it in a boiling water bath for 5 minutes. Remove the test tube, and place it in the room temperature water bath to cool. After 10 minutes, pipet 3.0 mL of 3% hydrogen peroxide into the tube and quickly stopper. Measure and record the volume of oxygen produced in 3 minutes.

NAME ____________________

SECTION __________ DATE __________

INSTRUCTOR ____________________

## *REPORT FOR EXPERIMENT 39*

# Enzymatic Catalysis—Catalase

**Data Table**

| Experiment | Temperature | Volume of Oxygen Produced |
|---|---|---|
| 1. Temperature Effect (a) | 0°C | |
| (b) | Room Temp. | |
| (c) | 37°C | |
| 2. Inhibitor Effect | Room Temp. | |
| 3. pH Effect HCl | Room Temp. | |
| NaOH | Room Temp. | |
| 4. Denaturation Effect | Room Temp. | |

**QUESTIONS AND PROBLEMS**

1. (a) What effect does changing from room temperature to 0°C have on the activity of the enzyme?

   (b) What effect does changing from room temperature to 37°C have on the activity of the enzyme?

2. (a) Does the addition of acid increase or decrease the activity of catalase?

   (b) Does the addition of base increase or decrease the activity of catalase?

3. What does copper(II) sulfate do to the enzyme? Explain.

4. Why did the heat-treated enzyme behave differently than the non-heated enzyme? Explain.

# EXPERIMENT 40

# A Partial Chemical Analysis of Urine

## MATERIALS AND EQUIPMENT

Solutions: Dilute (6 M) acetic acid ($CH_3COOH$), 10% acetic acid, dilute (6 M) nitric acid ($HNO_3$), dilute (6 M) hydrochloric acid (HCl), 0.2 M ammonium oxalate [$(NH_4)_2C_2O_4$], 0.1 M silver nitrate ($AgNO_3$), 0.2 M ammonium molybdate [$(NH_4)_2MoO_4$], 0.1 M barium chloride ($BaCl_2$), and Benedict's reagent. Fresh urine should be collected for each day's experiment.

## DISCUSSION

Our bodies are somewhat similar to industrial manufacturing plants. Both take in raw materials and produce products and waste. Human urine is a solution containing many different chemicals that are no longer required by the body. Excess chemicals and waste products from cells become dissolved in the blood. These substances are filtered out of the blood by the kidneys, mixed with water, and excreted as urine.

Medical personnel are very interested in the chemicals present in a patient's urine. Numerous diseases leave their mark in the compositon of urine. Often, a urine analysis either will directly indicate the existence of a specific disease or will suggest that other tests should be given.

In this experiment, we will determine the presence or absence of several chemicals in urine. Urea, $CO(NH_2)_2$, is the most abundant solute in urine. Chloride ion is the second most abundant solute. Since one consumes anywhere from 5 to 15 grams of salt (NaCl) daily in a normal diet it is not surprising to find a high chloride ion concentration in the urine. The chloride level varies considerably, depending on such factors as perspiration, diarrhea, and vomiting, which also eliminate chloride ions from the body.

Calcium intake usually amounts to 0.1 to 0.5 grams daily. Up to 50% of this is eliminated in the urine.

Phosphorus is ingested and eliminated in the form of phosphate. The absolute quantity of phosphate varies dramatically, depending on the type of diet. An average level of phosphate excretion in the urine is approximately 1.1 grams per day.

Nearly all the sulfur in the body comes from the sulfur-containing amino acids (cysteine and methionine) in proteins. Excess sulfur is excreted either as sulfate or as part of an organic molecule. Nearly 1.0 gram of sulfur is eliminated daily in the urine.

Amino acids usually are not found in the urine. However, in certain diseases, high levels of protein and amino acids are often present in the urine. These proteins can be detected by denaturing them. Denaturation usually results in the coagulation of the proteins and the formation of a floculant precipitate.

Low levels of carbohydrates are usually present in urine. Normal individuals generally excrete less than 0.7 gram of carbohydrate daily, mostly as reducing sugars. When a person has diabetes the cells absorb less glucose than they ordinarily do. As a result additional glucose is eliminated in the urine.

## PROCEDURE

**Wear protective glasses. Do not pipet by mouth.**

### A. Test for Chloride Ion

1. Pour 5 mL of urine into a clean test tube, and add 1 mL of dilute (6 M) nitric acid.
2. Add 5 drops of silver nitrate solution to the test tube and mix. A white precipitate indicates a positive test.

### B. Test for Calcium Ion

1. Pour 5 mL of urine into a clean test tube, and add 3 drops of dilute (6 M) acetic acid and mix.
2. Add 10 drops of ammonium oxalate solution to the test tube and mix. A white precipitate indicates a positive test.

### C. Test for Phosphate Ion

1. Pour 5 mL of urine into a clean test tube, and add 1 mL of dilute (6 M) nitric acid.
2. Add 1 mL of ammonium molybdate solution to the test tube and mix.
3. Heat the solution in a boiling water bath for about 2 minutes. A yellow precipitate indicates a positive test.

### D. Test for Sulfate Ion

1. Pour 5 mL of urine into a clean test tube and add 1 mL of dilute (6 M) hydrochloric acid.
2. Heat the solution in a boiling water bath for about 2 minutes.
3. Add 10 drops of barium chloride solution to the test tube and mix. A positive test is indicated by the formation of a finely divided white precipitate.

### E. Test for Protein

1. Filter about 12 mL of urine in order to remove any possible suspended material.
2. Pour 5 mL of this filtered urine into each of two clean test tubes.
3. Heat one of the test tubes in a boiling water bath for about 5 minutes.
4. Add 6 drops of 10% acetic acid, and heat for an additional 5 minutes.
5. Compare the heated sample to the unheated one. A positive test is indicated by turbidity (cloudiness) in the heated solution.

## F. Test for Carbohydrate

1. Pour 5 mLof Benedict's reagent into a clean test tube, and add 1 mLof urine.
2. Place the test tube in a boiling water bath, and allow the contents to heat for 10 minutes.
3. Remove the test tube, and note whether a reddish precipitate has formed. A positive test for reducing sugars is indicated by the formation of a red copper(I) oxide ($Cu_2O$).

NAME ____________________

SECTION __________ DATE __________

## *REPORT FOR EXPERIMENT 40*

INSTRUCTOR ____________________

# A Partial Chemical Analysis of Urine

**Data Table**

| Test | Observations | Conclusions (positive or negative) |
|---|---|---|
| A (Chloride ion) | | |
| B (Calcium ion) | | |
| C (Phosphate ion) | | |
| D (Sulfate ion) | | |
| E (Protein) | | |
| F (Carbohydrate) | | |

**QUESTIONS AND PROBLEMS**

1. Why do you filter the urine in the protein test?

2. What general property of proteins is the protein test dependent upon?

3. What is a reducing sugar, and which one would most likely be in urine at a high concentration?

4. A urine specimen gives a positive test for (a) chloride ion, (b) calcium ion, and (c) sulfate ion. Write the net ionic equations for the reactions that occurred.

   (a)

   (b)

   (c)

# EXPERIMENT 41

# Lipids

## MATERIALS AND EQUIPMENT

Solids: Cholesterol ($C_{27}H_{45}OH$), potassium bisulfate ($KHSO_4$), stearic acid [$CH_3(CH_2)_{16}COOH$], vegetable shortening (Spry, Crisco, etc.). Liquids: Acetic anhydride [$(CH_3CO)_2O$], chloroform ($CHCl_3$), glycerol ($CH_2OHCHOHCH_2OH$), ligroine (alkene free; boiling range 60° to 85°C), oleic acid [$CH_3(CH_2)_7CH{=}CH(CH_2)_7COOH$], concentrated sulfuric acid ($H_2SO_4$), vegetable oils (cottonseed, peanut, corn, soybean, etc.). Solutions: 5% bromine ($Br_2$) in 1,1,1-trichloroethane ($CCl_3CH_3$), 10% sodium hydroxide ($NaOH$).

## DISCUSSION

Lipids are naturally occurring substances that are arbitrarily grouped together on the basis of their solubility in fat solvents, such as ether, benzene, chloroform, and carbon tetrachloride, and their insolubility in water. Lipids are subdivided into classes based on structural similarities. Two important classes are the simple lipids and the steroids.

**Simple Lipids.** (a) Fats and oils: esters of fatty acids and glycerol. (b) Waxes: esters of high molecular weight fatty acids and high molecular weight alcohols.

**Steroids.** These are substances possessing a 17-carbon unit structure containing four fused rings known as the steroid nucleus. Cholesterol and several hormones are in this class. Steroids having an $-OH$ (alcohol) group attached to the ring are known as sterols. Cholesterol is an example of such a substance.

Fats (and oils) are one of the three general classes of foodstuffs–carbohydrates, proteins, and fats. The distinction between a fat and an oil is that a fat is a solid at room temperature while an oil is a liquid. Both have similar molecular structure. It is a triester (triacylglycerol) that may be considered as being derived from a glycerol molecule and three fatty acid molecules; thus

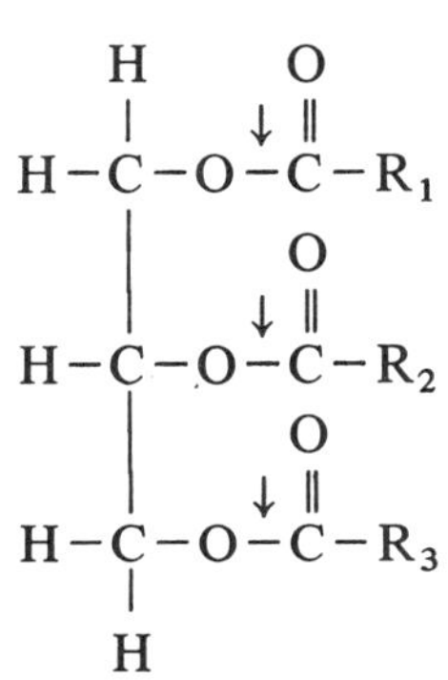

In this generalized formula, the part derived from glycerol is at the left, and the ester linkages are marked by small arrows. The portions to the right of the arrows are derived from three fatty acids. The fatty acids usually contain even numbers of carbon atoms ranging from 4 to 20 or more. The number of carbon-carbon double bonds in the carbon chains (represented by $R_1$, $R_2$, or $R_3$ in the generalized formula) usually varies from 0 to 4. Oleic acid, an 18-carbon acid with one double bond, is the most common fatty acid.

A fat with no carbon-carbon double bonds is saturated. If carbon-carbon double bonds are present, the fat is unsaturated. The term polyunsaturated means that the molecules of a particular product each contain several double bonds.

Halogens add readily to carbon-carbon double bonds.

$$X_2 + \sim CH=CH \sim \rightarrow \sim CHX-CHX \sim$$

A solution of bromine in 1,1,1-trichloroethane is used to detect and estimate the degree of unsaturation in fats. The reddish-brown color disappears when bromine adds to double bonds.

Fats are saponified when heated with a strong base, such as sodium hydroxide, yielding sodium salts of fatty acids (soaps) and glycerol. Fats are hydrolyzed to fatty acids and glycerol in the presence of fat splitting enzymes (lipases) or when heated with a strong acid.

$$\text{Saponification: Fat + Sodium hydroxide} \xrightarrow{\Delta} \underset{\text{(Soaps)}}{\text{Sodium salts of fatty acids}} + \text{Glycerol}$$

$$\text{Hydrolysis: Fat + Water} \xrightarrow{\text{Enzymes or } H^+} \text{Fatty acids + Glycerol}$$

Potassium bisulfate is used to distinguish glycerol esters from other lipids. It is both a strong acid and a powerful dehydrating agent. When $KHSO_4$ is heated with a fat, hydrolysis occurs and the glycerol produced is dehydrated to acrolein. Acrolein ($CH_2=CHCHO$) is an unsaturated aldehyde with a characteristic irritating odor.

Steroids are widely distributed in plants and animals and have a variety of functions. Examples of steroids are cholesterol, the sex hormones, ergosterol (which is a precursor of vitamin D) bile salts (which aid in the digestion of fats) and cortisone (a potent hormone useful in the treatment of allergies). Steroids have a 17-carbon skeletal structure consisting of four fused rings, numbered as shown.

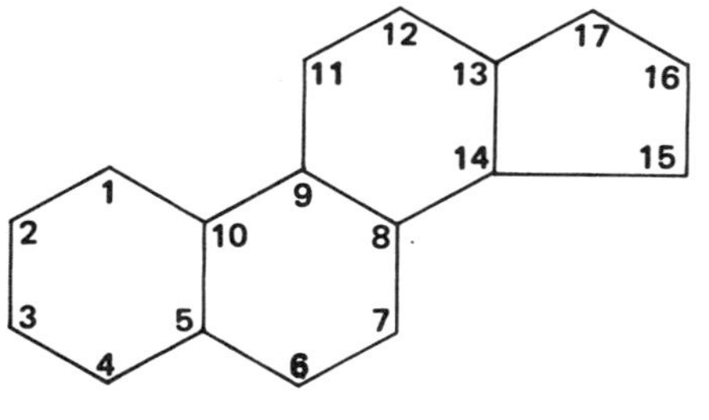


This hydrocarbon skeleton may have varying amounts of unsaturation and be substituted at various points, especially at positions 3 and 17.

Cholesterol is the most abundant steroid in the body and occurs in foods from animal sources such as eggs, butter, and cheese. Cholesterol is not found in plant tissues, but closely related steroids are present in plants.

Minute amounts of cholesterol and related steroids can be detected by the Lieberman-Burchard test. This test involves treating a chloroform solution of the steroid with acetic anhydride and concentrated sulfuric acid. The formation of a blue-green color is a positive test.

## PROCEDURE

**Wear protective glasses.**

NOTES:

1. Clean, dry test tubes are needed in this experiment and may be prepared as follows:
   (a) Clean each tube by brushing with hot water containing detergent.
   (b) Rinse thoroughly with water. For spotless tubes, rinse once with distilled water.
   (c) Wipe the outside of the tube, shake out excess water, grasp the tube with a test tube holder and heat over a nonluminous burner flame (without an inner cone) until dry.
   (d) Allow the tube to cool.
2. The tubes containing the reaction mixture from the acrolein test should be filled with hot water and allowed to soak for several minutes before cleaning by the method given in Note 1, above.
3. Dispose of mixtures containing chloroform and trichloroethane in the container provided.

### A. Acrolein Test

Place about 0.4 g (no more than 0.5 g) potassium bisulfate in a clean, dry test tube. Add 2 drops (or 0.05 to 0.1 g) of the material to be tested, and make sure that it is in contact with the potassium bisulfate. Now, using a test tube holder, heat the bottom of the tube over a hot burner flame. Incline the mouth of the tube at an angle of about 45° and constantly agitate the mixture by shaking the tube while heating, moving it in and out of the flame to control the rate of heating.

⚠ CAUTION: Keep the mouth of the tube pointed away from yourself and away from others while it is being heated; there is danger of spattering, and hot potassium bisulfate is an extremely disagreeable substance.

Continue heating until the potassium bisulfate is melted and a very slight darkening of the tube contents is visible. Stop heating and cautiously note the odor of the vapor coming from the tube by fanning it toward your nose from a distance of 3 to 6 inches. Acrolein produces a characteristic sharp irritating odor.

Do the acrolein test on the following materials in the order given. Describe and record qualitative differences, as well as intensity differences, among the odors produced.

1. Vegetable oil (cottonseed, corn, peanut, or safflower)
2. Oleic acid (or stearic acid)
3. Glycerol

### B. Reaction with Bromine

Place 7 clean, dry test tubes in a rack, and add 3 mL of ligroine to each. Ligroine is a mixture of saturated hydrocarbons and is used here as a solvent. Add the following substances to the ligroine in the test tubes.

| Tube No. | Material |
|---|---|
| 1 | Nothing (Control) |
| 2 | 3 drops vegetable oil (Peanut, cottonseed, or corn) |
| 3 | 3 drops of glycerol |
| 4 | 0.1 g stearic acid |
| 5 | 3 drops oleic acid |
| 6 | 0.1 g cholesterol |
| 7 | 3 drops melted vegetable shortening |

Mix the contents of each tube thoroughly for several seconds by firmly holding the top with one hand and tapping the side of the tube near the bottom with the forefinger of the other hand. **Do the following in the hood.** Add 2 drops of 5% bromine in 1,1,1-trichloroethane to Tube 1. Mix and note whether the bromine color has faded at 10 seconds and at 30 seconds following the bromine addition. Record your observations on the Report Sheet using this code: 0 = No fading, or barely noticeable fading, of color. 1 = Definitely noticeable color fading, but not complete. 2 = Complete, or nearly complete, disappearance of color.

Repeat the addition of bromine and the 10 and 30 second observations for each of the 6 remaining tubes in succession, and record your observations on the Report Sheet.

### C. Lieberman-Burchard Test

Place 5 clean, dry test tubes in a rack and add the materials to be tested as follows:

| Tube No. | Material |
|---|---|
| 1 | Stearic acid, pinhead sized quantity |
| 2 | Cholesterol, pinhead sized quantity |
| 3 | Glycerol, 1 drop |
| 4 | Vegetable oil, 1 drop (Cottonseed, peanut, soybean, or corn) |
| 5 | Melted vegetable shortening, 1 drop |

**Do the following in the hood.** Add 2 mL chloroform, 10 drops acetic anhydride, and 2 drops concentrated sulfuric acid to each tube, and mix thoroughly. After 5 minutes note the color present and its relative intensity in each solution. Record your data on the Report Form.

This is a sensitive test for the presence of sterols. Colors changing from pinkish, to blue, to green develop in the course of the reaction. The intensity of the color is roughly proportional to the amount of sterol present.

**D. Saponification**

Put 4 drops of vegetable oil or melted shortening into a test tube, and add 10 drops of 10% NaOH solution. Now using a test tube holder, incline the tube at an angle of about 45° and heat the contents over a low burner flame.

> ⚠ CAUTION: Be sure to wear your safety glasses and keep the tube pointed away from others, and from yourself, while it is being heated.

Shake the tube constantly while heating so that the liquid is continually sloshed about in the bottom third of the tube. Control the rate of heating by moving the tube in and out of the flame. Continue heating with constant agitation until the mixture in the tube is almost dry or until a solid begins to form. Allow the tube to cool for at least 2 minutes; add 5 mL of distilled water, stopper, and shake the tube vigorously for several seconds. Note and record the results.

To a second tube, add 4 drops of the vegetable oil and 10 drops of 10% NaOH solution. Do not heat this tube, but add 5 mL distilled water; stopper, and shake the tube vigorously for several seconds. Compare the results with those obtained with the first tube.

NAME ______________________

SECTION __________ DATE __________

INSTRUCTOR ______________________

## *REPORT FOR EXPERIMENT 41*

# Lipids

### A. Acrolein Test

1. Describe the odors produced by (a) vegetable oil, (b) oleic acid (or stearic acid), and (c) glycerol.

2. Which substance(s) gave a positive acrolein test?

### B. Reaction with Bromine

| Tube No. | Material tested | Color 10 sec. (0, 1, or 2) | Color 30 sec. (0, 1, or 2) | Conclusions (Saturated or unsaturated) |
|---|---|---|---|---|
| 1 | | | | |
| 2 | | | | |
| 3 | | | | |
| 4 | | | | |
| 5 | | | | |
| 6 | | | | |
| 7 | | | | |

**C. Lieberman-Burchard Test**

| Tube No. | Material tested | Color and relative intensity | Conclusions Positive or negative |
|---|---|---|---|
| 1 | | | |
| 2 | | | |
| 3 | | | |
| 4 | | | |
| 5 | | | |
| 6 | | | |

**D. Saponification**

Observations:

First tube:

Second tube:

## QUESTIONS AND PROBLEMS

1. Write the structural formula of a glyceryl tristearate molecule. Draw a dotted line enclosure about the portion(s) of the molecule that make it soluble in solvents such as benzene or ligroine.

2. Glycerol is a saturated trihydroxy alcohol. Acrolein is an unsaturated aldehyde having the formula, $CH_2{=}CHCHO$. Hot potassium bisulfate is a powerful dehydrating agent. Using structural formulas for organic substances, show how hot $KHSO_4$ converts glycerol to acrolein.

3. Explain why the fat, glyceryl tripalmitate, and cholesterol would be expected to give different results with the Acrolein Test.

4. Write an equation showing how bromine adds to oleic acid.

5. Does the vegetable oil (or oils) tested contain substances other than glycerides? Cite experimental results which justify your answer.

6. What was the evidence in Part D that the vegetable oil was saponified?

7. Write condensed structural formulas for the soaps produced in Part D if vegetable oil contained glycerides of lauric, palmitic, and oleic acids.

# EXPERIMENT 42

# Blood Cholesterol Levels

## MATERIALS AND EQUIPMENT

Liquids: Isopropyl alcohol [$CH_3CH(OH)CH_3$], acetone ($CH_3COCH_3$), and glacial acetic acid ($CH_3COOH$), concentrated sulfuric acid ($H_2SO_4$). Solutions: Whole blood (**certified AIDS virus free**), 1 : 1 ethanol ($C_2H_5OH$)–acetone solution, cholesterol standard solution ($C_{27}H_{45}OH$), digitonin solution ($C_{56}H_{92}O_{29}$), and iron reagent ($FeCl_3 \cdot 6H_2O$). Spectrophotometer, centrifuge and centrifuge tubes, 1 mL and 5 mL graduated pipets, protective gloves.

## DISCUSSION

Listen to the radio, watch television, or read the newspaper for a day and you will probably encounter some question or statement about cholesterol. The Office of the Surgeon General of the United States believes cholesterol can either directly or indirectly cause heart disease. Only animals produce and require cholesterol. Humans need a certain level of cholesterol to survive. Cholesterol is the principal steroid in vertebrates and is a starting molecule for the biosynthesis of some steroid hormones and bile acids. Cholesterol occurs both in blood and cells. A normal blood cholesterol level is in the range of 100-250 mg/100 mL. Problems may arise when cholesterol levels go significantly above this range. High levels of cholesterol indicate possible coronary disease, nephritis, or diabetes (among other possibilities). Since cholesterol is an alcohol, it can be esterified. A major portion of blood cholesterol exists as esters of fatty acids.

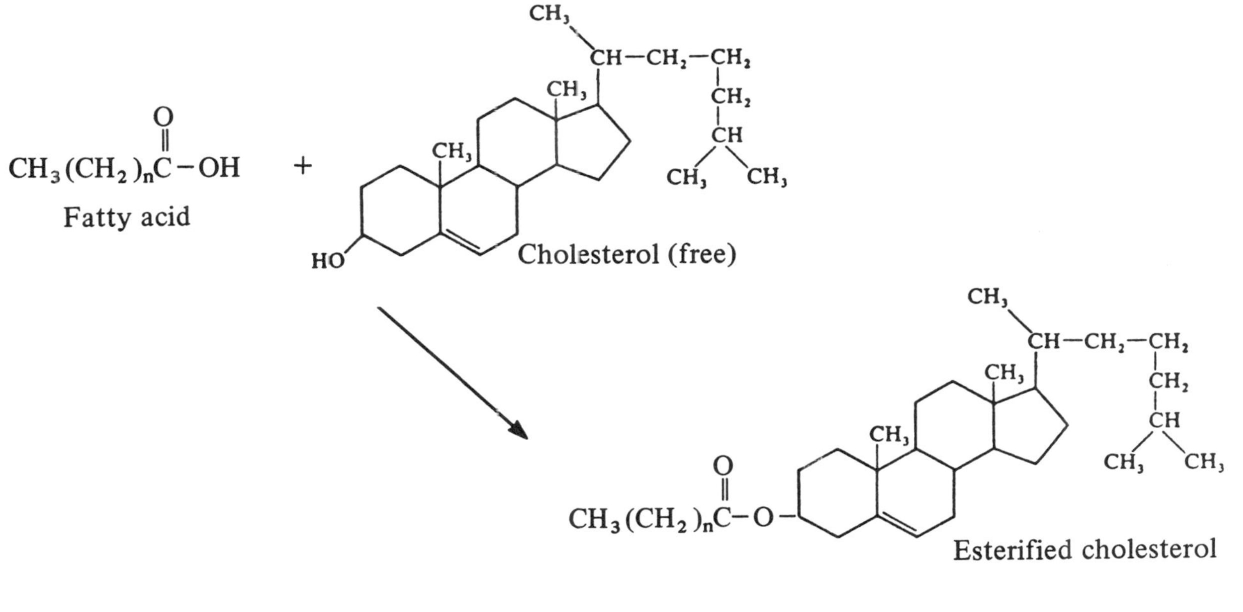

In this experiment, you will use a spectrophotometer to measure the concentration of both total cholesterol and free cholesterol by the Leffler method. By subtracting the free cholesterol from the total cholesterol, you obtain the amount of esterified cholesterol.

Whole blood consists primarily of plasma and red blood cells (RBC). The RBC's are removed by centrifugation. Then the cholesterols are extracted into an organic solvent. The cholesterols (both free and esterified) then are reacted with ferric chloride to produce a colored complex. This complex absorbs light at 560 nm. The concentration of total cholesterol in the blood will be determined by comparing its absorbance with that of a standard cholesterol solution.

Free cholesterol is selectively precipitated from solution by digitonin. Once the free cholesterol is separated, it is redissolved and analyzed.

## PROCEDURE

**Wear protective glasses.**

**Wear protective gloves.**

⚠ CAUTION: Use rubber suction bulb or water aspirator pump to draw liquids into your pipet. **Do not pipet by mouth.**

NOTE: To keep track of the various samples in this experiment, label four centrifuge tubes C-1 through C-4 and six test tubes T-1 through T-6. Use clean and dry tubes.

### A. Preparation of Plasma from Whole Blood

1. Put 1.5 to 2.0 mL of whole blood in centrifuge Tube C-1 and centrifuge for 10 minutes or until the red blood cells have sedimented to the bottom of the tube.
2. Cool Test Tube T-1 in an ice bath.
3. Carefully pipet the plasma (upper layer) from Centrifuge Tube C-1 and transfer to the cooled Test Tube T-1; keep this plasma cold.
4. Discard the RBC's in Tube C-1.

### B. Determination of Total and Free Cholesterol

1. Pipet 0.20 mL of plasma from Test Tube T-1 into Centrifuge Tube C-2.
2. Pipet 0.20 mL of the cholesterol standard (2.0 mg/mL) into Test Tube T-2.
3. Pipet 5.0 mL of isopropyl alcohol into C-2 and T-2 and **thoroughly** mix.
4. Centrifuge the plasma sample in C-2 for about 5 minutes to remove the precipitate.
5. Pipet 1.0 mL of the isopropyl-plasma supernatant from C-2 into T-3 and set aside for later use.
6. Pipet 1.0 mL of the isopropyl-cholesterol standard from T-2 into T-4 and set it aside for later use.
7. Pipet 0.20 mL of plasma from T-1 in the ice bath into C-3. Now pipet 3.0 mL of 1:1 ethanol-acetone solution into C-3. Thoroughly mix and centrifuge for about 5 minutes.
8. Pipet 2.0 mL of the supernatant from C-3 into C-4. Pipet 2.0 mL of the digitonin solution into C-4 and thoroughly mix.

9. Heat the mixture in C-4 in a water bath at approximately 55°C for 15 minutes and then cool in an ice bath for about one minute. Centrifuge, decant and discard the supernatant; **save** the precipitate.

10. Add 4 mL of acetone to the precipitate in C-4 and thoroughly mix; centrifuge for 15 minutes and decant the supernatant; save the precipitate.

11. Pipet 1.0 mL of isopropyl alcohol onto the precipitate in C-4; thoroughly mix and set aside for later use.

12. Pipet 1.0 mL of isopropyl alcohol into T-6. This is the blank required to zero the spectrophotometer. Save this blank.

13. Pipet 3.0 mL of glacial acetic acid into each of Tubes C-4, T-3, T-4, and T-6 and **thoroughly** mix. Be sure that no precipitate in C-4 remains undissolved. Pour the solution from Centrifuge Tube C-4 into Test Tube T-5.

14. Pipet 0.30 mL of the iron reagent into each of the four tubes, T-3 through T-6 and thoroughly mix.

15. **USING EXTREME CAUTION**, pipet 3.0 mL of concentrated sulfuric acid into each of the four tubes, T-3 through T-6, and thoroughly mix.

16. Cool the four samples to room temperature in a water bath.

17. Absorbancy measurements: Set the wavelength of the spectrophotometer to 560 nm. Transfer the solution from Tube T-6 (blank) to a cuvette and zero the instrument. Now measure the absorbancies of the solutions from Tubes T-3, T-4, and T-5. Make three measurements for each solution (zero the instrument between each measurement).

## C. Calculations

To obtain the cholesterol concentrations, use the following formulas:

$$\text{Total cholesterol} = \left(\frac{\text{Total cholesterol absorbance}}{\text{Std. cholesterol absorbance}}\right) \times \left(\begin{matrix}\text{Std. cholesterol}\\ \text{concentration}\end{matrix}\right)$$

$$\text{Free cholesterol} = \left(\frac{\text{Free cholesterol absorbance}}{\text{Std. cholesterol absorbance}}\right) \times \left(\begin{matrix}\text{Std. cholesterol}\\ \text{concentration}\end{matrix}\right) \times (0.31)$$

The concentration of the cholesterol standard is 2.0 mg/mL.

The factor 0.31 is a correction factor and is the ratio of the amount of the total cholesterol to the amount of free cholesterol measured, i.e., (1.0/5.2)/(2.0/3.2).

Total cholesterol is measured in Tube T-3 and free cholesterol in Tube T-5. Tube T-4 contains the cholesterol standard.

NAME ______________________

SECTION __________ DATE __________

INSTRUCTOR ______________________

## *REPORT FOR EXPERIMENT 42*

# Blood Cholesterol Levels

**Data Table**

| Tube No. | Absorbance | Average Absorbance |
|---|---|---|
| T-3<br>T-3<br>T-3 | | |
| T-4<br>T-4<br>T-4 | | |
| T-5<br>T-5<br>T-5 | | |

Calculated concentrations (show your calculations).

Concentration of total cholesterol in Tube T-3 = ______________________

Concentration of free cholesterol in Tube T-5 = ______________________

Concentration of esterified cholesterol = ______________________

## QUESTIONS AND PROBLEMS

1. Cholesterol will react with stearic acid in the presence of a small amount of mineral acid. Name the product and draw its structure.

2. Does the cholesterol level in your blood sample indicate the donor is possibly subject to cholesterol related diseases? Explain.

3. In the calculation for free cholesterol, we used a 0.31 factor. Show where this originates. Remember (1.0/5.2)/(2.0/3.2).

# STUDY AID 1

# Significant Figures

Every measurement that we make has some inherent error due to the limitations of the measuring instrument and the experimenter. The numerical value recorded for a measurement should give some indication of the reliability (precision) of that measurement. In measuring a temperature using a thermometer calibrated at one-degree intervals we can easily read the thermometer to the nearest one degree, but we normally estimate and record the temperature to the nearest tenth of a degree (0.1°C). For example, a temperature falling between 23°C and 24°C might be estimated at 23.4°C. There is some uncertainty about the last digit, 4, but an estimate of it is better information than simply reporting 23°C or 24°C. If we read the thermometer as "exactly" twenty-three degrees, the temperature should be reported as 23.0°C, not 23°C, because 23.0°C indicates our estimate to the nearest 0.1°C. Thus in recording any measurement, we retain one uncertain digit. The digits retained in a physical measurement are said to be significant, and are called **significant figures.**

Some numbers are exact and therefore have an infinite number of significant figures. Exact numbers occur in simple counting operations, such as 5 bricks, and in defined relationships, as 100 cm = 1 meter, 24 hrs = 1 day, etc. Because of their infinite number of significant figures, exact numbers do not limit the number of significant figures in a calculation.

**Counting Significant Figures.** Digits other than zero are always significant. Depending on their position in the number, zeros may or may not be significant. There are several possible situations:

1. All zeros between other digits in a number are significant. For example: 3.076, 4002, 790.2. Each of these numbers has four significant figures.

2. Zeros to the left of the first nonzero digit are used to locate the decimal point and are not significant. Thus 0.013 has only two significant figures.

3. Zeros to the right of the last nonzero digit, and to the right of the decimal point are significant, for they would not have been included except to express precision. For example, 3.070 has four significant figures; 0.070 has two significant figures.

4. Zeros to the right of the last nonzero digit, but to the left of the decimal, as in the numbers 100, 580, 37000, etc., may or may not be significant. For example, in 37000 the measurement might be good to the nearest 1000, 100, 10, or 1. There are two conventions which may be used to show the intended precision. If all the zeros are significant, then an expressed decimal may be added, as 580., or 37000. But a better system, and one which is applicable to the case when some but not all of the zeros are significant, is to express the number in exponential notation, including only the significant zeros. Thus for 300, if the zero following 3 is significant, we would write $3.0 \times 10^2$. For 17000, if two zeros are significant, we would write $1.700 \times 10^4$. The number we correctly expressed as 580. can also be correctly expressed as $5.80 \times 10^2$. With exponential notation there is no doubt as to the number of significant figures.

Many small numbers, such as 40°C, 250 mL, 700 mm, used in chemistry problems may appear to have only one or two significant figures since neither the expressed decimal nor the exponential form has been used to justify the significance of the zeros. It is customary to regard the zeros used in this way as being significant since the numbers do not really represent actual measurements but are merely values chosen for convenience.

**Addition or Subtraction.** The result of an addition or subtraction should contain no more digits to the right of the decimal point than are in that quantity which has the least number of digits to the right of the decimal point. Perform the operation indicated and then round off the number to the proper significant figure.

Example:

$$\begin{array}{r} 24.372 \\ 72.21 \\ 6.1488 \\ \hline 102.7308 \end{array} \quad (102.73)$$

Since the digit 1 in 72.21 is uncertain, the sum can have no digits beyond this point, so the sum should be rounded off to 102.73.

**Multiplication or Division.** In multiplication or division, the answer can have no more significant figures than the factor with the least number of significant figures. In multiplication or division the position of the decimal point has nothing to do with the number of significant figures in the answer.

Example: $3.1416 \times 7.5 \times 252 = 5937.624$ $(5.9 \times 10^3)$

The operations of arithmetic supply all the digits shown, but this does not make the answer precise to seven significant figures. Most of these digits are not realistic because of the limited precision of the number 7.5. So the answer must be rounded to two significant figures, 5900 or $5.9 \times 10^3$. It should be emphasized that in rounding-off the number you are not sacrificing precision, since the digits discarded are not really correct.

# STUDY AID 2

# Formulas and Chemical Equations

The atomic compositions of substances are expressed by formulas. In order to write correct chemical equations, it is necessary to know how to write correct formulas.

Ions are atoms or groups of atoms that have either a positive or a negative electrical charge. The relative electrical charge on an ion is indicated by a plus (+) or minus (−) sign together with a number. The number and the plus or minus sign are written as a superscript at the upper right corner of the symbol for the ion. Examples: $Na^+$ represents a sodium ion; it is a sodium atom with a +1 charge (the number 1 is omitted but understood to be present). $SO_4^{2-}$ represents a sulfate ion which has a −2 charge and is composed of one sulfur atom and four oxygen atoms.

In the examples given below, the ions shown are used to illustrate the formation of possible compounds. The first series of ions is hypothetical but will be used to illustrate the principle of combining positive and negative ions to write formulas of compounds.

| Ion | Name | Ion | Name | Ion | Name |
|---|---|---|---|---|---|
| $X^+$ | (Hypothetical) | $Na^+$ | Sodium | $Br^-$ | Bromide |
| $Y^{2+}$ | (Hypothetical) | $NH_4^+$ | Ammonium | $Cl^-$ | Chloride |
| $Z^{3+}$ | (Hypothetical) | $K^+$ | Potassium | $NO_3^-$ | Nitrate |
| $E^-$ | (Hypothetical) | $Ca^{2+}$ | Calcium | $O^{2-}$ | Oxide |
| $G^{2-}$ | (Hypothetical) | $Pb^{2+}$ | Lead(II) | $S^{2-}$ | Sulfide |
| | | $Mg^{2+}$ | Magnesium | $CrO_4^{2-}$ | Chromate |
| | | $Al^{3+}$ | Aluminum | $SO_4^{2-}$ | Sulfate |
| | | $Fe^{3+}$ | Iron(III) | $PO_4^{3-}$ | Phosphate |

Examples of compounds formed by combining positive and negative ions:

| Ions | Hypothetical Compound | Specific Compounds with Corresponding Ion Ratios |
|---|---|---|
| $X^+$, $E^-$ | XE | NaCl, $NH_4Cl$, KBr |
| $Y^{2+}$, $E^-$ | $YE_2$ | $CaCl_2$, $Mg(NO_3)_2$, $Pb(NO_3)_2$ |
| $X^+$, $G^{2-}$ | $X_2G$ | $Na_2O$, $(NH_4)_2SO_4$, $K_2CrO_4$ |
| $Y^{2+}$, $G^{2-}$ | YG | MgO, $PbCrO_4$, $CaSO_4$ |
| $Z^{3+}$, $E^-$ | $ZE_3$ | $AlCl_3$, $Fe(NO_3)_3$ |
| $Z^{3+}$, $G^{2-}$ | $Z_2G_3$ | $Al_2O_3$, $Fe_2(SO_4)_3$ |

When more than one atom of an element is used in the formula of a compound, it is indicated by a numerical subscript written to the right of the element; e.g., $CaCl_2$. The formula $CaCl_2$ indicates one $Ca^{2+}$ ion and two $Cl^-$ ions. If an ion contains more than one element; e.g., $NO_3^-$, and two or more of these ions are needed in the formula of a compound, the ion is enclosed in parentheses and the subscript is written after the parenthesis; e.g., $Cu(NO_3)_2$. This formula indicates one $Cu^{2+}$ ion and two $NO_3^-$ ions.

A chemical change or reaction results in the formation of substances whose compositions are different from the starting substances. A chemical equation is a shorthand expression for a chemical reaction. Substances in the equation are represented by their formulas. The equation indicates both the reactants (starting substances) and the products. The reactants are written on the left side and the products on the right side of the equation. An arrow (→) pointing to the products separates the reactants from the products. A plus sign (+) is used to separate one reactant (or product) from another.

Reactants ⟶ Products

**Example 1.** The combustion of carbon in oxygen or air

Word equation: Carbon + Oxygen ⟶ Carbon dioxide

Formula equation: $C + O_2 \longrightarrow CO_2$

**Example 2.** The combustion of magnesium in oxygen or air

Word equation: Magnesium + Oxygen ⟶ Magnesium oxide

Formula equation: $2\,Mg + O_2 \longrightarrow 2\,MgO$

**Balancing equations.** There is no detectable change in mass resulting from a chemical reaction. Therefore, the mass of the products must equal the mass of the reactants before the chemical change occurred. In representing the chemical change by an equation, this conservation of mass is attained by "balancing the equation." Because each atom has a particular mass, we balance an equation by adjusting the number of atoms of each kind of element to be the same on each side of the equation. Thus in Example 1 there are two oxygen atoms and one carbon atom on each side of the equation; therefore, the mass of the reactants and the products must be equal.

If the equation in Example 2 is written and left as

$$Mg + O_2 \longrightarrow MgO \qquad \text{(Unbalanced)}$$

it is not balanced since there are two oxygen atoms on the left side and only one oxygen atom on the right side. We can balance the oxygen atoms by placing a 2 in front of MgO, which gives us two oxygen atoms on each side of the equation.

$$Mg + O_2 \longrightarrow 2\,MgO \qquad \text{(Unbalanced)}$$

Now the magnesium atoms are unbalanced. We can balance the magnesium atoms and the whole equation by placing a 2 in front of Mg.

$$2\,Mg + O_2 \longrightarrow 2\,MgO \qquad \text{(Balanced)}$$

A correct formula of a substance may not be changed for the convenience of balancing an equation. Small whole numbers, as needed, are placed in front of the formulas to balance the equation. However, it is important to notice that when a number is placed in front of one formula to balance a particular element, as in 2 MgO above, it may unbalance another element in the equation.

A number placed in front of a formula multiplies every atom in the formula by that number. Thus 2 MgO means 2 Mg atoms and 2 O atoms; 3 $CaCl_2$ means 3 Ca atoms and 6 Cl atoms; 4 $H_2SO_4$ means 8 H atoms, 4 S atoms, and 16 O atoms; 3 $Cu(NO_3)_2$ means 3 Cu atoms, 6 N atoms, and 18 O atoms.

Many equations may be balanced by this "trial and error" or inspection method.

# STUDY AID 3

# Graphs

A graph is often the most convenient way to present or display a set of data. Various kinds of graphs have been devised, but the most common type uses a set of horizontal and vertical coordinates to show the relationship of two variables. It is called an *x*-*y* graph because the values of one variable are represented on the horizontal or *x*-axis (abscissa) and the values of the other variable are represented on the vertical or *y*-axis (ordinate). See Figure 1.

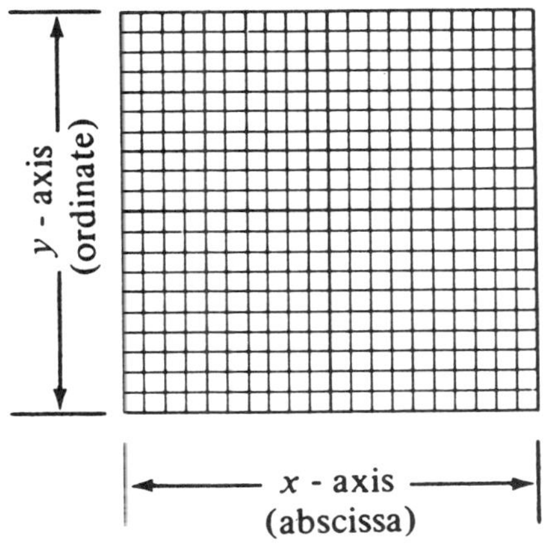


**Figure 1. Rectangular coordinate graph paper.**

As a specific example of how a graph is constructed, let us graph the relationship between Celsius and Fahrenheit temperature scales. Assume that we have the following data:

| °C | 0 | 50 | 100 |
|---|---|---|---|
| °F | 32 | 122 | 212 |

Also assume that we have a piece of graph paper which has 50 squares along the *x*-axis and 55 squares along the *y*-axis. We will plot °C along the *x*-axis, and °F along the *y*-axis.

## STEPS IN PREPARING THE GRAPH

1. Determine the scales, that is, how many units each division will represent. We have 50 divisions on the *x*-axis and data ranging from 0 to 100°C. Therefore it is convenient to let each division along this axis represent 2°C. The determination of scale along the *y*-axis is not so easy. Our data ranges from 32° to 212°F, a spread of 180°F. We have 55 divisions, which gives us a possible scale of 3.27°F/division (180°F/55 divisions). If we adopted this scale, we would use the entire *y*-coordinate, but the graph would be extremely awkward to plot and read. We can increase our scale to 4 or 5°F/division. Either scale would be acceptable. However, 5°F/division is more

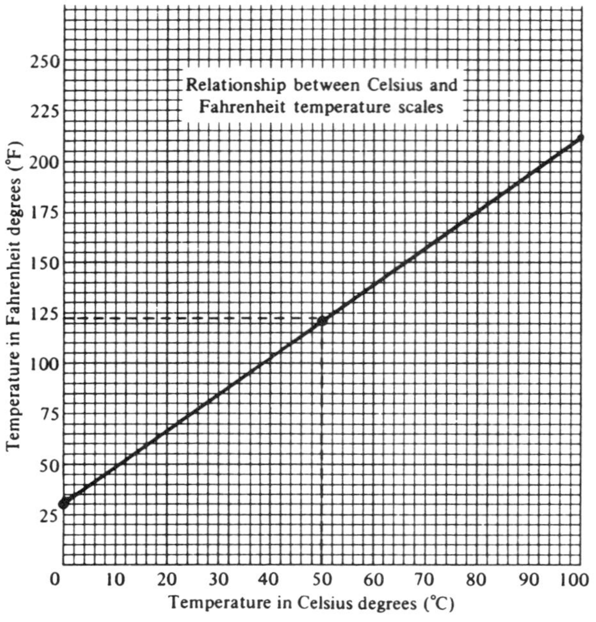


**Figure 2.**

convenient, especially if the graph paper has every fifth or tenth line darker as shown in Figure 2. For this reason we have chosen to use 5°F/division. The most convenient scales to use generally are 1/2, 1, 2, 5, or 10 units per division.

2. Determine the starting values for each coordinate. For our data the $x$-axis must begin at zero and go to 100, because having chosen the scale of 2°C/division, we will use the entire 50 divisions. Along the $y$-axis, we have a choice. We could begin numbering with 30°F (not 32°F, since we are using a scale of 5°F/division), which would use the lower 37 of the 55 available divisions. However, since there are so many extra divisions, we can also start numbering the Fahrenheit scale at zero, thus shifting the line upward and making a more attractive graph.

It is common for coordinates to be numbered starting with zero at the origin (lower left corner), but it is not required that numbering begin at zero. For example, if all the data for one variable were in the range of 200 to 250, it would be a poor choice of scale to make that coordinate range from 0 to 250. Judgment must be exercised when choosing scales; a good general rule is to choose a scale so that the data when plotted cover at least half of the length of each coordinate.

3. Number the major divisions along each axis. Because of our choice of scales, we will label the $x$-axis every 10°C, and the $y$-axis every 25°C. The axes need not be numbered every 5 divisions, but when the graph paper has darker lines every five spaces, it is convenient to number at these heavier lines. The numbered divisions must be on lines.

4. Label each axis. In our example, we label the $x$-axis as Temperature in Celsius degrees (°C) and the $y$-axis as Temperature in Fahrenheit degrees (°F). Shorter versions could have been used, but labels on the coordinates are absolutely essential.

5. Plot the points. Locate and plot the three points corresponding to the data. Here is how a point is located on the graph: using the 50°C-122°F data as an example, trace a vertical line up from 50°C on the $x$-axis and a horizontal line across from 122°F on the $y$-axis and mark the point where the two lines intersect. This process is called plotting. The other two points are plotted on the graph in the same way.

6. Draw a smooth line through the points. In our example, if the three points have been plotted correctly, they will lie on a straight line so that a straightedge can be used to draw the smooth line. When plotting data collected in the laboratory the best smooth line will not necessarily touch each of the plotted points. Thus some judgment must be exercised in locating the best smooth line, whether it be straight or curved.

7. Title the graph. Every graph should have a title which clearly expresses what the graph represents. Titles may be placed below the graph, above the graph, or on the upper part of the graph. The last choice, which is illustrated in Figures 2 and 3, is the most common for student laboratory reports. Of course the title must be placed so as not to interfere with the plot on the graph.

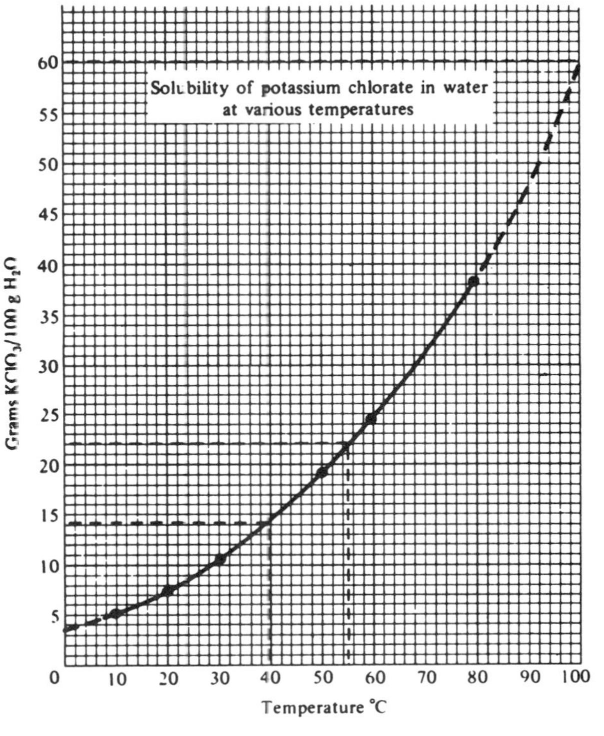


**Figure 3.**

## READING A GRAPH

Figure 3 is a graph showing the solubility of potassium chlorate in water at various temperatures. The solubility curve on this graph was plotted from the following data.

| Temperature °C | Solubility grams $KClO_3$/100 g water |
|---|---|
| 10 | 5.0 |
| 20 | 7.4 |
| 30 | 10.5 |
| 50 | 19.3 |
| 60 | 24.5 |
| 80 | 38.5 |

The graph in Figure 3 was constructed from experimentally determined solubilities at the six temperatures shown. These experimentally determined solubilities are all located on the smooth curve traced by the solid line portion of the graph. We are therefore confident that the solid line represents a very good approximation of the solubility data for potassium chlorate covering the temperature range from 10 to 80°C. All points on the plotted curve represent the composition of saturated solutions. Any point below the curve represents an unsaturated solution.

The dashed line portions of the curve are extrapolations; that is, they extend the curve above and below the temperature range actually covered by the plotted data. Curves such as this are often extrapolated a short distance beyond the range of the known data, although the extrapolated portions may not be highly accurate. Extrapolation is justified only in the absence of more reliable information.

The graph in Figure 3 can be used with confidence to obtain the solubility of $KClO_3$ at any temperature between 10 and 80°C, but the solubilities between 0 and 10°C and between 80 and 100°C are less reliable. For example, what is the solubility of $KClO_3$ at 40°C, at 55°C, and at 100°C? First draw a vertical line from each temperature to the plotted solubility curve. Now from each of these points on the curve draw a horizontal line to the solubility axis and read the corresponding solubility. The values that we read from the graph are

| | |
|---|---|
| 40°C | 14.2 g $KClO_3$/100 g water |
| 55°C | 22.0 g $KClO_3$/100 g water |
| 100°C | 60.0 g $KClO_3$/100 g water |

Of these solubilities, the one at 55°C is probably the most reliable because experimental points are plotted at 50° and 60°C. The 40°C solubility value is probably a bit less reliable because the nearest plotted points are at 30°C and 50°C. The 100°C solubility is the least reliable of the three values because it was taken from the extrapolated part of the curve, and the nearest plotted point is at 80°C. Actual handbook solubility values are 14.0 and 57.0 grams of $KClO_3$/100 grams of water at 40°C and 100°C respectively.

*STUDY AID 4*

# Use of the Mole in Chemical Calculations

It is often necessary to calculate the amount of product that can be obtained from a given amount of reactant or, conversely, to determine how much reactant is required to produce a stated amount of product. Calculations of this kind, based upon balanced chemical equations, are called **stoichiometry** (from Greek, meaning element measure).

In solving stoichiometric problems, the concept of the mole is very useful. In its broadest sense a mole is Avogadro's Number ($6.022 \times 10^{23}$) of any chemical species. Even though the word "mole" is used as a short expression for "molar mass," it is quite permissible to refer to moles of chemical species which are not really molecular in character. Reference may be made to moles of such diverse species as sulfur **atoms** (S), oxygen **atoms** (O), oxygen **molecules** ($O_2$), sulfuric acid **molecules** ($H_2SO_4$), sodium chloride **formula units** (NaCl), ammonium **ions** ($NH_4^+$), nitrate **ions** ($NO_3^-$)—or even to moles of **electrons** or of **protons.**

Since a mole, by definition, consists of Avogadro's Number of particles, it represents a mass in grams numerically equal to the molar mass of the particular species under consideration.

Consideration of Avogadro's famous hypothesis that "Equal volumes of all gases, at the same temperature and pressure, contain the same number of molecules" leads to the conclusion that one mole ($6.022 \times 10^{23}$ molecules) of any gas will have the same volume as one mole of any other gas at the same temperature and pressure. The volume of one mole of any gas is known to be 22.4 liters at 0°C (273°K) and one atmosphere (760 torr) pressure, i.e., at standard temperature and pressure (STP).

These three facets of the mole concept may be summarized:

1. One mole is $6.022 \times 10^{23}$ particles of the species under consideration.

2. One mole is a quantity in grams numerically equal to the molar mass of the substance under consideration.

3. If the substance is a gas, one mole is 22.4 liters of that gas at STP.

To use the mole concept to advantage in dealing with a particular problem, these four steps are generally needed:

**Step 1.** Write the balanced chemical equation for the reaction involved or check the equation given to make sure that it is balanced.

**Step 2.** Examine the problem statement and determine what is the **given substance** upon which to base the calculations. Then set up the calculation needed to convert the specified quantity of this substance to **moles of the given substance.**

**Step 3.** Obtain the chemically equivalent **moles of desired substance** by multiplying the moles of given substance (from Step 2) by the **mole ratio** obtained from the balanced equation:

$$\text{Mole ratio} = \frac{\text{Moles of desired substance (in the equation)}}{\text{Moles of given substance (in the equation)}}$$

The numbers in this mole ratio are the numbers (coefficients) in front of the respective substances in the balanced equation.

**Step 4.** Convert the moles of desired substance (obtained in Step 3) to whatever units are required in the problem by multiplying by the appropriate factor, i.e., molecules/mole, grams/mole, or liters/mole.

Consider the following solutions of problems related to the heating of potassium chlorate to produce potassium chloride and oxygen.

**Problem 1. How Many Grams of Oxygen Can Be Obtained From 15.0 g of $KClO_3$?**

SOLUTION:

**Step 1.** Write the balanced equation for the reaction.

$$2\,KClO_3 \longrightarrow 2\,KCl + 3\,O_2$$

Molar masses: $KClO_3$ = 122.6 KCl = 74.56 $O_2$ = 32.00

**Step 2.** $KClO_3$ is the given substance; convert the given quantity of $KClO_3$ to moles of $KClO_3$.

$$15.0\,\cancel{\text{g}\,KClO_3} \times \frac{1\text{ mol }KClO_3}{122.6\text{ g}\,\cancel{KClO_3}}$$

**Step 3.** Convert moles of $KClO_3$ to the chemically equivalent moles of $O_2$ by multiplying by the **mole ratio taken from the equation.**

$$15.0\,\cancel{\text{g}\,KClO_3} \times \frac{1\,\cancel{\text{mol}\,KClO_3}}{122.6\text{ g}\,\cancel{KClO_3}} \times \frac{3\text{ mol }O_2}{2\,\cancel{\text{mol}\,KClO_3}}$$

**Step 4.** Convert the moles of $O_2$ to grams of $O_2$.

$$15.0\,\cancel{\text{g}\,KClO_3} \times \frac{1\,\cancel{\text{mol}\,KClO_3}}{122.6\text{ g}\,\cancel{KClO_3}} \times \frac{3\,\cancel{\text{mol}\,O_2}}{2\,\cancel{\text{mol}\,KClO_3}} \times \frac{32.00\text{ g }O_2}{\cancel{\text{mol}\,O_2}} = 5.87\text{ g }O_2 \text{ (Answer)}$$

This answer was obtained by completing the indicated arithmetic, $\dfrac{15.0 \times 3 \times 32.00}{122.6 \times 2}$

**Problem 2. How Many Oxygen Molecules Can Be Obtained From 15.0 g of $KClO_3$?**

SOLUTION: The first three steps are exactly the same as in Problem 1, but the fourth step converts **moles** of $O_2$ to **molecules** of $O_2$.

$$15.0\,\cancel{\text{g}\,KClO_3} \times \frac{1\,\cancel{\text{mol}\,KClO_3}}{122.6\,\cancel{\text{g}\,KClO_3}} \times \frac{3\,\cancel{\text{mol}\,O_2}}{2\,\cancel{\text{mol}\,KClO_3}} \times \frac{6.022 \times 10^{23}\text{ molecules }O_2}{\cancel{\text{mol}\,O_2}}$$

$= 1.11 \times 10^{23}$ molecules $O_2$ (Answer).

**Problem 3. How Many Liters of Oxygen Gas, Measured at STP, Can Be Obtained From 15.0 g of $KClO_3$?**

SOLUTION: Here again the first three steps are identical to those of Problems 1 and 2. The final step converts **moles** of $O_2$ to **liters** of $O_2$.

$$15.0\,\cancel{g\ KClO_3} \times \frac{1\,\cancel{mol\ KClO_3}}{122.6\,\cancel{g\ KClO_3}} \times \frac{3\,\cancel{mol\ O_2}}{2\,\cancel{mol\ KClO_3}} \times \frac{22.4\ L\ O_2}{\cancel{mol\ O_2}} = 4.11\ L\ O_2 \quad \text{(Answer)}$$

Problems 1, 2, and 3 have dealt with questions relating to the determination of the amount of product obtainable from a given quantity of reactant. The following problems illustrate how the same general method may be used to predict the amount of reactant required to obtain a specified amount of product.

**Problem 4. How Many Grams of $KClO_3$ Must Be Decomposed to Produce 25.0 g of KCl?**

SOLUTION:

**Step 1.** Balance the equation for the reaction as shown in Problem 1.

**Step 2.** Consider KCl to be the given substance; convert to moles of KCl.

$$25.0\,\cancel{g\ KCl} \times \frac{1\ mol\ KCl}{74.56\,\cancel{g\ KCl}}$$

**Step 3.** Convert moles of KCl to chemically equivalent moles of $KClO_3$ by use of the mole ratio taken from the balanced equation.

$$25.0\,\cancel{g\ KCl} \times \frac{1\,\cancel{mol\ KCl}}{74.56\,\cancel{g\ KCl}} \times \frac{2\ mol\ KClO_3}{2\,\cancel{mol\ KCl}}$$

**Step 4.** Convert moles $KClO_3$ to grams of $KClO_3$.

$$25.0\,\cancel{g\ KCl} \times \frac{1\,\cancel{mol\ KCl}}{74.56\,\cancel{g\ KCl}} \times \frac{2\,\cancel{mol\ KClO_3}}{2\,\cancel{mol\ KCl}} \times \frac{122.6\ g\ KClO_3}{\cancel{mol\ KClO_3}} = 4.11\ g\ KClO_3 \quad \text{(Answer)}$$

**Problem 5. What Mass of $KClO_3$ Must Be Decomposed to Obtain 7.50 Liters of $O_2$ Measured at STP?**

SOLUTION: Oxygen is the given substance. Liters of $O_2$ are converted to grams of $KClO_3$ in the following manner:

$$\text{Liters } O_2 \longrightarrow \text{moles } O_2 \longrightarrow \text{moles } KClO_3 \longrightarrow \text{grams } KClO_3$$

$$7.50\,\cancel{L\ O_2} \times \frac{1\,\cancel{mol\ O_2}}{22.4\,\cancel{L\ O_2}} \times \frac{2\,\cancel{mol}\ KClO_3}{3\,\cancel{mol\ O_2}} \times \frac{122.6\ g\ KClO_3}{\cancel{mol\ KClO_3}} = 27.4\ g\ KClO_3 \quad \text{(Answer)}$$

# STUDY AID 5

# Organic Chemistry—An Introduction

Organic chemistry is known as the chemistry of the carbon compounds. All compounds classified as organic contain carbon. Organic compounds are found in all living matter, foodstuffs (fats, proteins, and carbohydrates), fuels of all kinds, plastics, fabrics, wood and paper products, paints and varnishes, dyes, soaps and detergents, cosmetics, medicinals, insecticides, refrigerants, etc. There are over two million known organic compounds.

To help study their properties, organic compounds are grouped into classes or series according to the similarity of their chemical makeup or structure. Some of the common classes are hydrocarbons, alcohols, aldehydes, ketones, acids, esters, ethers, amines, and amides.

The major reason for the large number of organic compounds is that carbon atoms have the ability to bond together, forming long chains and rings. Carbon atoms share electrons, forming covalent bonds. Between two carbon atoms, single, double, or triple covalent bonds may be formed by sharing one, two, or three pairs of electrons, respectively. These types of bonds are illustrated below:

| C : C | C::C | C:::C |
|---|---|---|
| C–C | C=C | C≡C |
| Single bond | Double bond | Triple bond |

A dash between carbon atoms indicates a covalent bond and represents one pair of electrons.

The names of the alkane series hydrocarbons are important to organic chemistry because they represent the basis for the systematic nomenclature of organic compounds. The first 10 members of this series and their molecular formulas are listed below:

| | | | | | | | |
|---|---|---|---|---|---|---|---|
| $CH_4$ | Methane | $C_4H_{10}$ | Butane | $C_6H_{14}$ | Hexane | $C_8H_{18}$ | Octane |
| $C_2H_6$ | Ethane | $C_5H_{12}$ | Pentane | $C_7H_{16}$ | Heptane | $C_9H_{20}$ | Nonane |
| $C_3H_8$ | Propane | | | | | $C_{10}H_{22}$ | Decane |

### Structural Formulas

A great many organic compounds are composed of carbon, hydrogen, and oxygen atoms. In these compounds we find these atoms bonded to each other in the following ways:

| | |
|---|---|
| Carbon to carbon | (C–C), (C=C), (C≡C) |
| Carbon to carbon to carbon | (C–C–C) |
| Carbon to hydrogen | (C–H) |
| Carbon to oxygen | (C=O) |
| Carbon to oxygen to hydrogen | (C–O–H) |
| Carbon to oxygen to carbon | (C–O–C) |

In chemical compounds, with some rare exceptions, a carbon atom will always have four bonds; a hydrogen atom, one bond; and an oxygen atom, two bonds. A bond consists of a pair of electrons shares between any two atoms.

Because of the different arrangements in which carbon atoms bond with each other, a single written molecular formula may represent more than one arrangement of the atoms, giving rise to more than one compound. For example, there are 2 different butanes ($C_4H_{10}$), 3 pentanes ($C_5H_{12}$), and 75 decanes ($C_{10}H_{22}$). To illustrate different compounds with the same molecular formula, we use **structural formulas.** Structural formulas show the order in which the atoms are bonded to each other, while molecular formulas show only the number and kind of each atom in a molecule. A few examples will illustrate:

Methane, $CH_4$

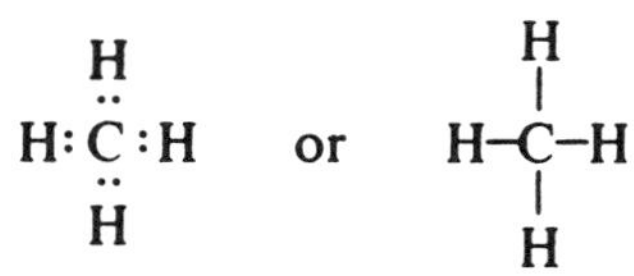

$CH_4$

Structural formula — Condensed structural formula

Ethane, $C_2H_6$

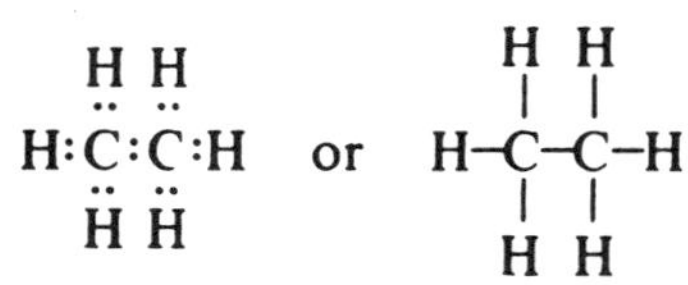

$CH_3{-}CH_3$ or $CH_3CH_3$

Structural formula — Condensed structural formula

Note that in the condensed structural formula, which is a convenient simplification of the structural formula, all of the atoms or groups attached to each carbon atom are written to the right of it.

Propane, $C_3H_8$

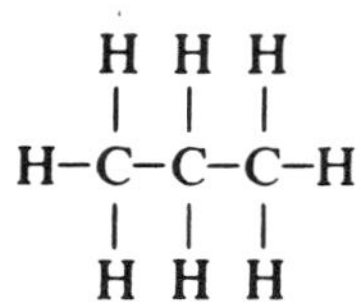

$CH_3CH_2CH_3$

Structural formula — Condensed structural formula

Methyl alcohol, $CH_3OH$

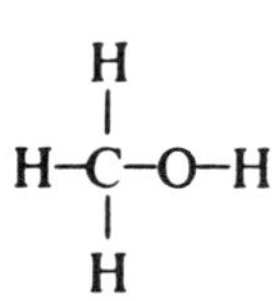

$CH_3OH$

Structural formula — Condensed structural formula

Butane, $C_4H_{10}$

(a) n-Butane (n = normal)

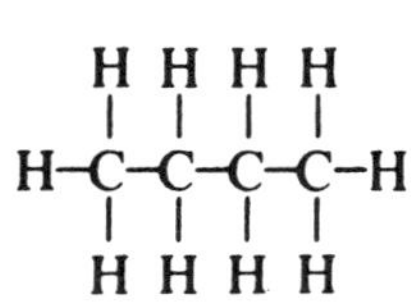

$CH_3CH_2CH_2CH_3$

Structural formula — Condensed structural formula

(b) Isobutane

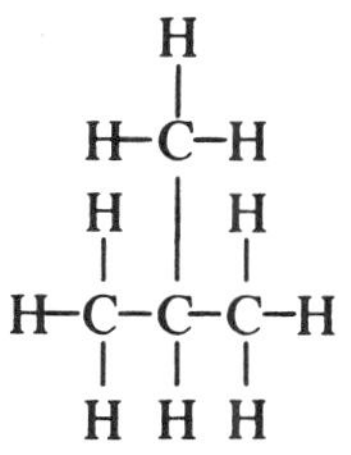

Structural formula

$$CH_3\overset{CH_3}{\overset{|}{C}}HCH_3 \quad \text{or}$$

$$CH_3CH(CH_3)CH_3 \quad \text{or}$$

$$CH_3CH(CH_3)_2$$

Condensed structural formula

## Isomerism

We have shown that there are two possible ways to bond 4 carbon atoms and 10 hydrogen atoms to form the structures of the two butanes. These butanes are indeed different compounds, each with its own physical and chemical properties. For example, n-butane boils at −0.5°C and isobutane boils at −11.7°C.

The phenomenon of two or more compounds having the same molecular formula is known as **isomerism.** The individual compounds are called **isomers**. Thus there are 2 isomers of butane, 3 of pentane, 18 of octane, and 75 of decane.

## Alkyl Groups

The nomenclature of organic chemistry is sprinkled with such terms as "methyl," "ethyl," and "isopropyl." These terms represent organic radicals and are derived from the names of the alkane hydrocarbons. An **alkyl group** represents an alkane hydrocarbon from which one hydrogen atom has been removed. For example, the methyl group $CH_3-$ is derived from methane by removing a hydrogen atom from $CH_4$. The name "methyl" is formulated by dropping the **ane** from methane and adding the letters **yl** to the stem **meth.** Other alkyl groups are derived in a similar manner—ethyl from ethane, etc. A few of the more common alkyl groups are listed below. The dash in the formula of the group indicates the carbon atom from which the hydrogen atom has been removed.

| | |
|---|---|
| $CH_3-$ | Methyl |
| $CH_3CH_2-$ | Ethyl |
| $CH_3CH_2CH_2-$ | n-Propyl (n = normal) |
| $CH_3\underset{\vert}{C}HCH_3$ | Isopropyl |
| $CH_3CH_2CH_2CH_2-$ | n-Butyl |

Note that two different propyl groups are formed, depending on whether the hydrogen atom removed is from an end carbon or a middle carbon atom. In a like manner four different butyl groups are formed (only one of which is shown). Alkyl groups are sometimes called alkyl radicals and are often designated by the letter R—. Thus RH represents an alkane hydrocarbon.

## Functional Groups

The formulas for many classes of organic compounds may be derived from the formulas of the alkane hydrocarbons by substituting a different group for one or more of the hydrogen atoms in the hydrocarbon chain. These groups are known as **functional groups** and characterize the classes of compounds that they represent.

The functional group of the alcohols is $-OH$. Two examples are methyl alcohol ($CH_3OH$) and ethyl alcohol ($CH_3CH_2OH$ or $C_2H_5OH$). In a like manner the formulas of an entire series of alcohols may be written by substituting an $-OH$ group for a hydrogen atom on the alkane chain or by combining the $-OH$ group with the alkyl groups given above. Thus n-propyl alcohol and iso- propyl alcohol are $CH_3CH_2CH_2OH$ and $\underset{\displaystyle OH}{\underset{|}{C}}H_3CHCH_3$, respectively.

Common functional groups, together with their classes of compounds, are given below.

| Functional Group | Class of Compound | Examples | |
|---|---|---|---|
| $-OH$ | Alcohol | $CH_3OH$ | Methyl alcohol (Methanol) |
| | | $CH_3CH_2OH$ | Ethyl alcohol (Ethanol) |
| $-C\begin{matrix}{\diagup\!\!\!\diagup} O \\ \diagdown OH\end{matrix}$ or $-COOH$ | Acid | $HCOOH$ | Formic acid (Methanoic acid) |
| | | $CH_3COOH$ | Acetic acid (Ethanoic acid) |
| $\overset{\displaystyle H}{\overset{|}{-C}}=O$ | Aldehyde | $H-\overset{\displaystyle H}{\overset{|}{C}}=O$ | Formaldehyde (Methanal) |
| | | $CH_3\overset{\displaystyle H}{\overset{|}{C}}=O$ | Acetaldehyde (Ethanal) |
| $R-\underset{\displaystyle O}{\underset{\parallel}{C}}-R$ | Ketone | $CH_3\underset{\displaystyle O}{\underset{\parallel}{C}}CH_3$ | Acetone (Propanone) |
| $-C\begin{matrix}{\diagup\!\!\!\diagup} O \\ \diagdown OR\end{matrix}$ | Ester | $H-C\begin{matrix}{\diagup\!\!\!\diagup} O \\ \diagdown OCH_3\end{matrix}$ | Methyl formate (Methyl methanoate) |
| | | $CH_3C\begin{matrix}{\diagup\!\!\!\diagup} O \\ \diagdown OCH_2CH_3\end{matrix}$ | Ethyl acetate (Ethyl ethanoate) |

NAME__________

SECTION______ DATE______

INSTRUCTOR__________

*EXERCISE 1*

# Significant Figures and Exponential Notation

1. How many significant figures are in each of the following numbers?

(a) 7.4 ______ (b) 462 ______ (c) 1.40 ______ (d) 16,000 ______

(e) 0.072 ______ (f) 0.0130 ______ (g) 0.0058 ______ (h) 0.1090 ______

2. Write each of the following numbers in proper exponential notation:

(a) 557 — (a) ______

(b) 0.064 — (b) ______

(c) 4,300 — (c) ______

(d) 382.0 — (d) ______

(e) 11,800,000 — (e) ______

(f) 0.007 — (f) ______

3. How many significant figures should be in the answer to each of the following calculations?

(a) $\begin{array}{r} 17.1 \\ +\ 0.77 \\ \hline \end{array}$

(b) $\begin{array}{r} 47.826 \\ -\ 9.4 \\ \hline \end{array}$

(c) $12.4 \times 5.3 =$

(d) $1.27 \times 3.1416 =$

(e) $\dfrac{0.517}{0.2742} =$

(f) $\dfrac{0.072}{4.36} =$

(g) $\dfrac{5.82 \times 760 \times 325}{723 \times 273} =$

(h) $\dfrac{0.37 \times 454 \times 5.620}{22.4} =$

(a) ______

(b) ______

(c) ______

(d) ______

(e) ______

(f) ______

(g) ______

(h) ______

4. For each of these problems, complete the answer with a 10 raised to the proper power. Note that each answer is expressed to the correct number of significant figures.

(a) $1.71 \times 10^3 \times 2.0 \times 10^2 = 3.4 \times$ ________ (a) ________

(b) $\frac{4.523 \times 10^3}{2.71 \times 10^1} = 1.67 \times$ ________ (b) ________

(c) $4.8 \times 10^1 \times 3.5 \times 10^4 = 1.7 \times$ ________ (c) ________

(d) $\frac{1.64 \times 10^{-2}}{1.2 \times 10^2} = 1.4 \times$ ________ (d) ________

(e) $\frac{3.7 \times 10^4}{8.42 \times 10^5} = 4.4 \times$ ________ (e) ________

5. Solve each of the following problems, expressing each answer to the proper number of significant figures. Use exponential notation for (c), (d), and (e).

(a)
```
  1.842
 45.2
+87.55
```

(b)
```
 714.3
- 18.56
```

(a) ________

(b) ________

(c) $1.83 \times 10^4 \times 7.55 \times 10^7 =$ (c) ________

(d) $3.3 \times 5{,}280 =$ (d) ________

(e) $\frac{5.07 \times 10^{-4} \times 8.51 \times 10^{-2}}{2.92 \times 10^4} =$ (e) ________

**Answers**

1. (a) 2, (b) 3, (c) 3, (d) 2, (e) 2, (f) 3, (g) 2, (h) 4.

2. (a) $5.57 \times 10^2$, (b) $6.4 \times 10^{-2}$, (c) $4.3 \times 10^3$, (d) $3.820 \times 10^2$, (e) $1.18 \times 10^7$, (f) $7 \times 10^{-3}$.

3. (a) 3, (b) 3, (c) 2, (d) 3, (e) 3, (f) 2, (g) 3, (h) 2.

4. (a) $10^5$, (b) $10^2$, (c) $10^6$, (d) $10^{-4}$, (e) $10^{-2}$.

5. (a) 134.6, (b) 695.7, (c) $1.38 \times 10^{12}$, (d) $1.7 \times 10^4$, (e) $1.48 \times 10^{-9}$.

NAME ______________________

SECTION __________ DATE __________

INSTRUCTOR ______________________

## EXERCISE 2

# Measurements

For each of the following problems show your calculation setup. In both your setup and answer show units and follow rules of significant figures. See Experiment 2 and the appendixes for any needed formulas or conversion factors.

1. Convert 88°F to degrees Celsius. ______________

2. Convert −15°C to degrees Fahrenheit. ______________

3. An object weighs 9.1 lbs. What is the mass in grams? ______________

4. A stick is 11.3 in. long. What is the length in centimeters? ______________

5. The water in a flask measures 452 mL. How many quarts is this? ______________

6. A piece of lumber measures 88.4 cm long. What is its length in:

   (a) Millimeters? ______________

   (b) Feet? ______________

7. A block is found to have a volume of 25.3 $cm^3$. Its mass is 21.7 g. Calculate the density of the block.

____________

8. A graduated cylinder was filled to 25.0 mL with liquid. A solid object was immersed in the liquid, raising the liquid level to 33.9 mL. The solid object was found to weigh 63.5 g. Calculate the density of the solid object.

____________

9. The density of the liquid in Problem 8 was 0.804 g/mL. What was the mass of the liquid in the graduated cylinder?

____________

10. Convert 66.2°F to degrees Celsius.

____________

11. A beaker contains 721 mL of water. The density of the water is 1.00 g/mL. Calculate:

(a) The volume of the water in liters.

____________

(b) The mass of the water in grams.

____________

12. The density of carbon tetrachloride, $CCl_4$, is 1.59 g/mL. Calculate the volume of 38.2 g of $CCl_4$.

____________

NAME ______________________

SECTION __________ DATE __________

INSTRUCTOR ______________________

*EXERCISE 3*

# Names and Formulas I

Give the names of the following compounds:

1. NaCl ______________________

2. $AgNO_3$ ______________________

3. $BaCrO_4$ ______________________

4. $Mg(OH)_2$ ______________________

5. $ZnSO_4$ ______________________

6. $K_2CO_3$ ______________________

7. $Al_2O_3$ ______________________

8. $CdF_2$ ______________________

9. $NH_4NO_2$ ______________________

10. $Fe(OH)_3$ (a) ______________________

(b) ______________________

11. $Zn_3(PO_4)_2$ ______________________

12. $KClO_3$ ______________________

13. CaS ______________________

14. $(NH_4)_2C_2O_4$ ______________________

 NAME____________________

Give the formulas of the following compounds:

| | | |
|---|---|---|
| 1. Barium chloride | 1. | ____________ |
| 2. Lead(II) nitrate | 2. | ____________ |
| 3. Titanium(III) iodide | 3. | ____________ |
| 4. Ammonium hydroxide | 4. | ____________ |
| 5. Potassium chromate | 5. | ____________ |
| 6. Cobalt(II) oxide | 6. | ____________ |
| 7. Magnesium perchlorate | 7. | ____________ |
| 8. Copper(II) sulfate | 8. | ____________ |
| 9. Sodium sulfite | 9. | ____________ |
| 10. Iron(III) chloride | 10. | ____________ |
| 11. Calcium cyanide | 11. | ____________ |
| 12. Copper(I) sulfide | 12. | ____________ |
| 13. Silver carbonate | 13. | ____________ |
| 14. Cadmium hypochlorite | 14. | ____________ |
| 15. Tin(IV) oxide | 15. | ____________ |
| 16. Sodium bicarbonate | 16. | ____________ |
| 17. Aluminum acetate | 17. | ____________ |
| 18. Nickel(II) phosphate | 18. | ____________ |

NAME ______________________

SECTION ________ DATE ____________

INSTRUCTOR ______________________

# EXERCISE 4

## Names and Formulas II

Give the names of the following compounds:

1. $(NH_4)_2S$ ______________________
2. $NiF_2$ ______________________
3. $Sb(ClO_3)_3$ ______________________
4. $Ca(OH)_2$ ______________________
5. HCl (aq) ______________________
6. $CrBr_3$ ______________________
7. $Cu_2CO_3$ (a) ______________________

   (b) ______________________
8. $K_2Cr_2O_7$ ______________________
9. $FeSO_4$ (a) ______________________

   (b) ______________________
10. $AgC_2H_3O_2$ ______________________
11. $H_2SO_4$ (aq) ______________________
12. $Cr_2O_3$ ______________________
13. $KBrO_3$ ______________________
14. $Cd(NO_3)_2$ ______________________

 NAME ______________________

Give the formulas of the following compounds:

1. Nitric acid 1. ____________
2. Sodium oxalate 2. ____________
3. Manganese(II) iodate 3. ____________
4. Zinc nitrite 4. ____________
5. Arsenic(III) iodide 5. ____________
6. Potassium permanganate 6. ____________
7. Titanium(IV) bromide 7. ____________
8. Sodium arsenate 8. ____________
9. Ammonium thiocyanate 9. ____________
10. Acetic acid 10. ____________
11. Manganese(IV) sulfide 11. ____________
12. Bismuth(III) arsenate 12. ____________
13. Sodium peroxide 13. ____________
14. Cobalt(II) chlorite 14. ____________
15. Stannous fluoride 15. ____________
16. Magnesium bicarbonate 16. ____________
17. Lead(II) acetate 17. ____________
18. Phosphoric acid 18. ____________

NAME ______________________

SECTION ________ DATE ____________

INSTRUCTOR ______________________

# EXERCISE 5

## Names and Formulas III

Give the names of the following compounds:

1. $CO_2$ ______________________
2. KSCN ______________________
3. $H_2O_2$ ______________________
4. $Ag_2CO_3$ ______________________
5. $Ni(MnO_4)_2$ ______________________
6. $CrF_3$ ______________________
7. $Co_3(AsO_4)_2$ ______________________
8. $Cu(OH)_2$ ______________________
9. KCN ______________________
10. $Bi(IO_3)_3$ ______________________
11. $H_2SO_3$ (aq) ______________________
12. $BaH_2$ ______________________
13. $SnS_2$ (a) ______________________

    (b) ______________________
14. $NaHSO_3$ ______________________
15. $HgC_2O_4$ ______________________
16. $As(NO_2)_5$ ______________________
17. $Pb(HCO_3)_2$ ______________________

EXERCISE 5 (continued) NAME ____________________

Give the formulas of the following substances:

1. Ammonium bicarbonate 1. ____________
2. Hydrogen sulfide 2. ____________
3. Barium hydroxide 3. ____________
4. Nitrous acid 4. ____________
5. Copper(II) bromide 5. ____________
6. Carbon tetrachloride 6. ____________
7. Nickel(II) perchlorate 7. ____________
8. Lead(II) nitrate 8. ____________
9. Ammonia 9. ____________
10. Chlorine 10. ____________
11. Chromium(III) sulfite 11. ____________
12. Sulfur dioxide 12. ____________
13. Carbonic acid 13. ____________
14. Copper(I) carbonate 14. ____________
15. Chloric acid 15. ____________
16. Barium arsenate 16. ____________
17. Calcium cyanide 17. ____________
18. Arsenic(III) oxide 18. ____________
19. Silver dichromate 19. ____________
20. Carbon disulfide 20. ____________
21. Aluminum fluoride 21. ____________
22. Manganese(IV) chloride 22. ____________

NAME ____________________

SECTION ________ DATE ____________

INSTRUCTOR ____________________

*EXERCISE 6*

# Equation Writing and Balancing I

Balance the following equations:

1. $Mg + O_2 \xrightarrow{\Delta} MgO$

2. $KClO_3 \xrightarrow{\Delta} KCl + O_2$

3. $Fe + O_2 \xrightarrow{\Delta} Fe_3O_4$

4. $Mg + HCl \longrightarrow MgCl_2 + H_2$

5. $Na + H_2O \longrightarrow NaOH + H_2$

Beneath each word equation write the formula equation and balance it. Remember that oxygen and hydrogen are diatomic molecules.

1. Sulfur + Oxygen $\xrightarrow{\Delta}$ Sulfur dioxide

2. Zinc + Sulfuric acid $\longrightarrow$ Zinc sulfate + Hydrogen

3. Carbon + Oxygen $\xrightarrow{\Delta}$ Carbon dioxide

4. Hydrogen + Oxygen $\xrightarrow{\Delta}$ Water

5. Aluminum + Hydrochloric acid $\longrightarrow$ Aluminum chloride + Hydrogen

 NAME ______________________

Balance the following equations:

1. $N_2 + H_2 \xrightarrow{\Delta} NH_3$

2. $CoCl_2 \cdot 6\,H_2O \xrightarrow{\Delta} CoCl_2 + H_2O$

3. $Fe + H_2O \xrightarrow{\Delta} Fe_3O_4 + H_2$

4. $F_2 + H_2O \longrightarrow HF + O_2$

5. $Pb(NO_3)_2 \xrightarrow{\Delta} PbO + NO + O_2$

Beneath each word equation write and balance the formula equation. Oxygen, hydrogen, and bromine are diatomic molecules.

1. Aluminum + Oxygen $\xrightarrow{\Delta}$ Aluminum oxide

2. Potassium + Water $\longrightarrow$ Potassium hydroxide + Hydrogen

3. Arsenic(III) oxide + Hydrochloric acid $\longrightarrow$ Arsenic(III) chloride + Water

4. Phosphorus + Bromine $\longrightarrow$ Phosphorus tribromide

5. Sodium bicarbonate + Nitric acid $\longrightarrow$ Sodium nitrate + Water + Carbon dioxide

NAME ____________________

SECTION ________ DATE ____________

INSTRUCTOR ____________________

*EXERCISE 7*

# Equation Writing and Balancing II

Complete and balance the following double displacement reaction equations (assume all reactions will go):

1. $NaCl + AgNO_3 \longrightarrow$

2. $BaCl_2 + H_2SO_4 \longrightarrow$

3. $NaOH + HCl \longrightarrow$

4. $Na_2CO_3 + HCl \longrightarrow$

5. $H_2SO_4 + NH_4OH \longrightarrow$

6. $FeCl_3 + NH_4OH \longrightarrow$

7. $Na_2SO_3 + HCl \longrightarrow$

8. $K_2CrO_4 + Mn(NO_3)_2 \longrightarrow$

9. $NaC_2H_3O_2 + HCl \longrightarrow$

10. $NaOH + NH_4NO_3 \longrightarrow$

11. $BiCl_3 + H_2S \longrightarrow$

12. $K_2C_2O_4 + HCl \longrightarrow$

13. $H_3PO_4 + Ca(OH)_2 \longrightarrow$

14. $(NH_4)_2CO_3 + HNO_3 \longrightarrow$

15. $K_2CO_3 + NiBr_2 \longrightarrow$

 NAME ____________________

Complete and balance the following equations. (Combination, 1-4; Decomposition, 5-8; Single displacement, 9-12; Double displacement, 13-16.)

1. $K + Cl_2 \longrightarrow$

2. $Zn + O_2 \longrightarrow$

3. $BaO + H_2O \longrightarrow$

4. $SO_3 + H_2O \longrightarrow$

5. $MgCO_3 \xrightarrow{\Delta}$

6. $NH_4OH \xrightarrow{\Delta}$

7. $Mn(ClO_3)_2 \xrightarrow{\Delta}$

8. $HgO \xrightarrow{\Delta}$

9. $Ni + HCl \longrightarrow$

10. $Pb + AgNO_3 \longrightarrow$

11. $Cl_2 + NaI \longrightarrow$

12. $Al + CuSO_4 \longrightarrow$

13. $KOH + H_3PO_4 \longrightarrow$

14. $Na_2C_2O_4 + CaCl_2 \longrightarrow$

15. $(NH_4)_2SO_4 + KOH \longrightarrow$

16. $ZnCl_2 + (NH_4)_2S \longrightarrow$

NAME ____________________

SECTION ________ DATE ____________

INSTRUCTOR ____________________

## *EXERCISE 8*

# Equation Writing and Balancing III

For each of the following situations, write and balance the formula equation for the reaction which occurs.

1. A strip of zinc is dropped into a test tube of hydrochloric acid.

2. Hydrogen peroxide decomposes in the presence of manganese dioxide.

3. Copper(II) sulfate pentahydrate is heated to drive off the water of hydration.

4. A piece of sodium is dropped into a beaker of water.

5. A piece of limestone (calcium carbonate) is heated in a bunsen burner flame.

6. A piece of zinc is dropped into a solution of silver nitrate.

7. Hydrochloric acid is added to a sodium carbonate solution.

8. Potassium chlorate is heated in the presence of manganese dioxide.

9. Hydrogen gas is burned in air.

10. Sulfuric acid solution is reacted with sodium hydroxide solution.

NAME ________________

SECTION ________ DATE ________

INSTRUCTOR ________________

# *EXERCISE 9*

# Graphical Representation of Data

## A. Reading a Graph

From the figure at the right, read values for the following:

1. The vapor pressure of ethyl ether at 20°C.

   ________________

2. The temperature at which ethyl chloride has a vapor pressure of 620 torr.

   ________________

3. The temperature at which ethyl alcohol has the pressure that ethyl chloride has at 2°C.

   ________________

Vapor Pressure-Temperature Curves

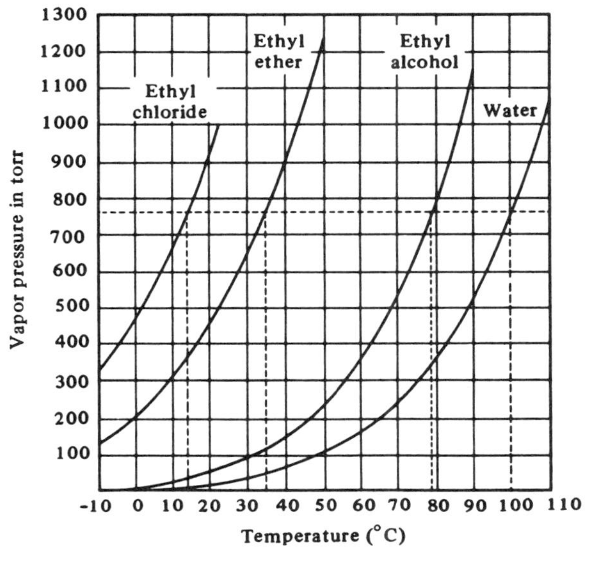


## B. Plotting Graphs

1. Plot the following pressure-temperature data for a gas on the graph below. Draw the best possible straight line through the data.

| Temperature, °C | 0 | 20 | 40 | 60 | 80 | 100 |
|---|---|---|---|---|---|---|
| Pressure, torr | 586 | 628 | 655 | 720 | 757 | 800 |

Plotted Pressure-Temperature Data for a Gas

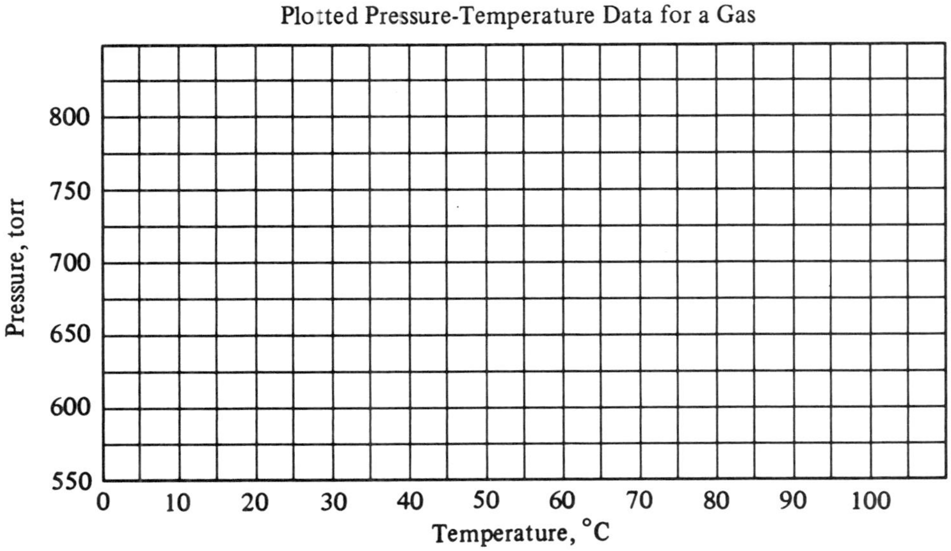

2. (a) Study the data given below; (b) determine suitable scales for pressure and for volume and mark these scales on the graph; (c) plot the eight points on the graph; (d) draw the best possible line through these points; (e) place a suitable title at the top of the graph.

Pressure-volume data for a gas

| Volume, mL | 10.70 | 7.64 | 5.57 | 4.56 | 3.52 | 2.97 | 2.43 | 2.01 |
|---|---|---|---|---|---|---|---|---|
| Pressure, torr | 250 | 350 | 480 | 600 | 760 | 900 | 1100 | 1330 |

Pressure, torr

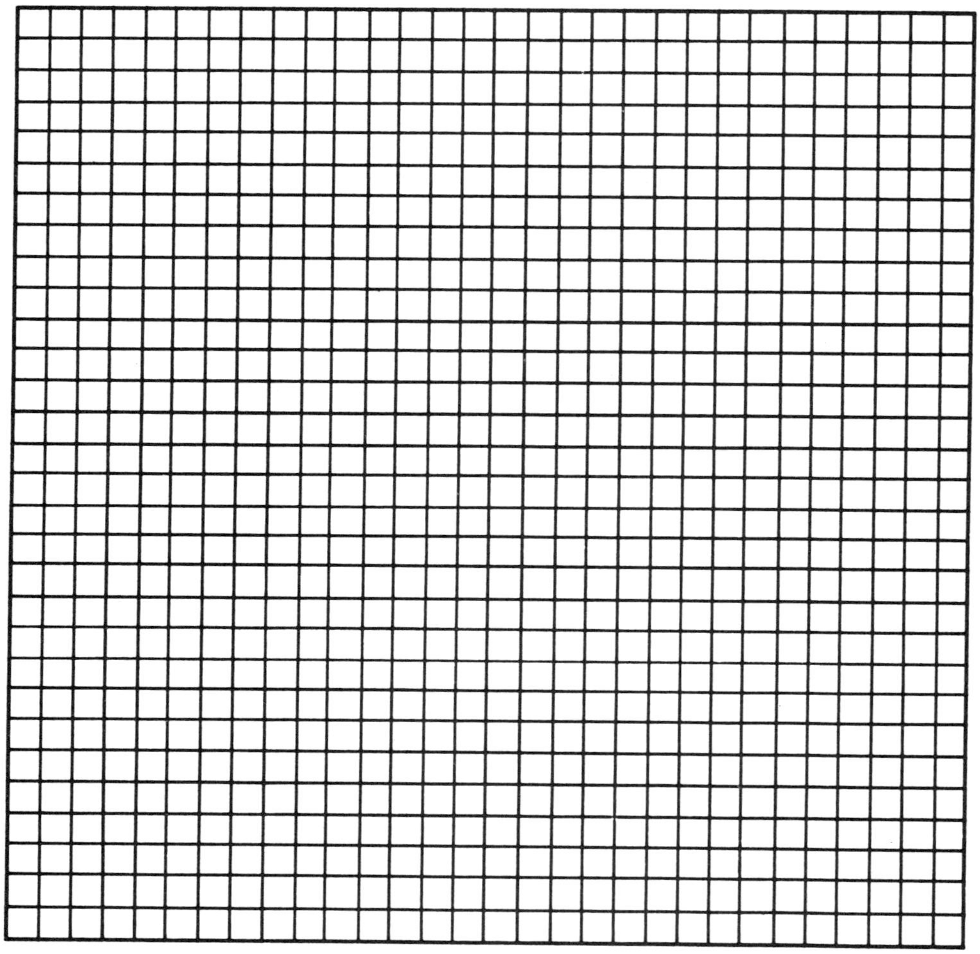

Volume, mL

Read from your graph:

(a) The pressure at 10.0 mL ____________________

(b) The volume at 700 torr ____________________

NAME ____________________

SECTION __________ DATE __________

INSTRUCTOR ____________________

## *EXERCISE 10*

# Moles

**Show calculation setups and answers for all problems.**

1. Find the molar mass of: (a) nitric acid, $HNO_3$; (b) sodium bicarbonate, $NaHCO_3$; and (c) cobalt(II) nitrate, $Co(NO_3)_2$.

(a) ______________

(b) ______________

(c) ______________

2. A sample of mercury(II) bromide, $HgBr_2$, weighs 9.42 g. How many moles are in this sample?

______________

3. What is the mass of 0.45 mol of ammonium carbonate, $(NH_4)_2CO_3$?

______________

4. How many molecules are contained in 6.53 mol of chlorine gas, $Cl_2$?

______________

5. Calculate the percent composition by mass of barium sulfite, $BaSO_3$.

Ba ______________

S ______________

O ______________

6. How many moles are present in 5.17 g of sodium chlorate, $NaClO_3$?

_______________

7. A sample of oxygen gas, $O_2$, weighs 30.0 g. How many molecules of $O_2$, and how many atoms of O are present in this sample?

_______________ **molecules** of $O_2$

_______________ **atoms** of O

8. A mixture of sand and salt is found to be 42 percent NaCl by mass. How many moles of NaCl are in 71 g of this mixture?

_______________

9. What is the mass of $2.2 \times 10^{23}$ molecules of carbon dioxide, $CO_2$?

_______________

10. A water solution of sulfuric acid has a density of 1.67 g/mL and is 75 percent $H_2SO_4$ by mass. How many moles of $H_2SO_4$ are contained in 500. mL of this solution?

_______________

NAME ____________________

SECTION __________ DATE __________

INSTRUCTOR ____________________

# EXERCISE 11

# Stoichiometry I

Show calculation setups and answers for all problems.

1. Use the equation given to solve the following problems:

   $Na_3PO_4 + 3\ AgNO_3 \longrightarrow Ag_3PO_4 + 3\ NaNO_3$

   (a) How many moles of $AgNO_3$ would be required to react with 1.0 mol of $Na_3PO_4$?

   ____________

   (b) How many moles of $NaNO_3$ can be produced from 0.30 mol of $Na_3PO_4$?

   ____________

   (c) How many grams of $Ag_3PO_4$ can be produced from 4.00 g of $AgNO_3$?

   ____________

   (d) If you have 8.44 g of $Na_3PO_4$, how many grams of $AgNO_3$ will be needed for complete reaction?

   ____________

   (e) When 25.0 g of $AgNO_3$ are reacted with excess $Na_3PO_4$, 17.7 g of $Ag_3PO_4$ are produced. What is the percentage yield of $Ag_3PO_4$?

   ____________

2. Use the equation given to solve the following problems:

$$2\,KMnO_4 + 16\,HCl \longrightarrow 5\,Cl_2 + 2\,KCl + 2\,MnCl_2 + 8\,H_2O$$

(a) How many moles of HCl are required to react with 45 g of $KMnO_4$?

____________________

(b) How many $Cl_2$ molecules will be produced using 5.0 mol of $KMnO_4$?

____________________

(c) To produce 55.0 g of $MnCl_2$, what mass of HCl will be reacted?

____________________

(d) How many moles of water will be produced when 7.0 mol of $KMnO_4$ are consumed?

____________________

(e) What is the maximum mass of $Cl_2$ that can be produced by reacting 35.0 g of $KMnO_4$ with 45.0 g of HCl?

____________________

NAME ____________________

SECTION __________ DATE __________

INSTRUCTOR ____________________

*EXERCISE 12*

# Gas Laws

**Show calculation setups and answers for all problems.**

1. A sample of nitrogen gas, $N_2$, occupies 3.0 L at a pressure of 3.0 atm. What volume will it occupy when the pressure is reduced to 1.0 atm and the temperature remains constant?

   ____________________

2. A sample of methane gas, $CH_4$, occupies 3.50 L at a temperature of 20.0°C. If the pressure is held constant, what will be the volume of the gas at 100.°C?

   ____________________

3. The pressure of hydrogen gas in a constant-volume cylinder is 2.25 atm at 0°C. What will the pressure be if the temperature is raised to 80°C?

   ____________________

4. A 325 mL sample of air is at 720. torr and 30°C. What volume will this gas occupy at 900. torr and 60.°C?

   ____________________

5. A sample of gas occupies 500. mL at STP. What volume will the gas occupy at 90.°C and 510. torr?

   ____________________

6. A quantity of oxygen occupies a volume of 17.7 L at STP. How many moles of oxygen are present?

____________

7. A 660. mL volume of hydrogen chloride gas, HCl, is collected at 27°C and 720. torr. What volume will it occupy at STP?

____________

8. What volume would 9.5 g of nitrogen gas, $N_2$, occupy at 200. K and 2.0 atm?

____________

9. Calculate the density of ammonia gas, $NH_3$, at STP.

____________

10. In a laboratory experiment, 259 mL of gas was collected over water at 24°C and 751 torr. Calculate the volume that the dry gas would occupy at STP.

____________

11. A volume of 50.0 mL of argon, Ar, is collected at 60°C and 820. torr. What does this sample weigh?

____________

NAME ____________________

SECTION __________ DATE __________

INSTRUCTOR ____________________

*EXERCISE 13*

# Solution Concentrations

**Show calculation setups and answers for all problems.**

1. What will be the percent composition by mass of a solution made by dissolving 12.0 g of zinc nitrate, $Zn(NO_3)_2$, in 45.0 g of water?

   **$Zn(NO_3)_2$** ____________________

   **$H_2O$** ____________________

2. **How many moles of sodium hydroxide, NaOH, are required to prepare 2.00 L of 0.380 M solution?**

   ____________________

3. What will be the molarity of a solution if 3.50 g of potassium hydroxide, KOH, are dissolved in water to make 150. mL of solution?

   ____________________

4. How many milliliters of 0.400 M solution can be prepared by dissolving 5.00 g of KOH in water?

   ____________________

5. How many grams of potassium bromide, KBr, could be recovered by evaporating 650. g of 15 percent KBr solution to dryness?

   ____________________

6. How many milliliters of 1.00 M HCl is needed to prepare 300. mL of 0.250 M HCl solution?

____________

7. A sample of potassium hydrogen oxalate, $KHC_2O_4$, weighing 0.717 g, was dissolved in water and titrated with 18.47 mL of an NaOH solution. Calculate the molarity of the NaOH solution.

____________

8. How many grams of hydrogen chloride are in 35 mL of concentrated (12 M) HCl solution?

____________

9. A sulfuric acid solution has a density of 1.73 g/mL and contains 80. percent $H_2SO_4$ by mass. What is the molarity of this solution?

____________

10. Sulfuric acid reacts with sodium hydroxide according to this equation:

$$H_2SO_4 + 2\,NaOH \longrightarrow Na_2SO_4 + 2\,H_2O$$

A 10.00 mL sample of the $H_2SO_4$ solution required 13.71 mL of 0.309 M NaOH for neutralization. Calculate the molarity of the acid.

____________

NAME____________________

SECTION________ DATE________

INSTRUCTOR____________________

## *EXERCISE 14*

# Stoichiometry II

Show calculation setups and answers for all problems.

1. Use the equation given to solve the following problems:

$$6\,KI + 8\,HNO_3 \longrightarrow 6\,KNO_3 + 2\,NO + 3\,I_2 + 4\,H_2O$$

(a) If 38 g of KI are reacted, how many grams of $KNO_3$ will be formed?

________________

(b) What volume of NO gas, measured at STP, will be produced if 37.0 g of $HNO_3$ are consumed?

________________

(c) If 0.500 mol of KI is to be reacted, how many milliliters of 6.00 M $HNO_3$ will be required?

________________

(d) When the reaction produces 6.0 mol of NO, how many molecules of $I_2$ will be produced?

________________

(e) How many grams of iodine can be obtained by reacting 45.0 mL of 0.350 M KI solution?

________________

2. Use the equation given to solve the following problems. All substances are in the gas phase.

$$N_2(g) + 3\,H_2(g) \longrightarrow 2\,NH_3(g)$$

(a) If 3.0 mol of $N_2$ react, how many moles of $NH_3$ will be formed?

____________

(b) When 3.50 mol of $N_2$ react, what volume of $NH_3$, measured at STP, will be formed?

____________

(c) If 9.0 L of $H_2$ are reacted, what volume of $NH_3$ will be formed? All volumes are measured at STP.

____________

(d) How many molecules of $NH_3$ will be formed when 20.0 L of $N_2$ at STP react?

____________

(e) What volume of $NH_3$, measured at 45°C and 710. torr, will be produced from 18.0 g of $H_2$?

____________

(f) If a mixture of 8.00 L of $N_2$ and 20.0 L of $H_2$ are reacted, what volume of $NH_3$ can be produced? Assume STP conditions.

____________

NAME ______________________

SECTION ______ DATE __________

INSTRUCTOR __________________

# EXERCISE 15

## Chemical Equilibrium

1. Consider the following system at equilibrium:

$$2\ CO_2(g) + 135.2\ \text{kcal} \rightleftarrows 2\ CO(g) + O_2(g)$$

Complete the following table. Indicate changes in moles and concentrations by entering I, D, N, or ? in the table. (I = increase, D = decrease, N = no change, ? = insufficient information to determine.)

| Change or stress imposed on the system at equilibrium | Direction of shift, left or right, to reestablish equilibrium | Change in number of moles | | | Change in molar concentrations | | |
|---|---|---|---|---|---|---|---|
| | | $CO_2$ | CO | $O_2$ | $CO_2$ | CO | $O_2$ |
| a. Add CO | | | | | | | |
| b. Remove $CO_2$ | | | | | | | |
| c. Decrease volume of reaction vessel | | | | | | | |
| d. Increase temperature | | | | | | | |
| e. Add catalyst | | | | | | | |
| f. Add both $CO_2$ and $O_2$ | | | | | | | |

2. Consider the reaction $PCl_5(g) \rightleftarrows PCl_3(g) + Cl_2(g)$

At 250°C, $PCl_5$ is 45% decomposed.

(a) If 0.110 mol of $PCl_5(g)$ is introduced into a 1.00 L container at 250°C, what will be the equilibrium concentrations of $PCl_5$, $PCl_3$, and $Cl_2$?

$PCl_5$ __________

$PCl_3$ __________

$Cl_2$ __________

(b) What is the value of $K_{eq}$ at 250°C?

$K_{eq}$ __________

3. For the reaction $H_2(g) + I_2(g) \rightleftarrows 2\ HI(g)$, $K_{eq} = 0.17$ at 500 K. What concentration of $I_2(g)$ will be in equilibrium with $H_2 = 0.040$ M, $HI = 0.015$ M?

______________________

4. $CaCO_3$ has a solubility in water of $6.9 \times 10^{-5}$ mol/L. Calculate the solubility product constant.

______________________

5. A 0.40 M HClO solution was found to have an $H^+$ concentration of $1.1 \times 10^{-4}$ M. Calculate the value of the ionization constant. The ionization equation is $HClO \rightleftarrows H^+ + ClO^-$.

______________________

6. Calculate (a) the $H^+$ ion concentration, (b) the pH, and (c) the percent ionization of a 0.40 M solution of $HC_2H_3O_2$ ($K_a = 1.8 \times 10^{-5}$).

(a) ______________________

(b) ______________________

(c) ______________________

NAME____________________

SECTION________ DATE__________

INSTRUCTOR__________________

# *EXERCISE 16*

## Oxidation-Reduction Equations I

Balance the following oxidation-reduction equations:

1. $P + HNO_3 + H_2O \longrightarrow H_3PO_4 + NO$

2. $H_2SO_4 + HI \longrightarrow H_2S + I_2 + H_2O$

3. $KBrO_2 + KI + HBr \longrightarrow KBr + I_2 + H_2O$

4. $Sb + HNO_3 \longrightarrow Sb_2O_5 + NO + H_2O$

5. $NO_2 + H_2O \longrightarrow HNO_3 + NO$

6. $Br_2$ + $NH_3$ → $NH_4Br$ + $N_2$

7. KI + $HNO_3$ → $KNO_3$ + NO + $I_2$ + $H_2O$

8. $H_2SO_3$ + $KMnO_4$ → $MnSO_4$ + $H_2SO_4$ + $K_2SO_4$ + $H_2O$

9. $K_2Cr_2O_7$ + $H_2O$ + S → $SO_2$ + KOH + $Cr_2O_3$

10. $KMnO_4$ + HCl → $Cl_2$ + KCl + $MnCl_2$ + $H_2O$

NAME ____________________

SECTION __________ DATE __________

INSTRUCTOR ____________________

# *EXERCISE 17*

## Oxidation-Reduction Equations II

Balance the following oxidation-reduction equations using the ion-electron method.

1. $MnO_4^- + \quad Cl^- + \quad H^+ \longrightarrow \quad Mn^{2+} + \quad Cl_2 + \quad H_2O$

2. $Ag_2S + \quad NO_3^- + \quad H^+ \longrightarrow \quad S + \quad NO + \quad Ag^+ + \quad H_2O$

3. $ClO_4^- + \quad I^- + \quad H^+ \longrightarrow \quad I_2 + \quad Cl^- + \quad H_2O$

4. $Br_2 + \quad H_2O \longrightarrow \quad BrO_3^- + \quad Br^- + \quad H^+$

5. $MnO_4^- + \quad HS^- + \quad H_2O \longrightarrow \quad S + \quad MnO_2 + \quad OH^-$

Balance the following oxidation-reduction equations using the ion-electron method.

6. $H_2O_2 + IO_3^- \longrightarrow I^- + O_2$ (acid solution)

7. $Cl_2 + SO_2 \longrightarrow SO_4^{2-} + Cl^-$ (acid solution)

8. $U^{4+} + MnO_4^- \longrightarrow Mn^{2+} + UO_2^{2+}$ (acid solution)

9. $Fe(CN)_6^{3-} + Cr_2O_3 \longrightarrow Fe(CN)_6^{4-} + CrO_4^{2-}$ (basic solution)

10. $Cr(OH)_3 + O_2^{2-} \longrightarrow CrO_4^{2-}$ (basic solution)

NAME ______________________

SECTION ________ DATE ____________

# *EXERCISE 18*

INSTRUCTOR ______________________

## Hydrocarbons

1. Write the structural formulas for the five isomers of hexane, all having the molecular formula $C_6H_{14}$.

2. Write the structural formulas for the six isomers of pentene, all having the formula $C_5H_{10}$.

3. Write the structural formulas for (a) acetylene, (b) isobutane, (c) benzene, (d) *n*-octane.

(a)

(b)

(c)

(d)

4. Write the formulas for the products of the following reactions:

(a) 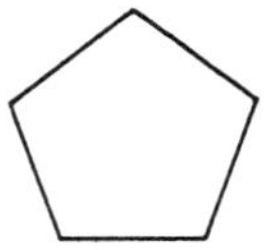 + $Br_2$ $\xrightarrow[\text{light}]{\text{UV}}$

(b) $CH_2{=}CH{-}CH_3$ + $Cl_2$ $\xrightarrow{\text{dark}}$

(c) 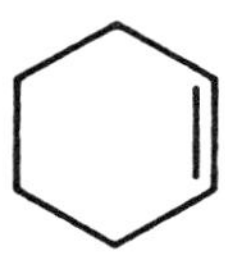 + $H_2$ $\xrightarrow[\text{and pressure}]{\text{catalyst, heat}}$

5. Complete and balance the following combustion reactions (assume complete combustion).

(a) $CH_3{-}CH_3$ + $O_2$ $\xrightarrow{\triangle}$

(b) $CH_3{-}C{\equiv}C{-}CH_3$ + $O_2$ $\xrightarrow{\triangle}$

(c) 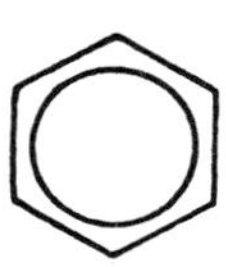 + $O_2$ $\xrightarrow{\triangle}$

NAME ______________________

SECTION ________ DATE __________

# EXERCISE 19

INSTRUCTOR ____________________

## Alcohols, Esters, Aldehydes, and Ketones

1. Write the structural formulas of eight isomers of pentanol, all having the formula $C_5H_{12}O$.

2. Write the structural formulas of the ketones having the formula $C_5H_{10}O$.

3. Write the structural formulas of the aldehydes having the formula $C_5H_{10}O$.

4. Write the formulas for the products of the following reactions:

(a) $CH_3CH_2CH_2OH + CuO \xrightarrow{\Delta}$

(b) 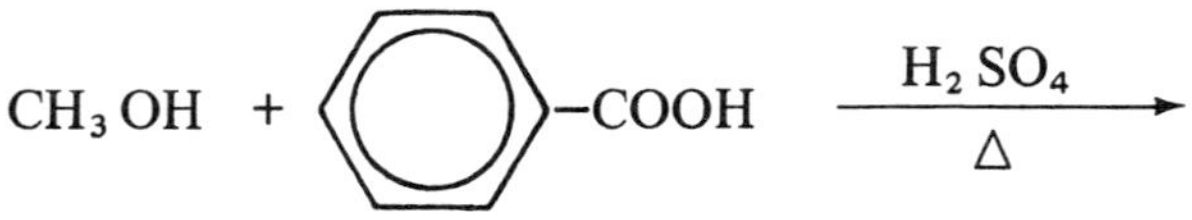

(c) $2\ CH_3CH_2COOH + HOCH_2CH_2CH_2OH \xrightarrow[\Delta]{H_2SO_4}$

(d) 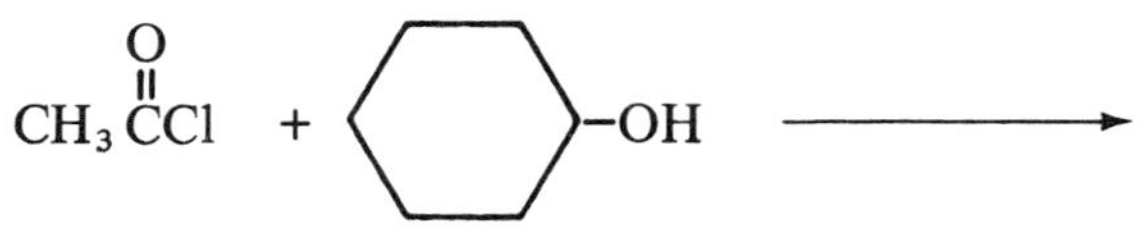

(e) 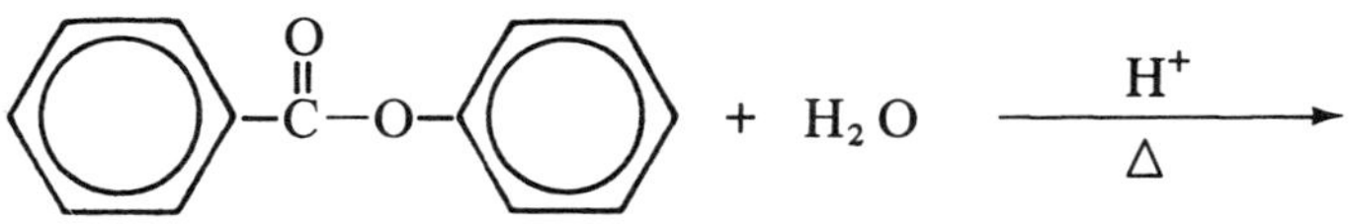

(f) $CH_3\overset{O}{\overset{\|}{C}}-O-\underset{CH_3}{\underset{|}{C}}HCH_3 + NaOH \xrightarrow[\Delta]{H_2O}$

5. Complete and balance the following combustion reactions (assume complete combustion).

(a) $CH_3-CH(OH)-CH_3 + O_2 \longrightarrow$

(b) 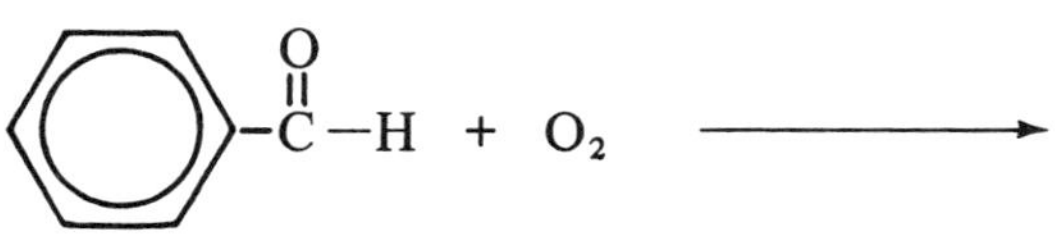

NAME ______________________

SECTION __________ DATE __________

INSTRUCTOR ______________________

## *EXERCISE 20*

# Functional Groups

1. Identify each functional group by name and circle the functional group on the formula.

(a) $CH_3-\underset{\displaystyle OH}{\underset{|}{CH}}-CH_3$

______________________

(b) 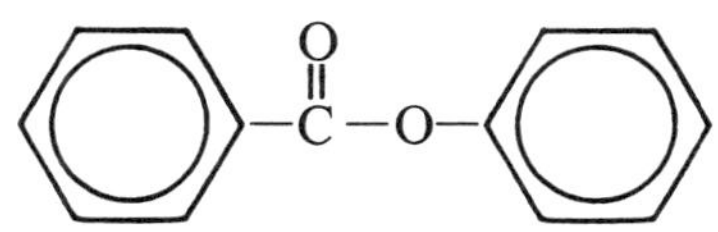

______________________

(c) $CH_3-\overset{\displaystyle O}{\overset{\|}{C}}-OH$

______________________

(d) $HO-CH_2-\overset{\displaystyle CH_3}{\overset{|}{\underset{\displaystyle CH_3}{\underset{|}{C}}}}-CH_3$

______________________

(e) $CH_2=CH-CH_3$

______________________

(f) $CH_3-\overset{\displaystyle O}{\overset{\|}{C}}-CH_2-CH_3$

______________________

(g) 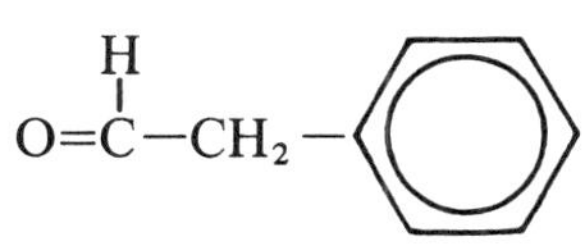

______________________

(h) $CH_3-CH_2-\overset{\displaystyle O}{\overset{\|}{C}}-O-\overset{\displaystyle O}{\overset{\|}{C}}-CH_2-CH_3$

______________________

(i) $CH_3-CH_2-NH_2$

______________________

(j) $CH_3-\overset{\displaystyle H}{\overset{|}{N}}-\underset{\displaystyle O}{\underset{\|}{C}}-CH_2-CH_3$

______________________

2. Name each of the compounds listed in Question 1.

(a) ____________________________________________

(b) ____________________________________________

(c) ____________________________________________

(d) ____________________________________________

(e) ____________________________________________

(f) ____________________________________________

(g) ____________________________________________

(h) ____________________________________________

(i) ____________________________________________

(j) ____________________________________________

NAME ______________________

SECTION __________ DATE __________

INSTRUCTOR ______________________

## EXERCISE 21

# Synthetic Polymers

1. Complete the following reactions showing two units of the polymers formed.

(a) $CH_3-CH=CH_2 \xrightarrow[\text{peroxide}]{\text{Organic}}$

(b) $CCl_2=CH_2 \xrightarrow[\text{peroxide}]{\text{Organic}}$

(c) 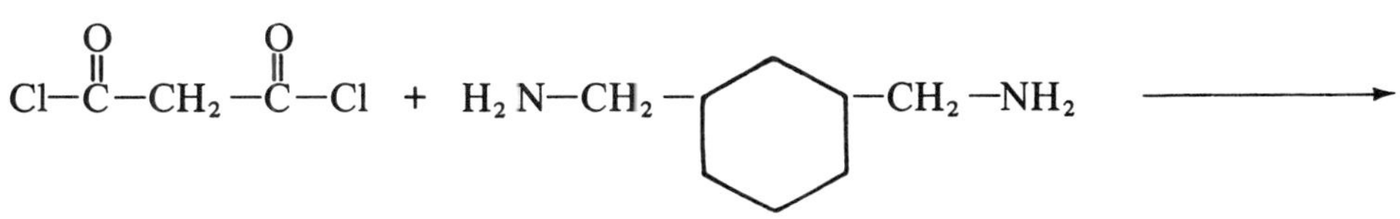

(d) 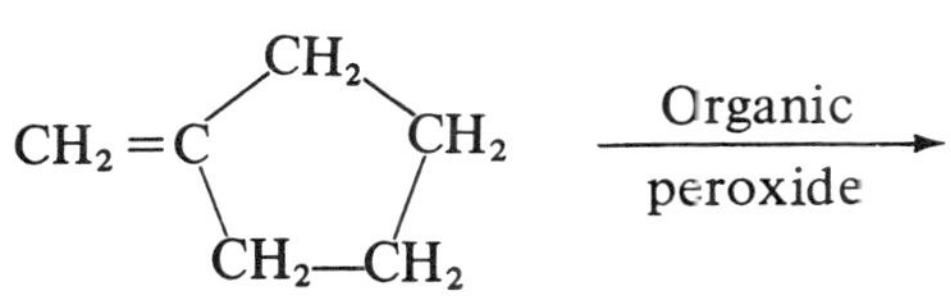

(e) 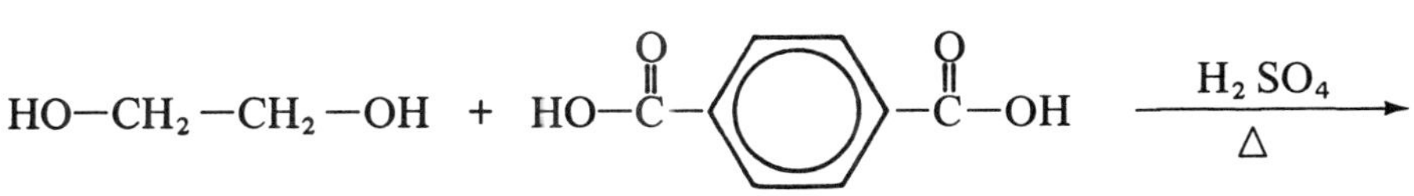

2. Why is a polymer usually a solid?

3. Write the structural formulas of the monomers from which the following polymers were obtained.

(a) $\sim\!\!\left(CH_2-CCl_2-CHCl-CH_2-CH_2-CCl_2-CH_2-CHCl\right)_n\!\!\sim$

(b) 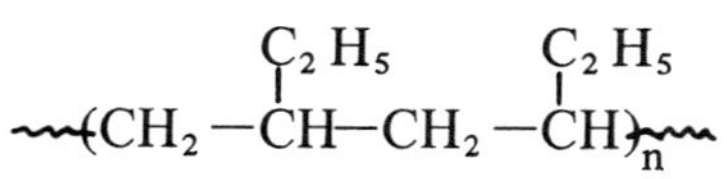

(c) 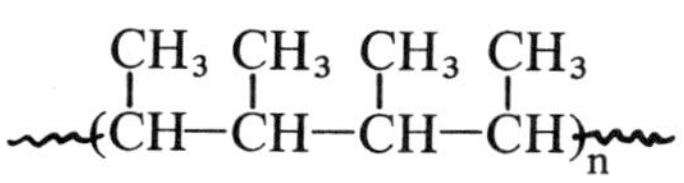

(d) 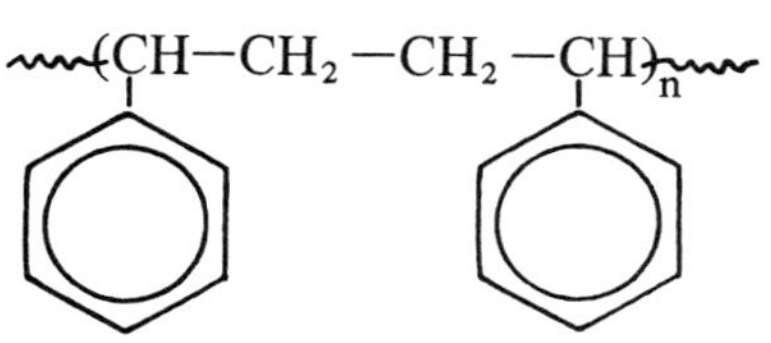

(e) 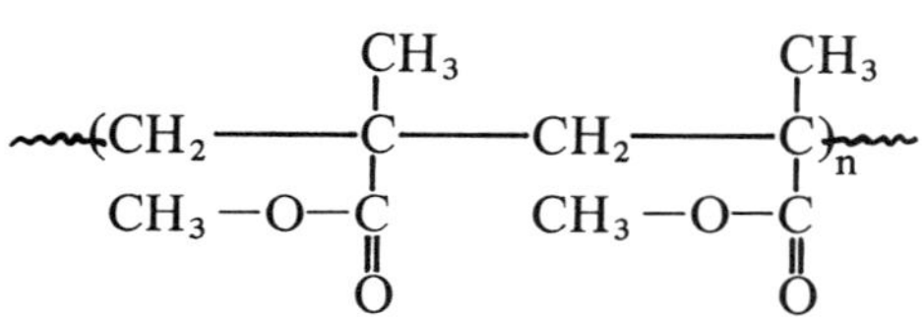

(f) 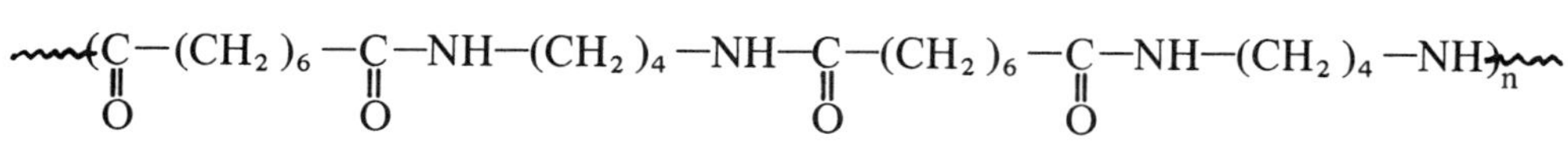

(g) $\sim\!\!\left(CH_2-O-CH_2-O\right)_n\!\!\sim$

NAME ____________________

SECTION ________ DATE ____________

INSTRUCTOR ____________________

## *EXERCISE 22*

# Carbohydrates

1. From the projection formulas given, write the Haworth perspective structures for α-D-glucopyranose and β-D-mannopyranose.

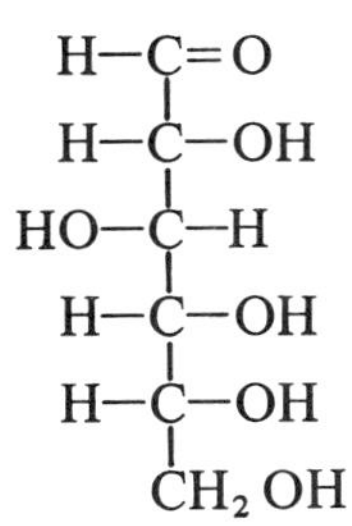

D-glucose

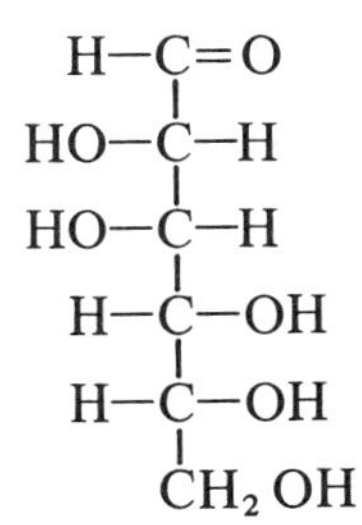

D-mannose

2. Complete and balance:

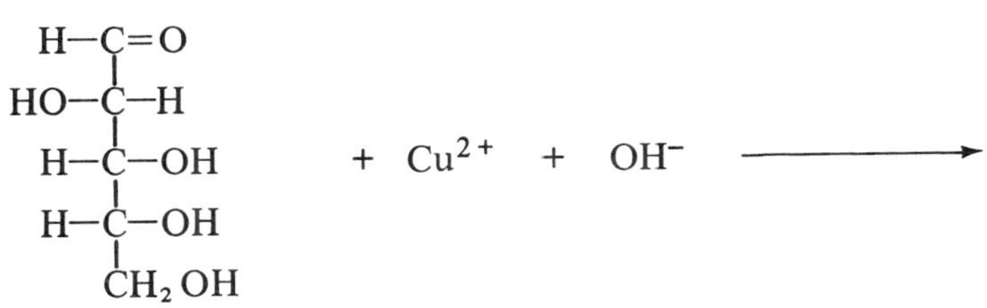

3. Write Fischer projection formulas for (a) a D-aldotetrose, (b) a D-ketopentose, and (c) an L-aldohexose.

4. What is the monosaccharide composition of

(a) Maltose

(b) Sucrose

(c) Lactose

(d) Cellulose

(e) Starch

(f) Glycogen

5. Circle the formulas of the reducing sugars:

(a)

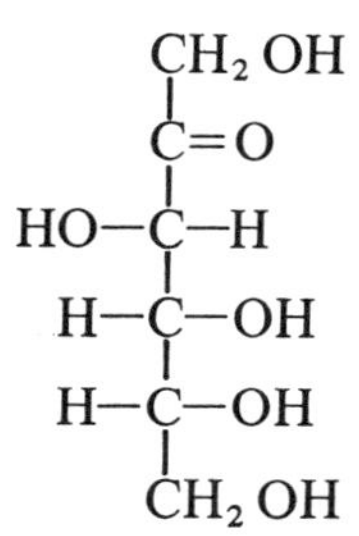

(b)

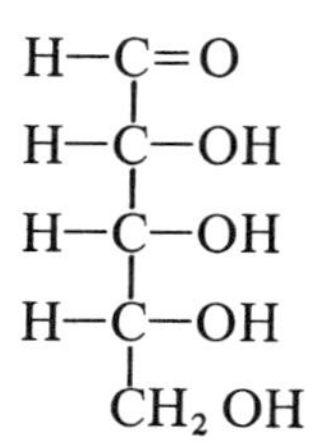

(c)

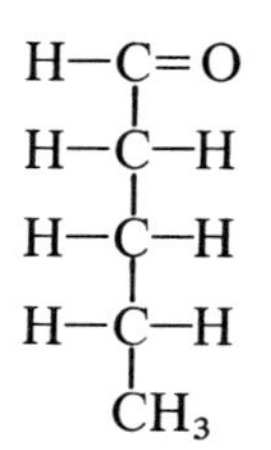

(d)

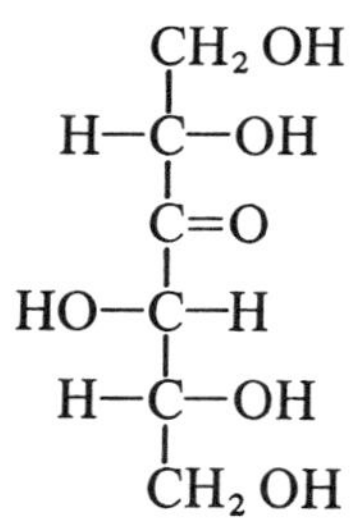

(e)

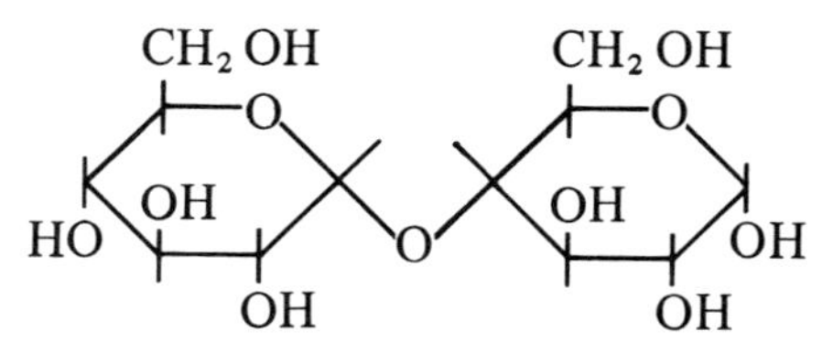

NAME ____________________

SECTION ________ DATE __________

*EXERCISE 23* INSTRUCTOR ______________

# Amino Acids and Polypeptides

1. Write structural formulas for:

   (a) Lysyltyrosine (Lys-Tyr)

   (b) Glycylvalylarginine (Gly-Val-Arg)

2. Nutrasweet artificial sweetener is a methyl ester of a dipeptide. The dipeptide is aspartylphenylalanine methyl ester. The methyl ester is on the carboxyl group of phenylalanine. Draw its structure.

3. Draw and name the hydrolysis products produced from the compound shown:

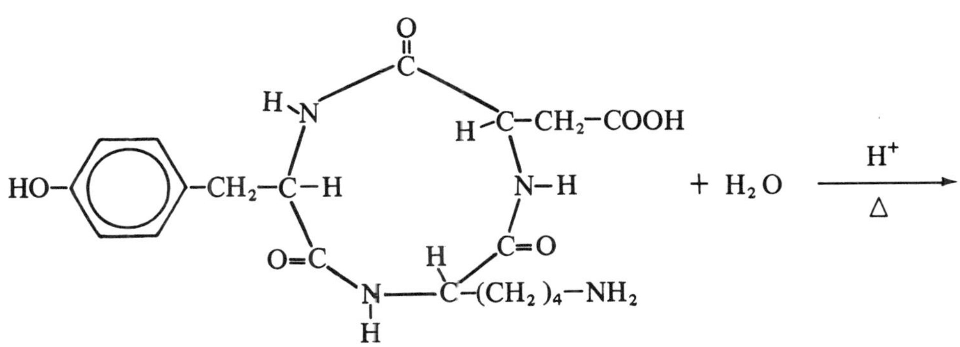

4. Indicate a positive result as (+), negative result as (—) for each of the following tests on polypeptides A and B:
(A) contains 3 alanine, 2 aspartic acid, 2 cysteine, 4 glycine, 2 histidine, 1 methionine, and 3 proline residues.
(B) contains 4 leucine, 5 glycine, 3 tryptophan, 3 valine, 2 serine, and 1 tyrosine residues.

| Test | Results on Polypeptide A | Results on Polypeptide B |
|---|---|---|
| Biuret | | |
| Ninhydrin | | |
| Xanthoproteic | | |
| Sulfur detection | | |

5. What is the sequence of amino acids in an octapeptide that contains one residue each of Glu, Arg, Thr, and Tyr, and two residues each of Phe and Gly, and hydrolyzes to the following fragments: (a) Arg-Gly-Phe, (b) Tyr-Thr, (c) Gly-Glu-Arg, and (d) Phe-Phe-Tyr?

NAME ____________________

SECTION __________ DATE __________

INSTRUCTOR ____________________

*EXERCISE 24*

# Lipids

1. Complete the following reactions:

(a)

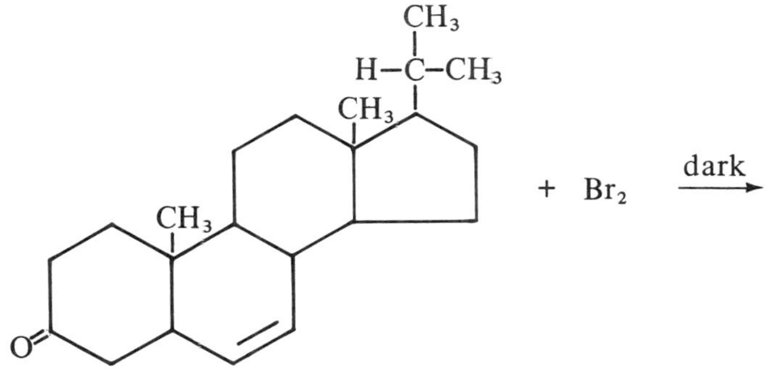

(b)

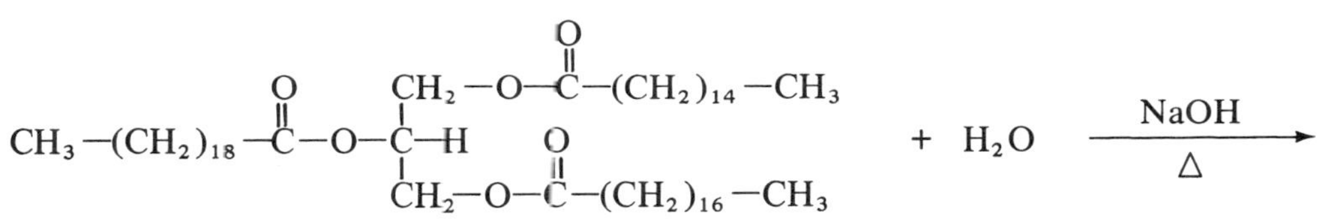

(c)

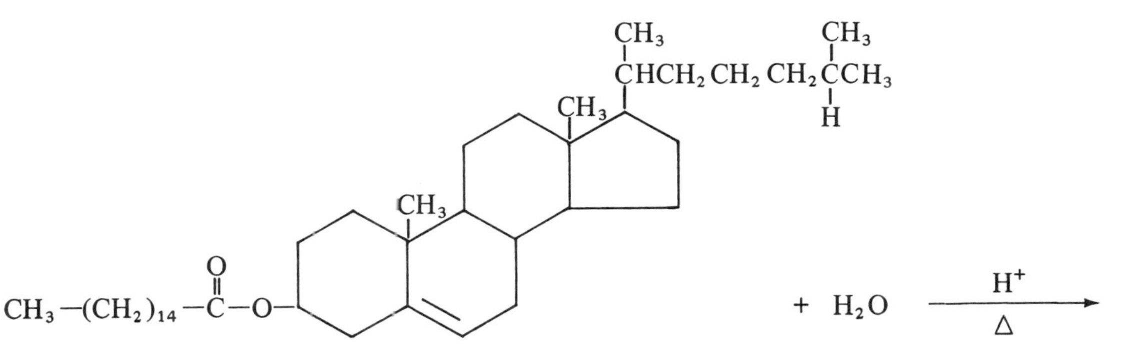

(d) $CH_2OH{-}CHOH{-}CH_2OH + 3\ CH_3(CH_2)_7CH{=}CH(CH_2)_7COOH \xrightarrow[\Delta]{H^+}$

(e) Phosphatidylserine, at pH 7, has the structure shown below. It is a major component of most membranes in cells. The R groups can be various fatty acid chains.

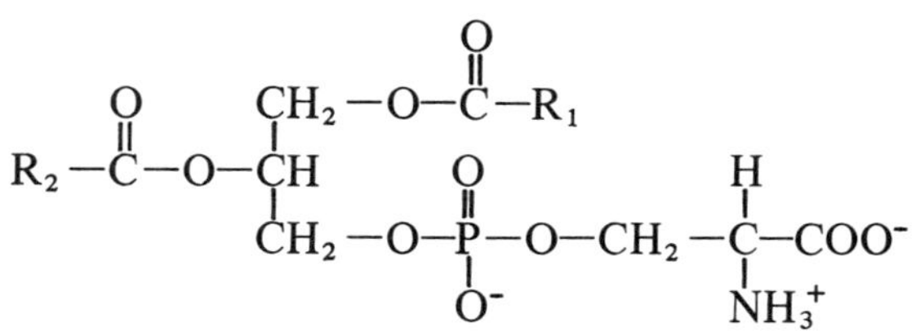

$+ H_2O \xrightarrow[\Delta]{H^+}$

NAME ____________________

SECTION __________ DATE __________

# *EXERCISE 25*

INSTRUCTOR ____________________

## Molecular Models and Isomerism

The objective of this exercise is to become familiar with the spacial arrangement of oragnic molecules and to observe the phenomenon of isomerism.

**Bonding rules:** Carbon must have 4 bonds to it; hydrogen 1 bond; oxygen 2 bonds; and chlorine 1 bond.

1. Construct four methane ($CH_4$) molecules using 1 C, 4 H, and 4 bonds for each molecule. Draw the structure of this $CH_4$ molecule.

2. Remove an H atom from each of two $CH_4$ molecules. You now have two methyl groups, $CH_3$—. Bond these two methyl groups together using a bond to give a model of ethane, $CH_3$— $CH_3(C_2H_6)$. Draw the structure of this ethane molecule.

3. Remove an H atom from the ethane molecule and an H atom from another methane molecule. You now have an ethyl, $CH_3CH_2$—, and a methyl, $CH_3$—, group. Bond these groups together with a bond to form a molecule of propane, $CH_3CH_2CH_3(C_3H_8)$. Draw the structure of propane.

4. Construct two ethyl groups and bond them together. You now have a model of butane, $CH_3CH_2CH_2CH_3(C_4H_{10})$. Draw the structure of this molecule.

5. Rearrange the butane model (from #4) to form isobutane, $C_4H_{10}$: (a) remove one methyl group from the butane model; (b) now remove one H from the central carbon atom of the remaining propyl, $CH_3CH_2CH_2$— group. Bond the methyl group to the central carbon atom and the hydrogen atom to the end carbon atom. The compound you now have is isobutane. Draw its structure.

6. Redraw the two butanes and compare their structures. This is an example of isomerism. **Isomerism** is the phenomenon of two or more compounds having the same molecular formula but different structural formulas.

7. Construct two molecules of propane, $CH_3CH_2CH_3$. On one molecule replace an H atom from an end carbon with a Cl atom. On the other molecule replace an H atom from the central carbon atom by a Cl atom. **This is another example of isomerism.** Both molecules have the formula $C_3H_7Cl$. The first molecule is named propyl chloride or 1-chloropropane; the second molecule is named isopropyl chloride or 2-chloropropane. Draw and label both structures.

8. Construct two different molecules, each having the molecular formula $C_2H_6O$. Follow the rules for bonding given earlier. One compound is called dimethyl ether and the other one is called ethyl alcohol or ethanol. Draw their structural formulas and give each its proper name.

9. If an H atom is replaced by a Cl atom in a molecule of butane, how many structural isomers of chlorobutane, $C_4H_9Cl$, can be formed? Draw the structures for these isomers.

10. (a) Suppose that two H atoms are replaced by two Cl atoms in a molecule of ethane. How many isomers can be formed? Draw their structural formulas. The molecular formula is $C_2H_4Cl_2$.

(b) Suppose that two Cl atoms replaced two H atoms in a molecule of propane. How many isomers can be formed? Draw their structural formulas. The molecular formula is $C_3H_6Cl_2$. Construct models if necessary.

11. Construct and draw models of the following alkyl groups:
methyl, ethyl, propyl, isopropyl, butyl, sec-butyl, isobutyl, and t-butyl

12. Construct and draw models of:

| pentane | 2-methylpentane | cyclopentane |
|---|---|---|
| $C_5H_{12}$ | $C_6H_{14}$ | $C_5H_{10}$ |

NAME ______________________

SECTION ________ DATE __________

INSTRUCTOR ________________

# EXERCISE 26

## Stereoisomerism—Optical Isomers

Stereoisomers are compounds that have the same structural formulas, but differ in their spacial arrangements. Two major types of stereoisomers are geometric isomers (cis-trans) and optical isomers. We will construct and study models of optical isomers in this exercise. The outstanding property of optical isomers is their optical activity; that is, the ability to rotate plane-polarized light. We will not study optical activity per se in this exercise, but we will observe the spacial structures and the criteria for compounds which have this property.

**Definitions**

**Asymmetric carbon atom:** A carbon atom that is bonded to four different atoms or groups of atoms.

**Superimposable molecules:** When one molecule is laid on top of another molecule and all the atoms of both molecules coincide exactly, the molecules are identical and are said to be superimposable on one another.

**Optical activity:** Molecules that have the ability to rotate plane polarized light are said to have optical activity.

**Mirror image isomers:** Non-superimposable molecules that appear as mirror images of one another (like the right and left hands).

**Enantiomers:** Non-superimposable molecules that appear as mirror images of one another are called enantiomers. Enantiomers have the property of optical activity.

**Chiral:** A molecule that is not superimposable on its mirror image is said to be chiral or to have chirality.

**Achiral:** A molecule that is superimposable on its mirror image is said to be achiral (it does not have chirality).

**Projection formula:** The three dimensional and projection formulas for 1-bromo-1-chloroethane, $CH_3CHBrCl$, are illustrated as follows:

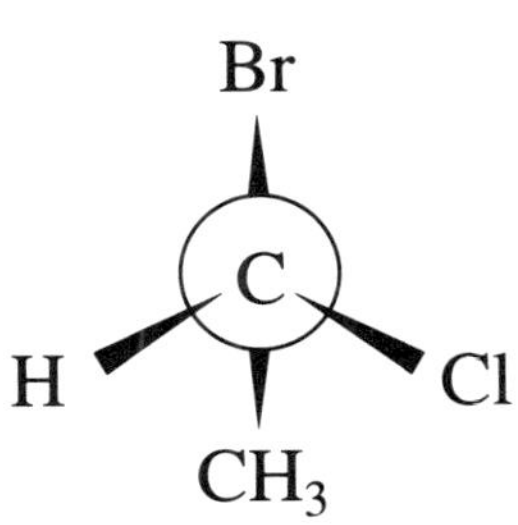

three-dimensional formula

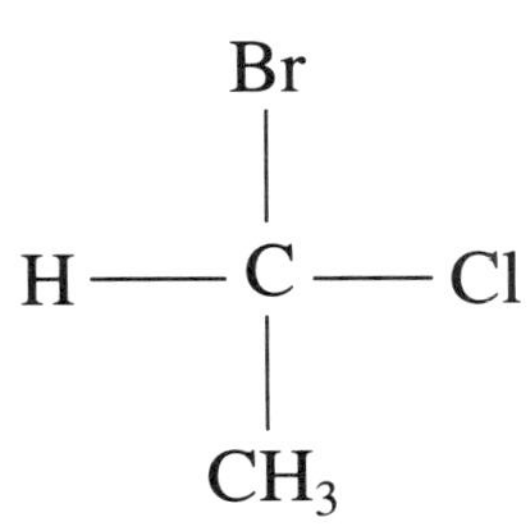

projection formula

C-1 is an asymmetric carbon atom with four different groups attached to it: $CH_3$, H, Cl, and Br. In the three dimensional formula the bonds from carbon to H and Cl are coming out of the plane of the paper toward the observer. The bonds from carbon to Br and $CH_3$ are projecting away from the observer. When we squash the molecule flat to illustrate it on a two-dimensional basis, this two-dimensional formula is known as a *projection formula*. In the projection formula it is understood that the H and Cl are coming out of the plane of the paper toward the observer and Br and $CH_3$ are going away from the observer.

**Procedure**

**A.** Construct two models of 1-bromo-1-chloroethane. Test to see if the two models are superimposable on one another. If they are superimposable, change the position of H and Cl on one model. Now you have two models that are not superimposable. These two are enantiomers, non-superimposable mirror image isomers. (Turn, twist, rotate, do anything that you want to these models except interchange the positions of the groups; they remain non-superimposable.) Position these enantiomers so that they appear as mirror images. Arrange them so that the Br is on top and the $CH_3$ is on the bottom of the asymmetric carbon atom, and the H and Cl are coming out of the plane toward you. Now draw the projection formulas for these enantiomers.

Since these two molecules are not superimposable they must indeed be different compounds, for only superimposable molecules are identical.

As you observe these two projection structures (and the two models) you should note that they are mirror image isomers. It is as if you had only one structure placed alongside a mirror which reflects its mirror image. With respect to the optical properties of these enantiomers, one isomer will rotate plane-polarized light to the right and is said to be *dextrorotatory;* the other isomer will rotate plane-polarized light to the left and is said to be *levorotatory*. The number of degrees that the light is rotated is not important to this exercise.

**B.** Construct two models of lactic acid.

```
  3    2    1
CH3–CH–C=O
      |    \
     OH    OH
```

C-2 is an asymmetric carbon atom.

Test to see if the two models are superimposable. If they are superimposable exchange the H and OH groups on C-2, which will make them non-superimposable. Once again the two non-superimposable isomers are enantiomers, and one compound will rotate plane-polarized light to the right and the other will rotate the light to the left.

Position the models so that they appear as mirror images. Arrange them so that the COOH group is on the top and the $CH_3$ is on the bottom of the asymmetric carbon atom, and the H and OH are coming toward you. Now draw projection structures of the two lactic acid enantiomers and observe that they are mirror images.

**Question 1.** One criteria for an organic compound to show optical activity is that the compound has at least one asymmetric carbon atom. Circle any of the following compounds that will show optical activity.

(a) $CH_2Cl_2$ (b) $CH_2ClBr$ (c) $CHClBrI$

(d) $CH_3CHCl_2$ (e) $CH_2Cl\text{-}CH_2Cl$ (f) $CH_3CHBrCl$

(g) $CH_2Cl\text{-}CH_2Br$ (h) $CH_3CH_2CH(CH_3)CH_2CH_3$ (i) $CH_3CH_2CH(CH_3)CH_2CH_2CH_3$

(j) $CH_3CBrCl\text{-}CH_2CH(CH_3)CH_3$ (k) $CH_2Cl\text{-}CHCl\text{-}CH_2Cl$

A compound that has only one asymmetric carbon atom can exist as two optically active isomers (enantiomers). In fact, the maximum number of optical isomers of a compound that may exist is related to the number of asymmetric carbon atoms and can be calculated from the formula $\mathbf{2^n}$, where n is the number of asymmetric carbon atoms in the compound. For example, when n = 1, two isomers exist; when n = 2, four isomers may exist; when n = 3, eight isomers may exist, etc. The formula $\mathbf{2^n}$ only gives you the maximum number of optical isomers that may exist. Other structural features of the compound may preclude some isomers from existing.

**Question 2.** What is the maximum number of stereoisomers for the following compounds:

(a) $CH_2ClCH_2Br$ ____ (b) $CHClBr\text{-}CHClBr$ ____ (c) $CH_2Cl\text{-}CHBr\text{-}CH_2Cl$ ____

(d) $CH_3CHClCHClCH_3$ ____ (e) $CH_2{=}CH\text{-}CHCl\text{-}CH_3$ ____ (f) $CH_3(CHCl)_3CH_2CH_3$ ____

(g) $CH_2Cl\text{-}CHCl\text{-}COOH$ ____ (h) $CH_2Cl\text{-}CHCl\text{-}CHBr\text{-}CH(OH)\text{-}CH_3$ ____

**C. (a)** Construct four models of the compound $\overset{1}{C}H_3\overset{2}{C}HCl\overset{3}{C}HBr\overset{4}{C}H_3$. Take two models and hold them with the two $CH_3$ groups away from you. Now arrange the Cl on C-2 and the Br on C-3 so that both groups are on your right on one model and both are on your left on the other model. Test to see if these models are superimposable or are enantiomers (mirror image isomers). Draw projection formulas for these models. If they are the same, only one structure needs to be drawn. If they are enantiomers, draw both structures as mirror image isomers.

**(b)** Take the other two models and arrange them as follows: with the two $CH_3$ groups away from you place the Cl on the right of C-2 and the Br on the left of C-3 in one model, and the Cl on the left of C-2 and the Br on the right of C-3 on the other model. Test to see if these two models are superimposable or enantiomers. Draw their projection formulas.

You now have two pairs of enantiomers in these four structures. This corresponds to the four stereoisomers for a compound with two asymmetric carbon atoms as indicated by the formula $\mathbf{2^n}$ ($2^2 = 4$). All four of these isomers are optically active. If we label these isomers as E-I and E-II for one set of enantiomers and E-III and E-IV for the other enantiomers, E-I and E-II are said to be diastereoisomers or diastereomers of E-III and E-IV. Thus, for example, E-I and E-III or E-IV are diastereomers. *Diastereomers* are stereoisomers of the same compound, but are not enantiomers.

**D. (a)** Construct four models of 2, 3-dichlorobutane $\overset{1}{C}H_3\overset{2}{C}HCl\overset{3}{C}HCl\overset{4}{C}H_3$. Take two of these models and hold them with the two $CH_3$ groups away from you. Place one Cl on the right of C-2 and the other Cl on the left of C-3 in one model. In the other model place one Cl on the left of C-2 and the other Cl on the right of C-3. Now test to see if these models are superimposable or enantiomers. Draw projection formulas for these structures.

(b) Take the other two models and arrange them as follows: with the two $CH_3$ groups away from you place one Cl on the right of C-2 and the other Cl on the right of C-3 in one model. In the other model, place both Cl groups on the left of C-2 and C-3. Test these two models to see if they are superimposable or enantiomers. If they are superimposable, draw one projection structure; if they are enantiomers, draw both projection structures as mirror image isomers.

**Question 3.**

(a) How many asymmetric carbon atoms are in 2,3-dichlorobutane?

(b) How many stereoisomers of 2,3-dichlorobutane did you find?

(c) Calculate the maximum number of stereoisomers possible for 2,3-dichlorobutane.

(d) Why is the number of isomers in Q3(c) different from that in Q3(b)?

In procedure D.(b) you found three stereoisomers of 2,3-dichlorobutane. The formula $2^n$ states that the maximum number of stereoisomers for a compound with two asymmetric carbon atoms is four. Why, then, are there only three isomers? The models constructed in Procedure D.(b) are superimposable. Although these models contain asymmetric carbon atoms, the molecule is symmetrical about a horizontal plane passing through its center, between C-2 and C-3. Thus the molecule shows no optical activity. Stereoisomers that contain asymmetric carbon atoms and are superimposable on their own mirror images are called *meso compounds*.

**Question 4.** Draw project formulas for (a) $CH_3CHBrCHBrCl$ and (b) $CH_2BrCHBrCHBrCH_2Br$. Label all pairs of enantiomers and meso compounds.

**Question 5.** Which of the structures shown are enantiomers of Compound **A** and which structures are the same as Compound **A**? For comparison purposes, projection formulas may be rotated 180°; may not be rotated 90°; and may not be lifted out of the plane of the paper.

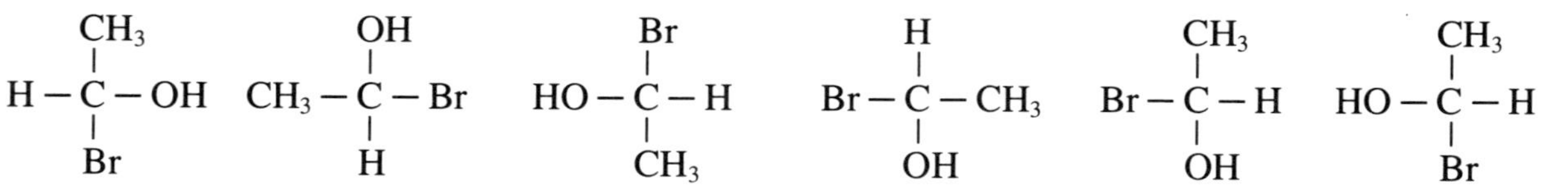

**A** **B** **C** **D** **E** **F**

# APPENDIX 1

# Suggested List of Equipment

## EQUIPMENT FOR STUDENT LOCKERS

1. 5 Beakers: 50,100,150,250,400 mL
2. 1 Burner, Tirrill (optional)
3. Ceramfab pad
4. 1 Clay triangle
5. 2 Crucibles, size 0
6. 2 Crucibles covers, size F
7. 1 Crucible tongs
8. 1 Evaporating dish, size 1
9. 1 File, triangular
10. 1 Filter paper (box)
11. 2 Flasks, Erlenmeyer, 125 mL
12. 2 Flasks, Erlenmeyer, 250 mL
13. 1 Flask, Florence, 500 mL
14. 5 Glass plates, 3 X 3 in.
15. 1 Graduated cylinder, 10 mL
16. 1 Graduated cylinder, 50 mL
17. 2 Litmus paper (vials), red and blue
18. 2 Medicine droppers
19. 1 Pipet, volumetric, 10 mL
20. 8 Rubber stoppers: 3 No. 1, solid; 1 No. 1, one-hole; 1 No. 4, one-hole; 1 No. 4, two-hole; 1 No. 5, solid; 1 No. 6, two-hole
21. 2 Rubber tubing (about 25 cm), 3/16 in. diameter
22. 1 Screw clamp
23. 1 Spatula
24. 1 Sponge
25. 12 Test tubes, 18 X 150 mm (or culture tubes)
26. 1 Test tube, ignition, 25 X 200 mm
27. 1 Test tube brush
28. 1 Test tube holder, wire
29. 1 Test tube rack
30. 1 Thermometer, 110°C
31. 1 Thistle top, plastic
32. 1 Utility clamp (single buret clamp)
33. 1 Wash bottle (plastic)
34. 2 Watch glasses, 4 in.
35. 5 Wide-mouth bottles, 8 oz
36. 1 Wing top
37. 1 Wire gauze

## AUXILIARY EQUIPMENT NOT SUPPLIED IN STUDENT LOCKERS

1. Aluminum foil (7 × 7 cm)
2. Balances
3. Beakers, 600 mL
4. Boyle's law apparatus
5. Buchner funnels and suction flasks
6. Burets, 25 or 50 mL
7. Buret clamps
8. Burners, Tirrill (if not individually supplied)
9. Capillary tubes (sealed at one end)
10. Centrifuges
11. Centrifuge tubes
12. Chromatography columns (polypropylene from Kontes)
13. Deflagration spoons
14. Erlenmeyer flasks, 500 mL
15. Filter paper, Whatman #1 (14 × 14 cm).
16. Glass rod, 5 or 6 mm
17. Glass tubing, 6 mm
18. Glass wool (pyrex)
19. Glass writing markers
20. Hair dryers
21. Magnetic stirrers with bars
22. Metric rulers
23. Oil baths
24. pH meters
25. Pipets, graduated, 1 mL, 5 mL, and 10 mL
26. Pipets, micro
27. Pipets, Pasteur
28. Pneumatic troughs
29. Protective gloves
30. Reflux and distillation equipment:
    100 mL or 200 mL round-bottom distilling flasks
    Distillation take-off heads
    Condensers
    200° or 250° thermometers
    250 mL separatory funnels
31. Ring stands
32. Ring supports, 4 to 5 in. diameter
33. Rubber bands cut from 3/16 inch rubber tubing
34. Spectrophotometers
35. Spray applicators
36. Suction bulbs
37. Wire stirrers for oil baths

# APPENDIX 2

# List of Reagents Required and Preparation of Solutions

## SOLIDS

Ammonium chloride, $NH_4Cl$

Barium chloride, $BaCl_2 \cdot 2\ H_2O$

Barium sulfate, $BaSO_4$

Benzoic acid, $C_6H_5COOH$

Benzophenone, $C_6H_5-CO-C_6H_5$

Benzoyl peroxide, $(C_6H_5COO)_2$

Boiling chips

Candles

Calcium carbide, $CaC_2$

Calcium hydroxide, $Ca(OH)_2$

Calcium oxide, CaO

Cholesterol, $C_{27}H_{45}OH$

trans-Cinnamic acid, $C_9H_8O_2$

Cobalt chloride paper

Copper turnings (light), Cu

Copper strips, Cu

Copper wire, #18, Cu

Copper wire, #24, Cu

Copper(II) sulfate pentahydrate, $CuSO_4 \cdot 5\ H_2O$

Cotton

Diphenylacetic acid, $C_{14}H_{12}O_2$

Diphenylacetic acid-cholesterol, 50:50

Glass wool, pyrex

Glucose, $C_6H_{12}O_6$

Ice

Iodine, $I_2$

Iron wire, #20-24, Fe

Iron(II) sulfide, FeS

Lead strips, Pb

Lead(II) chromate, $PbCrO_4$

Magnesium strips, Mg

Magnesium oxide, MgO

Magnesium sulfate, anhydrous, $MgSO_4$

Manganese dioxide, $MnO_2$

Marble chips

Marbles, about 20 mm diameter

Menthol, $C_{10}H_{20}O$

Potassium acid phthalate, $KHC_8H_4O_4$

Potassium bisulfate, $KHSO_4$

Potassium bromide, KBr

Potassium chlorate, C.P., $KClO_3$

Potassium chloride, C.P., KCl

Potassium chromate, C.P., $K_2CrO_4$

Potassium iodide, KI

Potassium nitrate, $KNO_3$

Potato (fresh)

Sand paper or emery cloth

Salicylic acid, $C_6H_4(COOH)(OH)$

Sodium, Na

Sodium bicarbonate, $NaHCO_3$

Sodium bisulfite, $NaHSO_3$

Sodium chloride (coarse crystals), NaCl

Sodium chloride (fine crystals), NaCl

Sodium peroxide, $Na_2O_2$

Sodium sulfate, $Na_2SO_4$

Sodium sulfite, $Na_2SO_3$

Stearic acid, $CH_3(CH_2)_{16}COOH$

Steel wool, Fe (Grade 0 or 1)

Sucrose, $C_{12}H_{22}O_{11}$

Sulfur, S

Tyrosine

Urea, $(NH_2)_2CO$

Urea-trans-cinnamic acid, 50:50

Vegetable shortening

Wood splints

Zinc, mossy, Zn

Zinc strips, Zn (0.01 inch thick).

## PURE LIQUIDS/COMMERCIAL MIXTURES

Acetic acid (glacial), $CH_3COOH$

Acetic anhydride, $(CH_3CO)_2O$

Acetone, $CH_3COCH_3$

Benzene, $C_6H_6$

Bromine, $Br_2$, (specimen sample in sealed vial).

n-Butyl alcohol (1-butanol), $C_4H_9OH$

Chloroform, $CHCl_3$

Ethyl alcohol (ethanol), 95%, $C_2H_5OH$

Glycerol, $C_3H_5(OH)_3$

Heptane (or low boiling petroleum ether), $C_7H_{16}$

Isoamyl alcohol, $C_5H_{11}OH$

Isopropyl alcohol (2-propanol), $C_3H_7OH$

Kerosene (alkene free)

Ligroine (60°-85°C boiling range)

Methyl alcohol (methanol), $CH_3OH$

Methyl methacrylate, $CH_2=CH(CH_3)COOCH_3$

Oleic acid, $CH_3(CH_2)_7CH=CH(CH_2)_7COOH$

Pentene (Amylene) $C_5H_{10}$

1,1,1-Trichloroethane, $CCl_3CH_3$

Vegetable oils (corn, cottonseed, peanut, soybean, etc.)

APPENDIX 2 (continued)

## SOLUTIONS

All solutions, except where otherwise directed, are prepared by dissolving the designated quantity of solute in distilled water, and diluting to 1 liter.

Acetic acid, concentrated (glacial); concentrated reagent

Acetic acid, dilute, 6 M; 355 mL concentrated acid/liter

Acetic acid, 10%; 100 mL concentrated acid/liter

Acetic acid-1-butanol-water (1:3:1 by volume); 200 mL $CH_3COOH$ + 600 mL $C_4H_9OH$/liter

Alanine, 0.2 M; 1.78 g alanine/ 100 mL

Alanine HCl, 0.1 M; 1.26 g alanine•HCl/ 100 mL

Alanine-aspartic acid-leucine-lysine solution (each 0.2 M); 1.78 g alanine + 2.66 g aspartic acid + 2.62 g leucine + 2.92 g lysine/100 mL

Adipoyl chloride, 0.4 M in cyclohexane; 18.3 g adipoyl chloride/250 mL cyclohexane

Albumin, 2%; 20 g albumin/liter (Make slurry with about 50 mL water, then add additional water slowly while stirring.)

Aluminum chloride, 0.10 M; 24.1 g $AlCl_3 \cdot 6\ H_2O$/ liter

Ammonium chloride, 0.1 M; 5.4 g $NH_4Cl$/liter

Ammonium chloride, saturated; 60 g $NH_4Cl$/ liter

Ammonium hydroxide, concentrated; concentrated reagent

Ammonium hydroxide, dilute, 6M; 400 mL concentrated/liter

Ammonium molybdate, 0.2 M; 3.53 g $(NH_4)_6Mo_7O_{24} \cdot 4\ H_2O$/100 mL

Ammonium oxalate, 0.2 M; 2.84 g $(NH_4)_2C_2O_4 \cdot H_2O$/100 mL

Arabinose, 1%; 10 g arabinose/liter

Arginine-tyrosine solution (each 0.1%); 100 mg arginine + 100 mg tyrosine/100 mL pH 6.0 phosphate buffer

Arsenomolybdate reagent; Nelson's arsenomolybdate reagent is commercially available from Sigma Chemical Co., St. Louis, Missouri

Aspartic acid, 0.2 M; 2.66 g aspartic acid/ 100 mL

Barfoed's reagent; dissolve 13.3 g $Cu(C_2H_3O_2)_2 \cdot H_2O$ in 200 mL $H_2O$. (Filter if necessary) and add 1.8 mL $HC_2H_3O_2$ (glacial). Cupric acetate is slow to dissolve.

Barium chloride, 0.10 M; 24.4 g $BaCl_2 \cdot 2\ H_2O$/liter

Barium hydroxide, saturated; 10 g $Ba(OH)_2 \cdot 8\ H_2O$/100 mL

Barium hydroxide, 0.2 M; 15.8 g $Ba(OH)_2 \cdot 8\ H_2O$/250 mL

Benedict's reagent; dissolve 86.5 g sodium citrate and 50 g anhydrous $Na_2CO_3$ in 400 mL water with heating. Dissolve 8.65 g $CuSO_4 \cdot 5\ H_2O$ in 50 mL water. Mix these two solutions slowly and add water to produce 500 mL of solution.

Bial's reagent; dissolve 1.5 g orcinol (5-methyl resorcinol) in 500 mL conc. HCl and add 1.5 mL of 10% aqueous $FeCl_3$

Blood, whole—as fresh as possible (certified AIDS free)

Bromine in 1,1,1-trichloroethane, 5% solution; 2.5 mL $Br_2$ plus 100 mL $CCl_3CH_3$

Bromine water, saturated; about 6 mL $Br_2$/liter

Buffer solution, standard pH 7.0; commercially available

Calcium chloride, 0.1 M; 14.7 g $CaCl_2 \cdot 2\,H_2O$/liter

Chlorine water; dilute 150 mL of 5.25% NaOCl (household bleach) to 1 liter. Add 15 mL concentrated HCl and mix gently.

Cobalt(II) chloride, 0.1 M; 23.8 g $CoCl_2 \cdot 6H_2O$/liter

Copper(II) nitrate, 0.1 M; 24.2 g $Cu(NO_3)_2 \cdot 3\,H_2O$/liter

Copper reagent; dissolve 24 g of anhydrous $Na_2CO_3$, 16 g of sodium potassium tartrate, 4 g of $CuSO_4 \cdot 5\,H_2O$, and 180 g of anydrous $Na_2SO_4$ in water and dilute to 1 liter.

Copper(II) sulfate, 0.1 M; 25.0 g $CuSO_4 \cdot 5\,H_2O$/liter

Copper(II) sulfate, 0.2 M; 50.0 g $CuSO_4 \cdot 5\,H_2O$/liter

Cysteine • HCl, 0.10 M; 1.56 g cysteine•HCl/100 mL

Dichlorofluorescein; dissolve 1 g of 2,7-dichlorofluorescein in 1 liter of 70% ethanol

Digitonin solution; dissolve 1.0 g digitonin in 50 mL ethanol

Dowex-50 slurry; The resin must be in the $H^+$ form. Newly purchased resin or used resin is washed sequentially with deionized water, acetone (use only for the new resin), 1 M NaOH, deionized water, 1 M HCl, and deionized water. The washed resin is suspended in the pH 6 phosphate buffer two or three times or until the slurry has a pH of 6. Store the slurry in the cold.

Ethanol-Acetone (1:1 by volume); 500 mL $CH_3CH_2OH$ + 500 mL $CH_3COCH_3$

Food colors, blue, green, and yellow; commercially available products

Formaldehyde, 10% solution; 25 mL formalin (40%) plus 75 mL $H_2O$

Fructose, 1% solution; 1.0 g fructose/100 mL

Gelatin, 2% solution; 10 g gelatin/500 mL (Dissolves slowly)

Glucose, 1% solution; 5.0 g $C_6H_{12}O_6$/500 mL $H_2O$

Glucose (dextrose) 10% solution; 10 g $C_6H_{12}O_6$ plus 90 mL water

Glucose, standard solutions; made up to contain 2.0, 5.0, 8.0, 12.0, 15.0, and 18.0 mg/100 mL

Glutamic acid•HCl, 0.10 M; 1.84 g glutamic acid•HCl/100 mL

Glycine•HCl, 0.10 M; 1.40 g glycine•HCl/100 mL

Glycine, 1% solution; 1.0 g glycine/100 mL

Hexamethylenediamine, 0.40 M in 0.40 M NaOH; dissolve 11.6 g $(CH_2)_6(NH_2)_2$/250 mL 0.40 M NaOH

Histidine•HCl, 0.10 M; 2.10 g histidine•HCl/100 mL

Hydrochloric acid, concentrated; concentrated reagent

Hydrochloric acid, dilute, 6 M; 500 mL concentrated acid/liter

Hydrochloric acid, 0.1 M; 8.33 mL concentrated acid/liter

Hydrochloric acid, 0.01 M; 0.83 mL concentrated acid/liter

Hydrogen peroxide, 3%; 3% reagent solution (or 100 mL 30%/ liter)

Hydrogen peroxide, 9%; 300 mL 30% $H_2O_2$/liter

Iodine in potassium iodide, 1%; 1.0 g $I_2$ + 2.0 KI/100 mL (Dissolve $I_2$ and KI in about 10 mL $H_2O$, then dilute)

Iodine in potassium iodide, 0.1 M; 25 g $I_2$ + 50 g KI/liter (Dissolve $I_2$ and KI in 50 mL $H_2O$, then dilute)

Iodine water, saturated; 5 g $I_2$/liter

Isopropyl alcohol-water (2:1 by volume); 667 mL $CH_3CH(OH)CH_3$/liter

Iron(III) chloride, 0.1 M; 27.1 g $FeCl_3 \cdot 6\,H_2O$ + 5 mL concentrated HCl/liter

Iron reagent; 2.5 g $FeCl_3 \cdot 6\,H_2O$/100 mL 85% phosphoric acid

Lead(II) acetate, 0.1 M; 3.25 g $Pb(C_2H_3O_2)_2$/100 mL

Lead(II) nitrate, 0.50 M; 165.6 g $Pb(NO_3)_2$/liter (for Experiment 15 only)

Lead(II) nitrate, 0.1 M; 33.1 g $Pb(NO_3)_2$/liter

Leucine, 0.2 M; 2.62 g leucine/100 mL

Lysine, 0.2 M; 2.92 g lysine/100 mL

Lysine•HCl, 0.10 M; 1.83 g lysine•HCl/100 mL

Magnesium sulfate, 0.1 M; 24.6 g $MgSO_4 \cdot 7\,H_2O$/liter

Maltose, 1% solution; 5.0 g maltose/500 mL

Mercury(I) nitrate, 0.1 M; dissolve 26.1 g $HgNO_3 \cdot H_2O$ in 50 mL concentrated $HNO_3$ and slowly dilute with water to 1 liter (Prepare in a well ventilated hood)

Milk, fat free (skim)

Millon's reagent (prepare fresh); mix 25 g Hg and 50 mL conc. $HNO_3$ and stir using a magnetic stirrer until all the Hg is reacted (10 to 15 minutes). Add 150 mL $H_2O$. Prepare in a well ventilated hood.

Mixed halide solution, chloride + iodide; 2.3 g NaCl + 1.7 g KI/liter

Mixed halide solution, chloride + bromide; 2.3 g NaCl + 4.8 g KBr/liter

Mixed halide solution, bromide + iodide; 4.8 g KBr + 1.7 g KI/liter

Molisch's reagent; dissolve 2.5 g α-naphthol in 50 mL 95% $C_2H_5OH$

Ninhydrin, 0.3%; 1.5 g ninhydrin/500 mL acetone

Ninhydrin, 0.2%; dissolve 0.2 g of ninhydrin in 100 mL of 1-butanol which is saturated with water

Nitric acid, concentrated; concentrated reagent

Nitric acid, dilute, 6 M; 375 mL concentrated acid/liter

Phenol, 1% solution; 5.0 g $C_6H_5OH$/500 mL

Phenolphthalein, 0.2% solution; dissolve 2 g phenolphthalein in 600 mL ethanol and dilute with water to 1 liter

Phosphoric acid, 85% reagent

Phosphoric acid, dilute, 3 M; 201 mL 85% solution/liter

Phosphate buffer, 0.2 M; 34.8 g $K_2HPO_4$/liter and adjust pH to 6.0 using a pH meter and dilute $H_3PO_4$ or KOH solution

Potassium chlorate, 0.1 M; 12.2 g $KClO_3$/liter

Potassium chloride, 0.1 M; 7.46 g KCl/liter

Potassium chromate, 0.1 M; 19.4 g $K_2CrO_4$/liter

Potassium dichromate, saturated; 250 g $K_2Cr_2O_7$/liter

Potassium nitrate, 0.1 M; 10.1 g $KNO_3$/liter

Potassium permanganate, 0.1 M; 15.8 g $KMnO_4$/liter

Potassium thiocyanate, 0.1 M; 9.7 g KSCN/liter

Seliwanoff's reagent; dissolve 0.50 g resorcinol in 1000 mL 4 M HCl (333 mL conc. HCl diluted to 1000 mL).

Silver nitrate, 0.10 M; 17.0 g $AgNO_3$/liter

Sodium arsenate, 0.1 M; 31.2 g $Na_2HAsO_4 \cdot 7\ H_2O$ + 4.0 g NaOH/liter

Sodium bicarbonate, 5% solution; 50 g $NaHCO_3$/liter

Sodium bicarbonate, saturated solution; 125 g $NaHCO_3$/liter

Sodium bromide, 0.1 M; 10.3 g NaBr/liter

Sodium carbonate, 0.1 M; 10.6 g $Na_2CO_3$/liter

Sodium chloride, 0.1 M; 5.85 g NaCl/liter

Sodium chloride, saturated solution; 60 g NaCl/liter

Sodium hydroxide, 10% solution; 111 g NaOH/liter

Sodium hydroxide, 0.1 M; 4.0 g NaOH/liter

Sodium iodide, 0.1 M; 15.0 g NaI/liter

Sodium nitrite, 0.1 M; 6.9 g $NaNO_2$/liter

APPENDIX 2 (continued)

Sodium phosphate, 0.1 M; 38.0 g $Na_3PO_4 \cdot 12\ H_2O$/liter

Sodium sulfate, 0.1 M; 14.2 g $Na_2SO_4$/liter

Starch, 1% solution; 5 g/500 mL (Make slurry and disperse in hot water.)

Sucrose, 1% solution; 10 g sucrose/liter (freshly prepared)

Sulfuric acid, concentrated; concentrated reagent

Sulfuric acid, dilute, 3 M; 167 mL concentrated acid/liter

Urine/fresh

Vinegar, commercial (colorless)

Xylose, 1% solution; 1.0 g xylose/100 mL

Zinc nitrate, 0.1 M; 29.8 g $Zn(NO_3)_2 \cdot 6\ H_2O$/liter

Zinc sulfate, 0.2 M; 14.4 g $ZnSO_4 \cdot 7\ H_2O$/250 mL

# APPENDIX 3

# Special Equipment or Preparations Needed

### Experiment 1. Laboratory Techniques

A small sample of lead(II) chromate is needed for comparison purposes only.

### Experiment 2. Measurements

An assortment of metal slugs or other solid objects are needed as unknowns for density determination. The diameter of the slugs should be such that they will fit into the 50 mL graduated cylinder. Suggested materials are aluminum, brass, magnesium, steel, etc. Slugs should be numbered for identification.

### Experiment 3. Preparation and Properties of Oxygen

Three demonstrations are suggested (see experiment for details).

### Experiment 4. Preparation and Properties of Hydrogen

For safety: (1) Instructor should dispense sodium metal (size of pieces should be no larger than a 4 mm cube); (2) it is advisable to instruct students in the handling and hazards of concentrated sulfuric acid (this is the first time it is used).

### Experiment 5. Water in Hydrates

An assortment of samples for unknowns for determination of percent water is needed. Samples can be issued in small coin envelopes. See Instructor's Manual for suggested list of samples.

### Experiment 6. Freezing Points—Graphing of Data

Slotted corks or stoppers.

### Experiment 11. Ionization—Acids, Bases, and Salts

Conductivity apparatus is needed for the demonstration (see experiment for details). The use of a magnetic stirrer greatly facilitates running the last part of the demonstration. The demonstration described is based on using the conductivity apparatus illustrated. Other types may be used without detracting from the results of the demonstration.

### Experiment 12. Identification of Selected Anions

Two unknown solutions (in test tubes) are to be issued to each student. Stock reagents used in the experiment are satisfactory for unknowns.

**Experiment 13. Identification of Halide Ions in Mixtures**

Two unknown solutions (in test tubes) are to be issued to each student. The three stock reagents plus other combinations of $Cl^-$, $Br^-$, and $I^-$ may be used for unknowns. Concentrations should be limited to 0.04 M $Cl^-$, 0.04 M $Br^-$, and 0.01 M $I^-$.

**Experiment 14. Properties of Lead(II), Silver, and Mercury(I) Ions**

An unknown solution containing one or more of the silver group cations is to be issued to each student.

**Experiment 15. Quantitative Precipitation of Chromate Ion**

Lead(II) nitrate solution is 0.50 M for this experiment.

**Experiment 16. Boyle's Law**

Boyle's law apparatus is needed. The apparatus described in the experiment is available from Central Scientific Company.

**Experiment 18. Gaseous Diffusion**

One glass tube-rubber stopper assembly is needed for every two students. See Figure 18.1 for details.

**Experiment 19. Liquids—Vapor Pressure and Boiling Points**

125 mL flasks containing acetone, methanol, ethanol, and water are needed for Part A. It is suggested that students work in pairs in Part B. A 1-gallon metal can is needed for the demonstration in Part C.

**Experiment 20. Neutralization—Titration I**

The following are needed by each student: A small vial or test tube containing about 4 grams of potassium hydrogen phthalate (KHP) (these vials are collected for reuse), one 25 or 50 mL buret, a buret clamp, a glass bead or pinch clamp (if Mohr buret is used), and 250 mL of unknown NaOH solution. The NaOH solution is used in Experiments 20 and 21. Buret tips prepared in Experiment 1 can be used here.

**Experiment 21. Neutralization—Titration II**

The following are needed by each student: A 10 mL volumetric pipet, one 25 or 50 mL buret, 50 mL of unknown acid solution, 50 mL of vinegar, and 125 mL of standard NaOH solution if Experiment 20 is not done.

### Experiment 22. Chloride Content of Salts

An unknown solution for the determination of chloride molarity is to be issued to each student. A solution of dichlorofluorescein indicator is needed.

### Experiment 24. Halogens

Provide a small specimen sample of bromine in a bottle for students to observe.

### Experiment 26. Heat of Reaction

Styrofoam cups are needed.

### Experiment 27. Boiling Points and Melting Points

Boiling point and melting point apparatus (see experiment for details), 200°-250°C. thermometers, wire stirrers, capillary melting point tubes, and unknown solids are required.

### Experiment 29. Alcohols, Esters, Aldehydes, and Ketones

Furnish No. 18 copper wire with five or six spiral turns at one end. Wire should be about 8 inches long overall.

### Experiment 30. Esterification—Distillation Synthesis of n-Butyl Acetate

Reflux and distillation equipment (see experiment for details), 200°-250°C. thermometer, and 250 mL separatory funnel are required.

### Experiment 31. Synthesis of Aspirin

Buchner funnel, suction flask, melting point apparatus, and capillary melting point tubes are needed.

### Experiment 32. Polymers—Macromolecules

Benzoyl peroxide is a shock and heat sensitive material. It should be dispensed to each student by the instructor or by qualified stock room personnel.

### Experiment 33. Carbohydrates

Pure fresh or frozen fruit juices, such as orange, lemon, lime, grapefruit, and apple are needed.

### Experiment 34. Glucose Levels in Whole Blood

Spectrophotometer, 10 mL graduated pipets, 20 mm diameter marbles, protective gloves, and whole blood (**certified AIDS virus free**) are needed.

### Experiment 36. Paper Chromatography

Five hundred mL Erlenmeyer flasks, 14 × 14 cm squares Whatman No. 1 filter paper, 7 × 7 cm squares of Al foil, micropipets and a spray applicator for ninhydrin are needed. An unknown amino acid or amino acid mixture is required for each student.

### Experiment 37. Ion-Exchange Chromatography

Chromatography columns (see experiment for details), Dowex-50 resin slurry, and 600 mL beakers must be provided.

### Experiment 38. Identification of an Unknown Amino Acid by Titration

pH meter, pH 7.0 buffer, burets, and magnetic stirrer are needed. An unknown amino acid is issued to each student or student pair.

### Experiment 39. Enzymatic Catalysis—Catalase

A potato is needed for each student or student pair.

### Experiment 40. A Partial Chemical Analysis of Urine

Fresh urine should be collected each day and stored in a refrigerator until used.

### Experiment 42. Blood Cholesterol Levels

Whole blood (**certified AIDS virus free**), spectrophotometer, centrifuge, centrifuge tubes, 1 mL and 5 mL graduated pipets, and protective gloves are needed.

## APPENDIX 4

# Units of Measurements

### Numerical Value of Prefixes Used with Units

| Prefix | Number | Power of 10 |
|---|---|---|
| Mega | 1,000,000 | $1 \times 10^{6}$ |
| Kilo | 1,000 | $1 \times 10^{3}$ |
| Hecto | 100 | $1 \times 10^{2}$ |
| Deca | 10 | $1 \times 10^{1}$ |
| Deci | 0.1 | $1 \times 10^{-1}$ |
| Centi | 0.01 | $1 \times 10^{-2}$ |
| Milli | 0.001 | $1 \times 10^{-3}$ |
| Micro | 0.000001 | $1 \times 10^{-6}$ |
| Nano | 0.000000001 | $1 \times 10^{-9}$ |

### Conversion of Units

| | | |
|---|---|---|
| 1 meter | = | 1000 mm |
| 1 cm | = | 10 mm |
| 2.54 cm | = | 1 in. |
| 453.6 gram | = | 1 lb |
| 1 gram | = | 1000 mg |
| 1 liter | = | 1000 mL |
| 1 mL | = | 1 $cm^3$ |
| 0.946 liter | = | 1 qt |

### Metric Abbreviations

| | |
|---|---|
| Meter | m |
| Centimeter | cm |
| Millimeter | mm |
| Liter | L |
| Milliliter | mL |
| Kilogram | kg |
| Gram | g |
| Milligram | mg |

### Temperature Conversion Formulas

$$°C = \frac{(°F - 32)}{1.8}$$

$$°F = 1.8\ °C + 32$$

# APPENDIX 5

## Solubility Table

| | $C_2H_3O_2^-$ | $AsO_4^{3-}$ | $Br^-$ | $CO_3^{2-}$ | $Cl^-$ | $CrO_4^{2-}$ | $OH^-$ | $I^-$ | $NO_3^-$ | $C_2O_4^{2-}$ | $O^{2-}$ | $PO_4^{3-}$ | $SO_4^{2-}$ | $S^{2-}$ | $SO_3^{2-}$ |
|---|---|---|---|---|---|---|---|---|---|---|---|---|---|---|---|
| $Al^{3+}$ | S | I | S | - | S | - | I | S | S | - | I | I | S | d | - |
| $NH_4^+$ | S | S | S | S | S | S | S | S | S | S | - | S | S | S | S |
| $Ba^{2+}$ | S | I | S | I | S | I | s | S | S | I | s | I | I | d | I |
| $Bi^{3+}$ | - | s | d | I | d | - | I | I | d | I | I | s | d | I | - |
| $Ca^{2+}$ | S | I | S | I | S | S | I | S | S | I | I | I | I | d | I |
| $Co^{2+}$ | S | I | S | I | S | I | I | S | S | I | I | I | S | I | I |
| $Cu^{2+}$ | S | I | S | I | S | I | I | - | S | I | I | I | S | I | - |
| $Fe^{2+}$ | S | I | S | s | S | - | I | S | S | I | I | I | S | I | s |
| $Fe^{3+}$ | I | I | S | I | S | - | I | - | S | S | I | I | S | I | - |
| $Pb^{2+}$ | S | I | I | I | I | I | I | I | S | I | I | I | I | I | I |
| $Mg^{2+}$ | S | d | S | I | S | S | I | S | S | I | I | I | S | d | s |
| $Hg^{2+}$ | S | I | I | I | S | s | I | I | S | I | I | I | d | I | - |
| $K^+$ | S | S | S | S | S | S | S | S | S | S | S | S | S | S | S |
| $Ag^+$ | s | I | I | I | I | I | - | I | S | I | I | I | I | I | I |
| $Na^+$ | S | S | S | S | S | S | S | S | S | S | S | S | S | S | S |
| $Zn^{2+}$ | S | I | S | I | S | I | I | S | S | I | I | I | S | I | I |

Key: S = Soluble in water

s = Slightly soluble in water

I = Insoluble in water (less then 1 g/100 g $H_2O$)

d = Decomposes in water

## APPENDIX 6

# Vapor Pressure of Water

| Temperature (°C) | Vapor Pressure (torr) | Temperature (°C) | Vapor Pressure (torr) |
|---|---|---|---|
| 0 | 4.6 | 26 | 25.2 |
| 5 | 6.5 | 27 | 26.7 |
| 10 | 9.2 | 28 | 28.3 |
| 15 | 12.8 | 29 | 30.0 |
| 16 | 13.6 | 30 | 31.8 |
| 17 | 14.5 | 40 | 55.3 |
| 18 | 15.5 | 50 | 92.5 |
| 19 | 16.5 | 60 | 149.4 |
| 20 | 17.5 | 70 | 233.7 |
| 21 | 18.6 | 80 | 355.1 |
| 22 | 19.8 | 90 | 525.8 |
| 23 | 21.2 | 100 | 760.0 |
| 24 | 22.4 | 110 | 1074.6 |
| 25 | 23.8 | | |

# APPENDIX 7

## Boiling Points of Liquids

| Liquid | Boiling Point (°C) |
|---|---|
| Acetone | 56.5 |
| Carbon disulfide | 46.3 |
| Carbon tetrachloride | 76.8 |
| Chloroform | 61.3 |
| Ethanol | 78.5 |
| Ether | 34.6 |
| Methanol | 64.6 |
| Water | 100.0 |

# PERIODIC TABLE OF THE ELEMENTS

Atomic number → 11
Symbol → Na
Atomic mass → 22.99

Atomic masses are based on carbon-12. Numbers in parentheses are mass numbers of most stable or best known isotopes of radioactive elements.

| Period | Group IA (1) | IIA(2) | IIIB(3) | IVB(4) | VB(5) | VIB(6) | VIIB(7) | VIII (8) | VIII (9) | VIII (10) | IB(11) | IIB(12) | IIIA(13) | IVA(14) | VA(15) | VIA(16) | VIIA(17) | Noble Gases (18) |
|---|---|---|---|---|---|---|---|---|---|---|---|---|---|---|---|---|---|---|
| 1 | 1<br>H<br>1.008 | | | | | | | | | | | | | | | | | 2<br>He<br>4.003 |
| 2 | 3<br>Li<br>6.941 | 4<br>Be<br>9.012 | | | | | | | | | | | 5<br>B<br>10.81 | 6<br>C<br>12.01 | 7<br>N<br>14.01 | 8<br>O<br>16.00 | 9<br>F<br>19.00 | 10<br>Ne<br>20.18 |
| 3 | 11<br>Na<br>22.99 | 12<br>Mg<br>24.31 | | | | | | | | | | | 13<br>Al<br>26.98 | 14<br>Si<br>28.09 | 15<br>P<br>30.97 | 16<br>S<br>32.06 | 17<br>Cl<br>35.45 | 18<br>Ar<br>39.95 |
| 4 | 19<br>K<br>39.10 | 20<br>Ca<br>40.08 | 21<br>Sc<br>44.96 | 22<br>Ti<br>47.90 | 23<br>V<br>50.94 | 24<br>Cr<br>52.00 | 25<br>Mn<br>54.94 | 26<br>Fe<br>55.85 | 27<br>Co<br>58.93 | 28<br>Ni<br>58.7 | 29<br>Cu<br>63.55 | 30<br>Zn<br>65.38 | 31<br>Ga<br>69.72 | 32<br>Ge<br>72.59 | 33<br>As<br>74.92 | 34<br>Se<br>78.96 | 35<br>Br<br>79.90 | 36<br>Kr<br>83.80 |
| 5 | 37<br>Rb<br>85.47 | 38<br>Sr<br>87.62 | 39<br>Y<br>88.91 | 40<br>Zr<br>91.22 | 41<br>Nb<br>92.91 | 42<br>Mo<br>95.94 | 43<br>Tc<br>98.91 | 44<br>Ru<br>101.1 | 45<br>Rh<br>102.9 | 46<br>Pd<br>106.4 | 47<br>Ag<br>107.9 | 48<br>Cd<br>112.4 | 49<br>In<br>114.8 | 50<br>Sn<br>118.7 | 51<br>Sb<br>121.8 | 52<br>Te<br>127.6 | 53<br>I<br>126.9 | 54<br>Xe<br>131.3 |
| 6 | 55<br>Cs<br>132.9 | 56<br>Ba<br>137.3 | 57*<br>La<br>138.9 | 72<br>Hf<br>178.5 | 73<br>Ta<br>180.9 | 74<br>W<br>183.9 | 75<br>Re<br>186.2 | 76<br>Os<br>190.2 | 77<br>Ir<br>192.2 | 78<br>Pt<br>195.1 | 79<br>Au<br>197.0 | 80<br>Hg<br>200.6 | 81<br>Tl<br>204.4 | 82<br>Pb<br>207.2 | 83<br>Bi<br>209.0 | 84<br>Po<br>(210) | 85<br>At<br>(210) | 86<br>Rn<br>(222) |
| 7 | 87<br>Fr<br>(223) | 88<br>Ra<br>226.0 | 89**<br>Ac<br>(227) | 104<br>Unq<br>(261) | 105<br>Unp<br>(262) | 106<br>Unh<br>(263) | 107<br>Uns<br>(262) | 108<br>Uno<br>(265) | 109<br>Une<br>(266) | | | | | | | | | |

←Transition Elements→

## Inner Transition Elements

| | | | | | | | | | | | | | | |
|---|---|---|---|---|---|---|---|---|---|---|---|---|---|---|
| * Lanthanide Series 6 | 58<br>Ce<br>140.1 | 59<br>Pr<br>140.9 | 60<br>Nd<br>144.2 | 61<br>Pm<br>(145) | 62<br>Sm<br>150.4 | 63<br>Eu<br>152.0 | 64<br>Gd<br>157.3 | 65<br>Tb<br>158.9 | 66<br>Dy<br>162.5 | 67<br>Ho<br>164.9 | 68<br>Er<br>167.3 | 69<br>Tm<br>168.9 | 70<br>Yb<br>173.0 | 71<br>Lu<br>175.0 |
| ** Actinide Series 7 | 90<br>Th<br>232.0 | 91<br>Pa<br>231.0 | 92<br>U<br>238.0 | 93<br>Np<br>237.0 | 94<br>Pu<br>(244) | 95<br>Am<br>(243) | 96<br>Cm<br>(247) | 97<br>Bk<br>(247) | 98<br>Cf<br>(251) | 99<br>Es<br>(252) | 100<br>Fm<br>(257) | 101<br>Md<br>(258) | 102<br>No<br>(259) | 103<br>Lr<br>(260) |

# Table of Atomic Masses (Weights) Based on Carbon-12

| Name | Symbol | Atomic No. | Atomic Mass |
|---|---|---|---|
| Actinium | **Ac** | 89 | (227)[a] |
| Aluminum | **Al** | 13 | 26.98154 |
| Americium | **Am** | 95 | (243)[a] |
| Antimony | **Sb** | 51 | 121.75 |
| Argon | **Ar** | 18 | 39.948 |
| Arsenic | **As** | 33 | 74.9216 |
| Astatine | **At** | 85 | (210)[a] |
| Barium | **Ba** | 56 | 137.34 |
| Berkelium | **Bk** | 97 | (247)[a] |
| Beryllium | **Be** | 4 | 9.01218 |
| Bismuth | **Bi** | 83 | 208.9804 |
| Boron | **B** | 5 | 10.81 |
| Bromine | **Br** | 35 | 79.904 |
| Cadmium | **Cd** | 48 | 112.40 |
| Calcium | **Ca** | 20 | 40.08 |
| Californium | **Cf** | 98 | (251)[a] |
| Carbon | **C** | 6 | 12.011 |
| Cerium | **Ce** | 58 | 140.12 |
| Cesium | **Cs** | 55 | 132.9054 |
| Chlorine | **Cl** | 17 | 35.453 |
| Chromium | **Cr** | 24 | 51.996 |
| Cobalt | **Co** | 27 | 58.9332 |
| Copper | **Cu** | 29 | 63.546 |
| Curium | **Cm** | 96 | (247)[a] |
| Dysprosium | **Dy** | 66 | 162.50 |
| Einsteinium | **Es** | 99 | (252)[a] |
| Erbium | **Er** | 68 | 167.26 |
| Europium | **Eu** | 63 | 151.96 |
| Fermium | **Fm** | 100 | (257)[a] |
| Fluorine | **F** | 9 | 18.99840 |
| Francium | **Fr** | 87 | (223)[a] |
| Gadolinium | **Gd** | 64 | 157.25 |
| Gallium | **Ga** | 31 | 69.72 |
| Germanium | **Ge** | 32 | 72.59 |
| Gold | **Au** | 79 | 196.9665 |
| Hafnium | **Hf** | 72 | 178.49 |
| Helium | **He** | 2 | 4.00260 |
| Holmium | **Ho** | 67 | 164.9304 |
| Hydrogen | **H** | 1 | 1.0079 |
| Indium | **In** | 49 | 114.82 |
| Iodine | **I** | 53 | 126.9045 |
| Iridium | **Ir** | 77 | 192.22 |
| Iron | **Fe** | 26 | 55.847 |
| Krypton | **Kr** | 36 | 83.80 |
| Lanthanum | **La** | 57 | 138.9055 |
| Lawrencium | **Lr** | 103 | (260)[a] |
| Lead | **Pb** | 82 | 207.2 |
| Lithium | **Li** | 3 | 6.941 |
| Lutetium | **Lu** | 71 | 174.97 |
| Magnesium | **Mg** | 12 | 24.305 |
| Manganese | **Mn** | 25 | 54.9380 |
| Mendelevium | **Md** | 101 | (258)[a] |
| Mercury | **Hg** | 80 | 200.59 |
| Molybdenum | **Mo** | 42 | 95.94 |
| Neodymium | **Nd** | 60 | 144.24 |
| Neon | **Ne** | 10 | 20.179 |
| Neptunium | **Np** | 93 | 237.0482[b] |
| Nickel | **Ni** | 28 | 58.71 |
| Niobium | **Nb** | 41 | 92.9064 |
| Nitrogen | **N** | 7 | 14.0067 |
| Nobelium | **No** | 102 | (259)[a] |
| Osmium | **Os** | 76 | 190.2 |
| Oxygen | **O** | 8 | 15.9994 |
| Palladium | **Pd** | 46 | 106.4 |
| Phosphorus | **P** | 15 | 30.97376 |
| Platinum | **Pt** | 78 | 195.09 |
| Plutonium | **Pu** | 94 | (244)[a] |
| Polonium | **Po** | 84 | (210)[a] |
| Potassium | **K** | 19 | 39.098 |
| Praseodymium | **Pr** | 59 | 140.9077 |
| Promethium | **Pm** | 61 | (145)[a] |
| Protactinium | **Pa** | 91 | 231.0359[b] |
| Radium | **Ra** | 88 | 226.0254[b] |
| Radon | **Rn** | 86 | (222)[a] |
| Rhenium | **Re** | 75 | 186.2 |
| Rhodium | **Rh** | 45 | 102.9055 |
| Rubidium | **Rb** | 37 | 85.4678 |
| Ruthenium | **Ru** | 44 | 101.07 |
| Samarium | **Sm** | 62 | 150.4 |
| Scandium | **Sc** | 21 | 44.9559 |
| Selenium | **Se** | 34 | 78.96 |
| Silicon | **Si** | 14 | 28.086 |
| Silver | **Ag** | 47 | 107.868 |
| Sodium | **Na** | 11 | 22.98977 |
| Strontium | **Sr** | 38 | 87.62 |
| Sulfur | **S** | 16 | 32.06 |
| Tantalum | **Ta** | 73 | 180.9479 |
| Technetium | **Tc** | 43 | 98.9062[b] |
| Tellurium | **Te** | 52 | 127.60 |
| Terbium | **Tb** | 65 | 158.9254 |
| Thallium | **Tl** | 81 | 204.37 |
| Thorium | **Th** | 90 | 232.0381[b] |
| Thulium | **Tm** | 69 | 168.9342 |
| Tin | **Sn** | 50 | 118.69 |
| Titanium | **Ti** | 22 | 47.90 |
| Tungsten | **W** | 74 | 183.85 |
| Unnilhexium | **Unh** | 106 | (263)[a] |
| Unnilpentium | **Unp** | 105 | (262)[a] |
| Unnilquadium | **Unq** | 104 | (261)[a] |
| Uranium | **U** | 92 | 238.029 |
| Vanadium | **V** | 23 | 50.9414 |
| Xenon | **Xe** | 54 | 131.30 |
| Ytterbium | **Yb** | 70 | 173.04 |
| Yttrium | **Y** | 39 | 88.9059 |
| Zinc | **Zn** | 30 | 65.38 |
| Zirconium | **Zr** | 40 | 91.22 |

[a]Mass number of most stable or best-known isotope

[b]Mass number of the isotope of longest half-life

# Names, Formulas and Charges of Common Ions

**Positive Ions (Cations)**

| 1+ | | 2+ | | 3+ | | 4+ | |
|---|---|---|---|---|---|---|---|
| Ammonium | $NH_4^+$ | Calcium | $Ca^{2+}$ | Aluminum | $Al^{3+}$ | Tin(IV) | $Sn^{4+}$ |
| Cesium | $Cs^+$ | Chromium(II) | $Cr^{2+}$ | Antimony(III) | $Sb^{3+}$ | (Stannic) | |
| Copper(I) | $Cu^+$ | Cobalt(II) | $Co^{2+}$ | Arsenic(III) | $As^{3+}$ | Lead(IV) | $Pb^{4+}$ |
| (Cuprous) | | Copper(II) | $Cu^{2+}$ | Bismuth(III) | $Bi^{3+}$ | (Plumbic) | |
| Francium | $Fr^+$ | (Cupric) | | Boron | $B^{3+}$ | Manganese(IV) | $Mn^{4+}$ |
| Gold | $Au^+$ | Iron(II) | $Fe^{2+}$ | Chromium(III) | $Cr^{3+}$ | | |
| Lithium | $Li^+$ | (Ferrous) | | Cobalt(III) | $Co^{3+}$ | | |
| Potassium | $K^+$ | Lead(II) | $Pb^{2+}$ | Iron(III) | $Fe^{3+}$ | | |
| Rubidium | $Rb^+$ | Manganese(II) | $Mn^{2+}$ | (Ferric) | | | |
| Silver | $Ag^+$ | Mercury(I) | $Hg_2^{2+}$ | Titanium(III) | $Ti^{3+}$ | | |
| Sodium | $Na^+$ | Mercury(II) | $Hg^{2+}$ | | | | |
| | | Nickel(II) | $Ni^{2+}$ | | | | |
| | | Tin(II) | $Sn^{2+}$ | | | | |
| | | (Stannous) | | | | | |
| | | Zinc | $Zn^{2+}$ | | | | |

**Negative Ions (Anions)**

| 1− | | 2− | | 3− | |
|---|---|---|---|---|---|
| Acetate | $C_2H_3O_2^-$ | Carbonate | $CO_3^{2-}$ | Arsenate | $AsO_4^{3-}$ |
| Bromide | $Br^-$ | Chromate | $CrO_4^{2-}$ | Borate | $BO_3^{3-}$ |
| Chlorate | $ClO_3^-$ | Dichromate | $Cr_2O_7^{2-}$ | Phosphate | $PO_4^{3-}$ |
| Chloride | $Cl^-$ | Hydrogen Phosphate | $HPO_4^{2-}$ | Phosphite | $PO_3^{3-}$ |
| Chlorite | $ClO_2^-$ | (Biphosphate) | | Phosphide | $P^{3-}$ |
| Cyanide | $CN^-$ | Oxalate | $C_2O_4^{2-}$ | | |
| Dihydrogen phosphate | $H_2PO_4^-$ | Oxide | $O^{2-}$ | | |
| Fluoride | $F^-$ | Peroxide | $O_2^{2-}$ | | |
| Hydride | $H^-$ | Sulfate | $SO_4^{2-}$ | | |
| Hydrogen carbonate | $HCO_3^-$ | Sulfide | $S^{2-}$ | | |
| (Bicarbonate) | | Sulfite | $SO_3^{2-}$ | | |
| Hydrogen oxalate | $HC_2O_4^-$ | Silicate | $SiO_3^{2-}$ | | |
| Hydrogen sulfate | $HSO_4^-$ | | | | |
| (Bisulfate) | | | | | |
| Hydrogen sulfide | $HS^-$ | | | | |
| Hydrogen sulfite | $HSO_3^-$ | | | | |
| (Bisulfite) | | | | | |
| Hydroxide | $OH^-$ | | | | |
| Hypochlorite | $HClO^-$ | | | | |
| Iodide | $I^-$ | | | | |
| Nitrate | $NO_3^-$ | | | | |
| Nitrite | $NO_2^-$ | | | | |
| Perchlorate | $ClO_4^-$ | | | | |
| Permanganate | $MnO_4^-$ | | | | |
| Thiocyanate | $SCN^-$ | | | | |